# Species

# Species

## New Interdisciplinary Essays

edited by Robert A. Wilson

A Bradford Book
The MIT Press
Cambridge, Massachusetts
London, England

This book was set in Palatino by Asco Typesetters, Hong Kong and was printed and bound in the United States of America.

Library of Congress Cataloging-in-Publication Data

Species: new interdisciplinary essays / edited by Robert A. Wilson.
p. cm.
"A Bradford book."
Includes bibliographical references and index.
ISBN 978-0-262-73123-2
1. Species—Philosophy. I. Wilson, Robert A. (Robert Andrew)
QH83.S725 1999
576.8'6—dc21

98-34713
CIP

# Contents

# Acknowledgments

Initial work on this volume was supported by SSHRC grant number 410-96-0497 awarded by the Social Sciences and Humanities Research Council of Canada while I was at Queen's University; I completed the volume with the assistance of a Research Board Grant from the University of Illinois, Urbana-Champaign. I would like to thank Allison Dawe for her research assistance with the project in its early stages; Peter Asaro for his assistance in its latter stages; and Linda May for her masterly, efficient, and patient work in formatting the manuscript.

Although it should almost go without saying, I would especially like to thank all the contributors to the volume, not simply for collectively being the *sine qua non* for the volume, but for making the editorial aspects of the project less chorelike than they might have been. (Sometimes I even had the illusion that I was having fun.) Apart from providing the essays themselves, many contributors also read and offered comments on other contributors' essays and gave general advice, as needed, about the constitution of the volume as a whole.

# Introduction

This volume of twelve specially commissioned essays about species draws on the perspectives of prominent researchers from anthropology, botany, developmental psychology, the philosophy of biology and science, protozoology, and zoology. The concept of species has played a focal role in both evolutionary biology and the philosophy of biology, and the last decade has seen something of a publication boom on the topic (e.g., Otte and Endler 1989; Ereshefsky 1992b; Paterson 1994; Lambert and Spence 1995; Claridge, Dawah, and Wilson 1997; Wheeler and Meier 1999; Howard and Berlocher 1998). *Species: New Interdisciplinary Essays* is distinguished from other recent collections on species and the species problem in two ways.

First, by attempting to be more explicitly integrative and analytical, this volume looks to go beyond both the exploration of the detailed implications of any single species concept (cf. Lambert and Spence 1995, and Wheeler and Meier 1999) and the survey of the ways in which species are conceptualized by researchers in various parts of biology (cf. Claridge, Dawah, and Wilson 1997). As a whole, it takes a step back from much of the biological nitty-gritty that forms the core of these recent books on species in order to gain some focus on general claims about and views of species. Authors for the current volume were explicitly encouraged to address some subset of five general themes that tied their particular discussions to broader issues about species with a philosophical edge to them. Half the contributors have their primary training in philosophy. The volume is thus deliberately more philosophical in its orientation and in the content of the essays. Yet the biological detail in *Species: New Interdisciplinary Essays* is, I believe, rich enough for the volume as a whole to contribute both to the philosophy of biology and to evolutionary biology itself.

Second, the volume adds historical and psychological dimensions typically missing from contemporary discussions of species. The historical slant is reflected in essays that consider the Linnaean hierarchy (e.g., Ereshefsky) and the Modern Synthesis (e.g., Nanney), as well as in those essays that draw on more general considerations from the philosophy of science (e.g., Boyd) and in those that offer particular solutions to the species problem (e.g., de Queiroz). Although the principal purpose of these essays is not to contribute to the history of biology, they are often able to appeal to that history in order to enrich our understanding of species and the biological world. The psychological perspective is most explicit in the essays by Atran and by Keil and Richardson, but also underlies central arguments in several other papers (e.g., by Wilson and by Griffiths). Together, these two features of the volume provide for a broad perspective on species and on the issues in the philosophy of biology and in biology proper to which species are central.

The papers have been organized into five sections that seemed to me to represent the most cohesive clusters of views and the most interesting sequence of papers to read from beginning to end. Those sections are: "Monism, Pluralism, Unity and Diversity"; "Species and Life's Complications"; "Rethinking Natural Kinds"; "Species in Mind and Culture"; and "Species Begone!" The rest of this introduction mainly provides an overview of the papers in the order that they appear. There are, of course, other thematic commonalities, shared perspectives, and oppositions that this organization (or any single artifactual classification scheme, such as a table of contents) will obscure. One alternative way of thematically locating particular papers in the volume and of viewing the orientation of the volume as a whole is to consider the five themes that authors were invited to address and the pair of themes each paper concentrates on most intensely. Those themes, ranked in order from those that feature in the highest number of papers to those that feature in the smallest number, together with some accompanying questions, are:

### 1 Unity, Integration, and Pluralism

Given the proliferation of species concepts in recent years, how should these concepts be viewed? In what ways do they compete with one another? Which proposals should be seen as the main contenders for "the" species concept, and by which criteria should they be evaluated? What are the prospects for developing an integrated species concept? Should one be a *pluralist* about species? [Dupré, Hull, de Queiroz, Boyd, Wilson, Atran, Mishler]

### 2 Species Realism

What sort of realism, if any, should one adopt with respect to species? In what ways does our answer to this question both reflect and influence our view of other elements in the Linnaean hierarchy? What interplay is there between a stance on the realism issue and broader issues in both the philosophy of biology and the philosophy of science more generally? [Dupré, Sterelny, Boyd, Wilson, Griffiths, Keil and Richardson, Ereshefsky]

### 3 Practical Import

In what ways are answers to the questions asked under the other four themes important for the practice of evolutionary biology and related sciences? Should we view the resolution of the cluster of issues often called "the species problem" as foundational in some way? To what extent is the species problem (merely) definitional? What is the relationship between the species problem and empirical practice within the biological (and other) sciences? [Hull, de Queiroz, Nanney, Sterelny, Griffiths, Mishler]

## 4 Historical Dimensions

In what ways are the views of major historical figures or movements in evolutionary biology of significance for contemporary views of species? Is our own view of important historical episodes (e.g., formation of the Linnaean hierarchy, the Modern Synthesis) skewed in important ways? How can we shed light on contemporary discussions by reflecting on the recent history of evolutionary biology? [Nanney, Ereshefsky]

## 5 Cognitive Underpinnings

To what extent do the literature on children's naive biology and anthropological work in cross-cultural psychology support nativist and universalist views of species? What fruitful interplay exists between explorations of the mental representation of biological knowledge and the philosophy of biology as it has been traditionally circumscribed? [Atran, Keil and Richardson]

The summaries of the sections and essays indicate that many other issues are raised in *Species: New Interdisciplinary Essays*, including the plausibility of the individuality thesis about species, the death of essentialism, the interplay between ecology and evolution, the relationship between common sense and scientific taxonomies, and the challenge that recent developmental systems theory poses to taxonomy in terms of evolutionary homologies. *Species: New Interdisciplinary Essays* advances debate about all of these issues. Between the overviews of contemporary debates and the novel insights provided in many of the essays, the volume should prove invaluable for professionals working in the contributing fields and useful for advanced undergraduate and graduate courses in either the foundations of evolutionary biology or the philosophy of biology.

Let me turn more directly to the individual essays and the sections into which they are organized, beginning with "Monism, Pluralism, Unity, and Diversity", containing papers by John Dupré, David Hull, and Kevin de Queiroz. As the title of his essay ("On the Impossibility of a Monistic Account of Species") suggests, Dupré argues for the rejection of monism about species. He claims, moreover, that this conclusion is the proper one to draw from the complete assimilation of Darwin's insights about the organization of the biological world. There are no perfectly sharp boundaries between preexisting natural kinds—species—that would allow for a monistic account of species. Rather, what we find when we investigate the biological world is diversity, and our schemes of classification should reflect both this diversity and our various theoretical and practical ways of exploring the biological world. As well as recounting familiar objections to the numerous attempts to provide a monistic account of species, Dupré also offers novel responses to some putative problems facing pluralism. However, he tempers his pluralism, and the acknowledgment that our taxonomic system is the product of a highly contingent process, with a concession to monism: because one of the points of biological taxonomy is to facilitate communication between scientists, we ought to view species as the basal unit in one overarching taxonomic hierarchy. Thus dispensing with overlapping taxonomies, this view represents a less radical version of pluralism than Dupré himself has advocated in the past (e.g., Dupré 1993; cf. Kitcher 1984).

Hull is more sceptical about the prospects for pluralistic accounts of species in his essay, "On the Plurality of Species: Questioning the Party Line." After sketching some broad issues that arise more generally with respect to pluralist views, he turns to examine some of the prominent expressions of pluralism by Kitcher (1984), Ereshefsky (1992a), and Stanford (1995). Hull then turns to compare his own (1997) attempts to classify and evaluate the plethora of species concepts with Mayden's attempts (1997). Whereas Hull reached the "grudging conclusion," as he calls it here, in 1997 that no *one* species concept won out within the criteria he proposed, Mayden arrived at a form of monism. Returning to one of the earlier themes of his essay, Hull attributes the difference here in part to the fact that as a practicing scientist, Mayden has to make more definitive theoretical commitments than a philosopher who stands outside the practice of science and surveys options. Stances on the pluralism issue typically reflect social and institutional facts about the advocates of those stances, rather than bias-free views argued out from first principles.

The third essay in this section, de Queiroz's "The General Lineage Concept of Species and the Defining Properties of the Species Category," develops a solution to the species problem that de Queiroz has recently (1998) defended: what he calls the *general lineage* concept of species. This view equates species with segments of population lineages, and de Queiroz argues not only that it underlies "virtually all modern ideas about species," but that it illuminates a wide range of issues about species, including debates about speciation, the individuality thesis, and species realism. de Queiroz also proposes that it allows one to dissolve the debate between monists and pluralists. He continues by tracing the history of the population lineage concept from Darwin through the early part of the Modern Synthesis in the work of Huxley, Mayr, Dobzhansky, and Wright, to the more explicitly lineage-focused concepts of Simpson, Hennig, and Wiley. Although this historical sketch constitutes a minor part of de Queiroz's wide-ranging essay, it serves to buttress his proposals about the ways in which the population lineage concept underlies many apparent disagreements between advocates of different species concepts. The paper concludes with a philosophical diagnosis of why this underlying unity has been largely unrecognized in contemporary debates.

The two papers in the next section, "Species and Life's Complications," look at very different issues that arise for particular species concepts. David Nanney's "When Is a Rose?: The Kinds of *Tetrahymena*," probes Mayr's biological species concept (BSC) and the notions that it employs, such as a closed gene pool, from the perspective of a longtime practicing ciliate biologist. Nanney conveys interesting information about the ciliates (*Tetrahymena* in particular)—such as the relative independence of genetic and morphological subdivisions, and the clonal propagation of these ancient protists (some of which include asexually reproducing populations)—that pose problems for the BSC; he also reveals enough of the history of protozoology to suggest why the field has a strained relationship to the Modern Synthesis and concepts forged during it. One striking conclusion of the essay is that microbiology, having essentially bypassed the Modern Synthesis, awaits a new synthesis that focuses on more than the most recent snapshot of a history of life that stretches back almost four billion years.

Kim Sterelny's essay, "Species as Ecological Mosaics," offers a defense of a form of realism about species committed neither to universalism about any species concept or definition nor to any type of species selection. Some (but not all) species form what Sterelny calls *ecological mosaics*; which are made up of ecologically diverse populations of organisms. As structured and diverse metapopulations, such mosaics are subject to evolutionary change when there is an ecological or geographic fracturing of the metapopulation, but they are also stabilized by what he calls *Mayr's Brake*, the mechanisms of reproductive isolation central to Mayr's well-known account of speciation. Sterelny explores this idea through a discussion of Vrba's and Eldredge's views of evolutionary change in which he argues, amongst other things, that those views should be divorced from their authors' own fondness for Paterson's (1985) recognition concept of species. Sterelny's scepticism about universalism and thus monism draws on the claim that, like organisms, species and the complex, ecological organization they possess were invented at some point in evolutionary time, forming a grade of biological organization that, like organismal individuality, only some clusters of biological entities have.

The essays by Richard Boyd, Paul Griffiths, and Rob Wilson in the next section, "Rethinking Natural Kinds," revive the ontological issue of whether species are natural kinds or individuals by offering a reexamination of the notion of a natural kind. Boyd's paper, "Homeostasis, Species, and Higher Taxa," develops the conception of *homeostatic property cluster* (HPC) kinds that he briefly introduced in earlier work (1988, 1991). Here, Boyd both provides the broader philosophical context in which that conception functions and shows how it applies to several issues concerning species. More specifically, he defends the idea that HPC kinds are an integral part of an overall, realist view of science that accommodates the inexactitude, natural vagueness, and historicity of many sciences, including the biological sciences. He then argues that species and at least some higher taxa are HPC kinds, and indicates how his view makes plausible a form of pluralistic realism. A passing theme in the essay is that in the HPC conception of natural kinds, the contrast between natural kinds and individuals is of less importance than it is in a traditional notion of natural kinds, thus deflating the significance of the individuality thesis about species defended by Hull (1978) and Ghiselin (1974, 1997), and the subsequent debate over it.

My own contribution—"Realism, Essence, and Kind: Resuscitating Species Essentialism?"—takes its cue from Boyd's earlier work on HPC kinds. After outlining how both the individuality thesis about species taxa and pluralism about the species category have been developed because of problems with traditional realism, I use two examples from the taxonomy of neural states to suggest that there is more than merely conceptual space for a view closer to traditional realism than either of these fairly radical proposals. This middle-ground position is a version of the HPC view of natural kinds, and in contrast to Boyd's own development of this view, I argue that this position is incompatible with both the individuality thesis and pluralistic realism. This essay thus steps outside of the philosophy of biology to the philosophy of psychology and neuroscience to shed some light on natural kinds more generally and on realism and pluralism about species in particular.

Griffiths's "Squaring the Circle: Natural Kinds with Historical Essences" looks at the treatment of the notion of natural kinds by a variety of researchers across the biological sciences, including systematists (regarding species taxa) and process structuralists (regarding developmental biology). Griffiths defends the idea that natural kinds can have historical essences, using this idea to address the claim that there are no (or few) laws of nature in the biological sciences. For Griffiths, concepts of taxa and of parts and processes in biology can be based on the idea of an evolutionary rather than a distinctly structural or developmental homology. Griffiths sees phylogenetic inertia and its basis in the developmental structure of organisms as a mechanism for producing what Boyd calls the "causal homeostasis" of natural kinds.

The two papers in the next section, "Species in Mind and Culture," present perspectives on the issues surrounding the psychological and cultural representations of central biological concepts, such as the species concept. In "The Universal Primacy of Generic Species in Folkbiological Taxonomy: Implications for Human Biological, Cultural, and Scientific Evolution", Scott Atran draws on recent cross-cultural experimental research with the Maya in Guatemala and with midwestern urban college students that probes the strength of inductive inferences across various levels of biological categories. Atran has found surprising similarities across these forest-dwelling and urbanized populations that cry out for psychological explanation. He argues for the universality across cultures of what he calls *generic species*, a level of organization in the biological world that doesn't distinguish the Linnaean species and genus categories; he proposes a domain-specific representation of this category and explores its relationship to essence-based habits of the mind and the cultural development of various species concepts in Western science. Atran concludes his paper with some thoughts about recent views of pluralism and species and about what these views imply about the relation between common sense and science.

Frank Keil and Daniel Richardson discuss the psychological representation of species and of biological knowledge more generally in their essay, "Species, Stuff, and Patterns of Causation." They argue that the substantial developmental literature on biological knowledge often presents a misleading conception of what intuitive or folkbiology must be like in order for species and other biological categories to have the distinctive psychological features that they do, suggesting several new lines of empirical research. By exploring what has been called "psychological essentialism" about biological kinds and its relationship to essentialism in the philosophy of biology, Keil and Richardson call for more careful empirical examination of the nature of our mental representation of the biological world and identify a number of cognitive biases that contribute to what they call the "vivid illusion of species." They claim that although species do seem to have a distinctive psychological representation, the specific form that representation takes remains largely an open empirical question.

The concluding section—"Species Begone!"—contains two essays that, in their own ways, express some skepticism about the *special* reality of species that is the focus of biological and philosophical controversy regarding species (as in "the species problem"). Both authors feel that species are as real as higher taxa, but no more than the genuses, families, orders, and so on that those species constitute. Marc Ereshefsky's "Species and the Linnaean Hierarchy" offers a review of our current thinking about the species category, advocating a replacement of the entire Linnaean system of classification. Ereshefsky questions the distinctive reality of the species category by pointing to the problems in drawing the distinction between species and higher taxa and by using the critiques of monistic accounts of species that motivate pluralism to suggest the heterogeneity of the species category. Because the point of the Linnaean hierarchy and the distinctions that it draws (e.g., between species and higher taxa) has been lost through the Darwinian revolution, our current taxonomic practice creates problems that alternative systems of classification may avoid. Ereshefsky concludes by examining two such systems, though he acknowledges that any change should not be made lightly.

In "Getting Rid of Species?" Brent Mishler explores the application of phylogenetics to species taxa. Like Ereshefsky, Mishler views the Linnaean hierarchy as outdated, and like de Queiroz (1992; cf. de Queiroz, chapter 3 in this volume), he thinks that phylogenetic schemes of classification are necessary. Mishler argues that taxa at all levels, including the least inclusive, should be recognized because of evidence for monophyly. He believes that the failure of the various species concepts to uniquely define the species rank in the phylogenetic hierarchy reflects reality, thus highlighting the need to get rid of the species rank altogether. Thus, a rank-free phylogenetic taxonomy should be applied consistently to all taxa, including the least inclusive. Mishler concludes by reflecting on the implications of his proposed reform on our ecological thinking about biodiversity and conservation.

## REFERENCES


Boyd, R. (1988). How to be a moral realist. In G. Sayre-McCord, ed., *Essays on moral realism*. Ithaca, N.Y.: Cornell University Press.

Boyd, R. (1991). Realism, anti-foundationalism, and the enthusiasm for natural kinds. *Philosophical Studies* 61, 127–148.

Claridge, M., H. Dawah, and M. Wilson (1997). *The units of biodiversity*. London: Chapman and Hall.

de Queiroz, K. (1992). Phylogenetic definitions and taxonomic philosophy. *Biology and Philosophy* 7, 295–313.

de Queiroz, K. (1998). The general lineage concept of species, species criteria, and the process of speciation: A conceptual unification and terminological recommendations." In Howard and Berlocher.

Dupré, J. (1993). *The disorder of things: Philosophical foundations for the disunity of science*. Cambridge, Mass.: Harvard University Press.

Ereshefsky, M. (1992a). Eliminative pluralism. *Philosophy of Science*, 59, 671–690.

Ereshefsky, M., ed. (1992b). *The units of evolution: Essays on the nature of species*. Cambridge, Mass.: MIT Press.

Ghiselin, M. (1974). A radical solution to the species problem. *Systematic Zoology* 23, 536–544. Reprinted in Ereshefsky 1992b.

Ghiselin, M. (1997). *Metaphysics and the origins of species*. Albany, N.Y.: SUNY Press.

Howard, D. J., and S. H. Berlocher, eds. (1998). *Endless forms: Species and speciation*. Oxford: Oxford University Press.

Hull, D. (1976). Are species really individuals? *Systematic Zoology* 25, 174–191.

Hull, D. (1978). A matter of individuality. *Philosophy of Science* 45, 335–360. Reprinted in Ereshefsky 1992b.

Hull, D. (1997). The ideal species definition and why we can't get it. In Claridge, Dawah, and Wilson.

Kitcher, P. (1984). Species. *Philosophy of Science* 51, 308–333. Reprinted in Ereshefsky 1992b.

Lambert, D., and H. Spence, eds. (1995). *Speciation and the recognition concept*. Baltimore: Johns Hopkins University Press.

Mayden, J. (1997). A hierarchy of species concepts: The denouement in the saga of the species problem. In Claridge, Dawah, and Wilson.

Otte, D., and J. Endler, ed. (1989). *Speciation and its consequences*. Sunderland, Mass.: Sinauer.

Paterson, H. (1985). The recognition concept of species. In E. Vrba (ed.) Species and Speciation. Pretoria: Transvial Museum Monograph No. 4. Reprinted in Ereshefsky 1992b and Paterson 1994.

Paterson, H. (1994). *Evolution and the recognition concept of species: Collected writings*. Baltimore: Johns Hopkins University Press.

Stanford, K. (1995). For pluralism and against monism about species. *Philosophy of Science* 62, 70–91.

Wheeler, Q., and R. Meier, eds. (1999). *Species concepts and phylogenetic theory: A debate*. New York: Columbia University Press.

# Contributors

**Scott Atran** holds appointments at the Centre National de la Recherche Scientifique (CNRS-CREA) in Paris and the University of Michigan, Ann Arbor. He is the author of *Cognitive Foundations of Natural History* (Cambridge University Press, 1990) and the editor, with Douglas Medin, of *Folkbiology* (MIT Press, 1999).

**Richard Boyd** is a professor of philosophy at Cornell University. He has authored papers on scientific and moral realism, and is the editor, with Philip Gasper and J. D. Trout, of *The Philosophy of Science* (MIT Press, 1991).

**Kevin de Queiroz** is an associate curator and zoologist in the Department of Vertebrate Zoology at the National Museum of Natural History of the Smithsonian Institution. He was previously a Tilton Postdoctoral Fellow at the California Academy of Sciences. He has published both empirical papers on the evolution and systematics of vertebrates and theoretical papers on the methods and philosophy of systematic biology, including several previous papers on species concepts. He is currently studying the phylogeny of *Holbrookia* and *Anolis* lizards and is developing a phylogenetic approach to biological nomenclature.

**John Dupré** holds appointments at Birkbeck College, London, and the University of Exeter, having taught previously at Stanford University. In 1993, he published *The Disorder of Things: Metaphysical Foundations of the Disunity of Science* (Harvard University Press).

**Marc Ereshefsky** is an associate professor in the Department of Philosophy at the University of Calgary. He has published a number of papers on species and is the editor of *The Units of Evolution: Essays on the Nature of Species* (MIT Press, 1992).

**Paul E. Griffiths** is senior lecturer in the Unit for History and Philosophy of Science, University of Sydney, Australia. He previously taught at Otago University in New Zealand. His publications include *What Emotions Really Are: The Problem of Psychological Categories* (University of Chicago Press,

1997) and, with Kim Sterelny, *Sex and Death: An Introduction to the Philosophy of Biology* (University of Chicago Press, 1999).

**David L. Hull** is the Dressler Professor in the Humanities in Weinberg College at Northwestern University, Evanston, Illinois. He is the author of *Darwin and His Critics* (Harvard University Press, 1973; reprinted by the University of Chicago Press, 1983), *Philosophy of Biological Science* (Prentice-Hall, 1974), *Science as a Process* (University of Chicago Press, 1988), *The Metaphysics of Evolution* (State University of New York Press, 1989), and *Science and Selection* (Cambridge University Press, 1998). He is editor of the series *Science and Its Conceptual Foundations* at University of Chicago Press.

**Frank C. Keil** is a professor in the Department of Psychology at Yale University. He taught for twenty years at Cornell University, is the author of *Semantic and Conceptual Development: An Ontological Perspective* (Harvard University Press, 1979) and *Concepts, Kinds, and Cognitive Development* (MIT Press, 1989), and is the coeditor, with Robert Wilson, of *The MIT Encyclopedia of the Cognitive Sciences* (MIT Press, 1999).

**Brent D. Mishler** is a professor in the Department of Integrative Biology and Director of the University and Jepson Herbaria at the University of California, Berkeley. He received his Ph.D. from Harvard University in 1984 and was in the faculty at Duke University until 1993, when he moved to UC Berkeley. He is a systematist specializing in mosses and has thus grappled with species both in practice and in theory. He has published extensively on species concepts as well as on phylogenetics at many scales, from moss species to the overall relationships of the green plants.

**David L. Nanney** became a professor in the School of Life Science at the University of Illinois in Urbana-Champaign in 1959. He became an emeritus professor in the Department of Ecology, Ethology, and Evolution in 1991. He received his graduate training under Tracy Sonneborn at Indiana University (Ph.D., 1951) and taught at the University of Michigan before moving to Illinois. His studies have focused on the domestication of the ciliated protozoan *Tetrahymena thermophila* as a laboratory instrument and have included genetic, developmental, and evolutionary mechanisms.

**Daniel C. Richardson** is a graduate student in the Department of Psychology at Cornell University.

**Kim Sterelny** is an Australian now living in exile in New Zealand. Starting as a philosopher of language and mind, he has become increasingly interested in the philosophy of biology. He retains his original interests, but now with a strongly evolutionary spin. He is coauthor, with Paul Griffiths, of *Sex and Death: An Introduction to the Philosophy of Biology*, due to appear with the University of Chicago Press in 1999.

**Robert A. Wilson** is an associate professor in the Department of Philosophy and a member of the Cognitive Science Group at the Beckman Institute at the University of Illinois, Urbana-Champaign. He taught previously at Queen's University, Canada. He is the author of *Cartesian Psychology and Physical Minds: Individualism and the Sciences of the Mind* (Cambridge University Press, 1995), and is the general editor, with Frank Keil, of *The MIT Encyclopedia of the Cognitive Sciences* (MIT Press, 1999).

# I Monism, Pluralism, Unity, and Diversity

# 1 On the Impossibility of a Monistic Account of Species

John Dupré

*[i]f we can once and for all lay the bogey of the existence of true relationship and realize that there are, not one, but many kinds of relationship—genealogical relationship, morphological relationship, cytological relationship, and so on—we shall release ourselves from the bondage of the absolute in taxonomy and gain enormously in flexibility and adaptability in taxonomic practice.*

—J. S. L. Gilmour, "The Development of Taxonomic Theory Since 1851"

*By the classification of any series of objects, is meant the actual, or ideal, arrangement of those which are like and the separation of those which are unlike; the purpose of this arrangement being to facilitate the operations of the mind in clearly conceiving and retaining in the memory, the characters of the objects in question. Thus there may be as many classifications of any series of natural, or of other, bodies, as they have properties or relations to one another, or to other things; or, again, as there are modes in which they may be regarded by the mind.*

—T. H. Huxley, *Introduction to the Classification of Animals*

Most of the philosophical difficulties that surround the concept of species can be traced to a failure to assimilate fully the Darwinian revolution. It is widely recognized that Darwin's theory of evolution rendered untenable the classical essentialist conception of species. Perfectly sharp discontinuities between unchanging natural kinds could no longer be expected. The conception of sorting organisms into species as a fundamentally classificatory exercise has nevertheless survived. Indeed, the concept of a species traditionally has been the paradigmatic unit of classification. Classification is centrally concerned with imposing conceptual order on diverse phenomena. Darwin's theory, as the title of his most famous work indicates, is about the origins of diversity, though, so it is no surprise that the dominant task in post-Darwinian taxonomy has been to connect classificatory systems to the received, Darwinian, account of the origin of diversity. Attractive though this task undoubtedly is, it has proved unsuccessful. The patterns of diversity that evolution has produced have turned out to be enormously diverse, and in many cases the units of evolutionary analysis have proved quite unsuitable for the basic classificatory aims of taxonomy. Or so I argue.

Why do we classify organisms? A natural and ancient explanation—expressed clearly by, for example, Locke (1689, bk. 3, chap. 5, sec. 9) and

Mill (1862)—is that we do so to facilitate the recording and communication of information. If I tell you some animal is a fox, I immediately convey a body of information about its physiology, habits, and so on. The more you know about animals or mammals or foxes, the more information about that particular animal I convey. If organisms came in sharply distinguished natural kinds, internally homogeneous and reliably distinguishable from the members of any other kind, then the identification of such kinds would be the unequivocal aim of taxonomy. A classificatory system that recognized such natural kinds would be unequivocally the best suited to the organization and dissemination of biological information. But this is just what Darwin has shown us we cannot expect (see e.g., Hull 1965). In a domain of entities characterized, in part, by continuous gradation of properties and varyingly sharp and frequent discontinuities, matters are much less clear. It is this fact about the biological world that makes attractive the idea of taxonomic pluralism—the thesis that there is no uniquely correct or natural way of classifying organisms and that a variety of classificatory schemes will be best suited to the various theoretical and practical purposes of biology.

Many biologists and philosophers appear to think that pluralistic accounts of species will lead us to Babel (see e.g., Ghiselin 1997, 117–121). Biologists, they suppose, will be unable to communicate with one another if they are working with different species concepts. In this paper, I argue that species pluralism is nevertheless unavoidable. However, I also defend a kind of minimal monism: to serve the traditional epistemic goals of classification, it is desirable to have one general set of classificatory concepts. However, this general taxonomy will need to be pragmatic and pluralistic in its theoretical bases. For specialized biological purposes, such as the mapping of evolutionary history, it may often be necessary to adopt specialized classificatory systems. My monism is merely semantic: I suggest it would be best to reserve the term *species*—which is, as I have noted, the traditional philosophical term for classificatory concepts—for the base-level categories of this general, pragmatic, taxonomy. Such an antitheoretical concept of species will discourage the conspicuously unsuccessful and controversial efforts to find a solution to the "species problem," and leave it to working biologists to determine the extent to which they require specialized classificatory schemes for their particular theoretical projects.

Monists, needless to say, disagree about which actual species concept biologists should accept. The cheapest way to buy monism might be with a radically nominalistic phenetic concept, as conceived by numerical taxonomists (Sneath and Sokal 1973). If biological classification could be conceived as merely an exercise in recording degrees of objective similarity, then some particular degree of similarity could be defined as appropriate to the species category. But few people now think this can be done. Philosophically, attempts to construe a notion of objective similarity founder on the fact that indefinitely many aspects of difference and of similarity can be discovered between any two objects. Some account of what makes a property biologically interesting is indispensable: there can be no classification wholly innocent of theoretical contamination. Without wishing to deny that phenetic approaches to classification have provided both theoretical insights and practical benefits, I restrict my attention in this essay to more theoretically laden routes to species monism. My conclusions, however, leave entirely open the possibility that a version of pheneticism, modified by some account of what kinds of properties might be most theoretically interesting, may be appropriate for important domains of biology. The classification of bacteria is a likely example (see e.g., Floodgate 1962 and further discussion below).

In the section "Troubles with Monism," I trace some of the difficulties that have emerged in attempting to provide monistic accounts of taxonomy motivated by central theses about the evolutionary origins of diversity. I thereby hope to substantiate my claim that as more has been learned about the diversity of the evolutionary process, the hopes of grounding therein a uniform account of taxonomy in general, or even the species category in particular, have receded. In the final section, I outline my more constructive proposal for responding to this situation.

## TROUBLES WITH MONISM

The potential conflict between two main goals of classification has long been recognized. The first and most traditional goal is to facilitate the communication of information or to organize the vast quantities of detailed biological information. From this point of view, a taxonomy should be constructed so that knowing the taxon to which an organism belongs should tell us as much as possible about the properties of that organism. This goal must, of course, be qualified by pragmatic considerations. Indefinite subdivision of classifications can provide, theoretically, ever more detailed information about the individuals classified: assignment to a subspecies or a geographical race will presumably give more information than mere assignment to a species. As the basal taxonomic unit, the species should be defined, therefore, to classify organisms at a level at which the gains from finer classification would be outweighed by the costs of learning or transmitting a more complicated set of categories. If organisms varied continuously with no sharp discontinuities, this balancing of costs and benefits would present a largely indeterminate problem. By happy chance for many kinds of organisms there appear to be sharp discontinuities at a relatively fine classificatory level that are much sharper than any discontinuities at any lower level. To the extent that this is the case, the selection of the appropriate level for assignment of organisms to species appears unproblematic.

In recent years, this goal of organizing biological information has been emphasized much less than a second, that of mapping the currents of the evolutionary process. A recent anthology of biological and philosophical essays on the nature of species carried the title *The Units of Evolution* and the subtitle *Essays on the Nature of Species* (Ereshefsky 1992). Though the idea that, by definition, species should be the units of evolution is not uncontroversial, it is widely held. What is a unit of evolution? Evolutionary change is not change in the properties of any individual organism, but change over time in the distribution of properties within some set of organisms. (We need not worry here whether these properties are conceived as genetic or phenotypic.) A unit of evolution is the set of organisms in which changes in the distribution of properties constitute a coherent evolutionary process.

Because an evolutionary change is one with the potential to be maintained in future organisms, it is easy to see that the temporal dimension of a unit of evolution must be defined by relations of ancestry. As long as we are concerned with biological evolution in which properties are transmitted genetically (and ignore some complexities of gene exchange in bacteria), then evolution will be constrained within sets of organisms defined temporally by parent-offspring relations. We must then consider what determines the synchronic extent of a unit of evolution. A natural and attractive idea is that a species should include all and only those organisms with actual or potential reproductive links to one another. This condition would determine the set of organisms among whose descendants a genetic change in any member of the set might possibly be transmitted. To the extent that the biological world is characterized by impenetrable barriers to genetic exchange, then there will be distinct channels down which evolutionary changes can flow. The sets of organisms flowing down these channels, then, will be the units of evolution.

Here, of course, is the great appeal of the so-called *biological species concept* (BSC)—until recently the dominant conception of the nature of species. According to this view, a species is conceived as a group of organisms with actual or potential reproductive links to one another and reproductively isolated from all other organisms. Recalling for a moment my brief discussion of classification as mere ordering of information, one might also suppose that the sharp discontinuities that (sometimes) determine the optimal level for making base-level discriminations should correspond to lines of reproductive isolation. The flow of evolutionary change down reproductively isolated channels, after all, should be expected to lead to ever-growing morphological distinctness. Thus, the goals of representing the evolutionary process and of optimally ordering biological phenomena would turn out to coincide after all.

Unfortunately, however, the biological world proves much messier than this picture reveals. Certainly, there are cases in which species can be identified with discoverable lowest-level sharp discontinuities marked by reproductive barriers. But such cases are far from universal, and the appealing picture drawn thus far has a range of important complications to which I now turn.

### Asexual Species

A familiar objection to the BSC is that it has nothing to say about asexual species. A fully asexual organism is reproductively isolated from everything except its direct ancestors and descendants. The leading proponent of the BSC, Ernst Mayr, has concluded that there are, strictly speaking, no species of asexual organisms (Mayr 1987). But asexual species still require classification, and indeed some asexual species are more sharply distinguishable from related species than are some sexual species. Moreover, asexual organisms evolved just as surely as did sexual species. Thus, whichever view we take of the fundamental goal of assigning organisms to species, the exclusion of asexual organisms should lead us to see the BSC as at best one species concept among two or several concepts necessary for encompassing biological reality. A more radical attempt to save the BSC is suggested by David Hull (1989): in asexual organisms, the species are simply organism lineages—that is, an organism and its descendants (p. 107).[1] I take it that although Hull's proposal is attractive theoretically, it will divorce the identification of species in these cases from any practical utility in classification. It should also be noted that even this radical move may not work to give the biological species concept universal applicability. In bacteria, although reproduction is asexual, various mechanisms are known by which bacteria exchange genetic material. The pattern of relationships between bacteria is thus netlike, or reticulated, rather than treelike.[2] Although I suppose that one might hope to identify a new species as originating at each node in the net, such an identification would imply the existence of countless species, many lasting only a few minutes or even seconds. The impracticality of this idea suggests that we would be better abandoning the idea of applying the BSC, or indeed any evolutionarily based species concept, to bacteria. Many bacterial taxonomists (see Nanney, this volume) indeed seem to have this inclination.[3]

## Gene Flow beyond Sharp Discontinuities

A second familiar difficulty with the biological species concept is that apparently well-distinguished species frequently do, in fact, exchange genetic material. The classic illustration is American oaks (see Van Valen 1976). Various species of oaks appear to have coexisted in the same areas for millions of years while exchanging significant amounts of genetic material through hybridization. Ghiselin (1987) is quite happy to conclude that these oaks form a large and highly diversified species. Two responses should be offered to this conclusion. First, and most obviously, the need to make such a move illustrates the divergence between this kind of theoretically driven taxonomy and the pragmatic goal of providing a maximally informative ordering of nature. This divergence may not much bother the theoretically inclined, but it does illustrate one of the ways in which we cannot both have our cake and eat it in the way indicated in the most optimistic explication of the BSC.[4] Second, such examples throw serious doubt on the central motivation for the BSC, which is that genetic isolation is a necessary condition for a group of organisms to form a coherent unit of evolution. The example shows that different species of oaks have remained coherent and distinct vehicles of evolutionary change and continuity for long periods of time. Ghiselin's conclusion looks like nothing more than an epicycle serving solely to protect the BSC from its empirical inadequacy.

### The Absence of Sharp Discontinuities

In some groups of plants and of microorganisms, and very probably in other kinds of organisms, there is considerable variation, but no apparent sharp discontinuities. It is even tempting to suggest that within certain plant genera there are no species. A good example would be the genus *Rubus*, blackberries and their relatives. Because *Rubus* lacks sharp differentiation between types, but admits great variation within the genus as a whole, it seems unlikely that there could be any consensus on its subdivision into species.[5] If we assume that this lack of sharp differentiations is due, in part, to gene flow, the option is again open to call *Rubus* a single and highly polymorphic species. Though less objectionable than in the case where there are sharply distinguished types, as with oaks, this move again separates theory-driven taxonomy from the business of imposing useful order on biological diversity.

### Lack of Gene Flow within Sharply Differentiated Species

A somewhat less familiar point is that a considerable amount of research has shown that often there is surprisingly little genetic flow within well-differentiated species (Ehrlich and Raven 1969), most obviously in the case of species that consist of numbers of geographically isolated populations, but that nevertheless show little or no sign of evolutionary divergence. Even within geographically continuous populations, however, it appears that genetic interchange is often extremely local. This kind of situation puts great weight on the idea of *potential* genetic flow in defending the BSC. If populations are separated by a distance well beyond the physical powers of an organism to traverse, should their case nevertheless be considered one of potential reproduction, on the grounds that if, *per impossibile*, the organisms were to find one another, they would be interfertile? The alternative, paralleling Ghiselin's line on oaks, would be to insist that such apparent species consisted of numbers of sibling species, differentiated solely by their spatial separation. Again, one is led to wonder what the point of either maneuver would be. Clearly, to the extent that species retain their integrity despite the absence of genetic exchange, it must be concluded that something other than gene interchange explains the coherence of the species. Contenders for this role in cases like either of the kinds just considered include the influence of a common selective regime and phyletic or developmental inertia. I might finally note that although I do not know whether any systematic attempt has been made to estimate the extent of gene flow in the genus *Rubus*, in the likely event that the flow is quite spatially limited, the claim that the whole complex group with its virtually worldwide distribution can be seen as reproductively connected is tenuous to say the least.

The conclusion I want to draw at this point is that the BSC will frequently lead us to distinguish species in ways quite far removed from traditional Linnaean classification and far removed from the optimal organization of taxonomic information. Moreover, the theoretical motivation for the BSC seems seriously deficient. The sorts of criticisms I have been enumerating above have led, however, to a decline in the extent to which the BSC is now accepted, and this decline has been accompanied by increasing interest in a rather different approach to evolutionarily centered taxonomy that can be broadly classified under the heading of the *phylogenetic species concept* (PSC). (The definite article preceding the term should not be taken too seriously here, as there are several versions of the general idea.)

The central idea of all versions of the PSC is that species—and, in fact, higher taxa as well—should be monophyletic. That is, all the members of a species or of a higher taxon should be descended from a common set of ancestors. An appropriate set of ancestors is one that constitutes a new branch of the phylogenetic tree. Such a group is known as a *stem species.* The important distinction between versions of PSC is whether a taxon is merely required to contain *only* descendants of a particular stem species or to contain *all* and only such descendants. The latter position is definitive of cladism, whereas the former, generally described as evolutionary taxonomy, requires some further criterion for deciding which are acceptable subsets of descendants.[6] Two issues arise in explicating a more detailed account of the PSC. First, what constitutes the division of a lineage into two distinct lineages and hence qualifies a group as a stem species? Second, what constitutes a lineage and its descendants as a species (or, indeed, as any other taxonomic rank)?

The traditional answer to the first question is that a lineage has divided when two components of it are reproductively isolated from one another, but the difficulties raised in connection with the BSC suggest that this answer is inadequate. Examples such as oaks suggest that reproductive isolation is not necessary for the division of a lineage, and worries about the lack of gene flow within apparently well-defined species suggest that it is not sufficient either. An illuminating diagnosis of the difficulty here is provided by Templeton (1989), who distinguishes *genetic* exchangeability, the familiar ability to exchange genetic material between organisms, and *demographic* exchangeability, which exists between two organisms to the extent that they share the same fundamental niche (p. 170). The problem with asexual taxa and with a variety of taxa for which gene exchange is limited is that the boundaries defined by demographic exchangeability are broader than those defined by genetic exchangeability. Conversely, for cases in which well-defined species persist despite gene exchange, the boundaries defined by genetic exchangeability are broader than those defined by demographic exchangeability (p. 178).

In the light of these considerations, Templeton proposes the *cohesion species concept* (CSC). It is not entirely clear how this concept should be interpreted. In the conclusion of his paper, he writes that species should be defined as "the most inclusive group of organisms having the potential for genetic and/or demographic exchangeability" (p. 181). If we assume that the connective "and/or" should be interpreted as inclusive disjunction, this definition would suggest that the "syngameon" of oaks—that is, the set of distinct but hybridizing species—should be treated as a species. But it is clear from earlier discussion that such an application is not what Templeton intends. Earlier, he defines the CSC as "the most inclusive population of individuals having the potential for phenotypic cohesion through intrinsic cohesion mechanisms" (p. 168). A central and convincing motivation for this definition is the claim that a range of such mechanisms promotes phenotypic cohesion, of which genetic exchange and genetic isolation are only two. Equally important are genetic drift (cohesion through common descent), natural selection, and various ecological, developmental, and historical constraints. The basic task, according to Templeton, is to "identify those mechanisms that help maintain a group as an evolutionary lineage" (p. 169).

What, then, is an evolutionary lineage? The significance of the conflicting criteria of genetic and demographic exchangeability is that they show it to be impossible to define that lineage in terms of any unitary theoretical criterion. Rather, lineages must first be identified as cohesive groups through which evolutionary changes flow, and only then can we ask what mechanisms promote this cohesion, and to what extent the identified groups exhibit genetic or demographic exchangeability. Presumably, this initial identification of lineages must be implemented by investigation of patterns of phenotypic innovation and descent over time. With the abandonment of any general account of speciation or any unitary account of the coherence of the species, it appears that species will be no more than whatever groups can be clearly distinguished from related or similar groups. This approach may seem theoretically unsatisfying, but to the extent that it reflects the fact that there are a variety of mechanisms of speciation and a variety of mechanisms whereby the coherence of the species is maintained, it would also seem to be the best concept we can hope for.

This conclusion makes pressing the second question distinguished above: How do we assign taxonomic rank, especially species rank, to a particular lineage or set of lineages? A prima facie advantage of the BSC is that it provides a clear solution to this problem: a species is the smallest group of individuals reproductively connected (or at least potentially connected) one to another and reproductively isolated from all other individuals. The difficulty is that this definition would leave one with species ranging from huge and diverse syngameons to clonal strains with a handful of individuals. Apart from the theoretical difficulties discussed above, any connection between the theoretical account of a species and a practically useful classification would surely be severed.

The question that must be faced, then, is whether from the PSC point of view the idea that the species is the basal taxonomic unit—where taxonomy is conceived as providing a practically useful classification—can be maintained. Abandoning the BSC will take care of species that look unsatisfactorily large by allowing a variety of cohesion mechanisms apart from reproductive isolation, but it will tend to imply the presence of disturbingly small species. Frequently there are clearly distinguishable groups of organisms—subspecies, varieties, geographical races—below the species level. There is no reason to suppose that these groups are not monophyletic and no reason to suppose that they are not, at least for the moment, evolving independently. There is no doubt that such groups are often clearly distinguishable, and indeed for many purposes classification at this level is the most important. Stebbins (1987, 198) notes, for instance, that foresters are often more concerned with geographic races than species and indeed can be hampered in their work by the confusing attachment of the same specific name to trees with quite distinct ecological properties and requirements. A judge at a dog show is not much concerned with the criteria that identify something as *Canis familiaris*. Such groups may go extinct, they may merge with other subgroups in the species, or they may be destined to evolve independently into full-blown species or higher taxa. Their evolutionary significance is thus unknown and unknowable. The same, of course, could be said of groups recognized as full species, though the second alternative (merging with other groups) may be rare.

## THE CASE FOR PLURALISM

An evolutionarily based taxonomy appears to be faced at this point with only two possible options. The first is to consider species as by definition the smallest units of evolution. Leaving aside the insurmountable difficulty of detecting such units in many cases, my argument so far has been that this option will provide a fundamental classification that is often much too fine to be useful for many of the purposes for which taxonomies have traditionally been used.[7] Mishler and Donoghue (1982) suggest that this proposal is also conceptually confused. They argue that "there are many evolutionary, genealogical units within a given lineage ... which may be temporally and spatially overlapping" (1982, 498). They suggest, therefore, that it is an error to suppose that there is any such thing as a unique basal evolutionary unit and that the particular evolutionary unit one needs to distinguish will depend on the kind of enquiry with which one is engaged. If there is no unique basal unit, then there is no privileged unit and, from an evolutionary point of view, no theoretical reason to pick out any particular group as the species. Mishler and Donoghue therefore propose the second option, to "[a]pply species names at about the same level as we have in the past, and decouple the basal taxonomic unit from notions of 'basic' evolutionary units" (p. 497). This process involves seeing species on a par with genera and higher taxa—that is, as ultimately arbitrary levels of organization, chosen on a variety of pragmatic grounds.[8]

Although Mishler and Donoghue see the species as an ultimately arbitrary ranking criterion, they do maintain a version of the PSC and, hence, do not see it as arbitrary from the point of view of grouping. In fact, they endorse the strong, cladistic concept of monophyly as a condition on a group constituting a species (or, for that matter, a taxon at any other level). Their "pluralism," however, entails that "comparative biologists must not make inferences from a species name without consulting the systematic literature to see what patterns of variation the name purports to represent" (p. 500). But given this degree of pluralism, and the rejection of the attempt to equate the basal taxonomic unit with any purportedly fundamental evolutionary unit, one may reasonably wonder why it is desirable to insist nevertheless on the requirement of monophyly. I suspect that part of the motivation for this requirement is the idea that there must be *some* answer to the question what a species *really* is. It was once, no doubt, reasonable to suppose that evolution had produced real, discrete species at approximately the classificatory level of the familiar Linnaean species. Perhaps this supposition was an almost inevitable consequence of the transition from an essentialist, creationist view of nature to an evolutionary view. Acceptance of evolutionary theory would require that it more or less serve to explain biological phenomena as theretofore understood. Nevertheless, a further century of development of the evolutionary perspective has given us a radically different picture of biological diversity. The sharpness of differentiation between kinds and the processes by which such differentiation is produced and maintained have proved to be highly diverse. There is no reason to suppose that evolution has provided any objectively discoverable and uniquely privileged classification of the biological world.

Why, then, should we continue to insist that evolution should provide a necessary condition, namely monophyly, on any adequate biological taxon? I can think of only three possible answers. First, it might be held that a better understanding of evolution is so overwhelmingly the most important biological task that any taxonomy should be directed at improving this understanding. Second, it might be thought that an evolutionarily based taxonomy, despite its problems, would provide the best available taxonomy, or at least a perfectly adequate taxonomy, for any biological project even far removed from evolutionary concerns. Or third—and this, I suspect, is the most influential motivation—it may be held on general methodological grounds that a central concept such as the species must be provided with a unitary definition. This third motivation might be grounded either in a general commitment to unification as a scientific desideratum or on the fear that failure to provide a unified account of the species category will lead to massive confusion as biologists attempt to communicate with one another. I argue, however, that none of these proposed justifications of the demand for monophyly stand up to much critical scrutiny.

The first answer can be quickly dismissed. Even as distinguished an evolutionist as Ernst Mayr (1961) has emphasized the distinction between *evolutionary* and *functional* biology, the former being concerned with questions about ultimate causation (how did a trait come to exist?), the latter with questions of proximate causation (how does the trait develop or function in particular individuals?). Following Kitcher (1984), I prefer to distinguish these types of questions as historical and structural. It is clear that questions about the ontogeny of the human eye, say, or about the processes by which it provides the individual with information about the environment, have little to do with questions about how humans came to have the kinds of eyes they have. Of course, just noting this fact doesn't show that we need a taxonomy based specifically on structural aspects of organisms, but it does remind us that there is more to biology than evolution. A particularly salient domain, about which I say a bit more below, is ecology.

We should turn, then, to the second, and more promising, line of thought. The fact that a great variety of kinds of investigation takes place within biology certainly does not show that one scheme of classification, based on phylogenetic methods, might not be adequate to all these purposes. To some degree, it should be acknowledged that this question is purely empirical: only the progress of biological enquiry can determine whether different overlapping schemes of classification may be needed. This point needs to be stated carefully. There is no doubt at all that interesting structural or physiological properties crosscut any possible phylogenetically based classification. An investigation into the mechanics of flight, for instance, will have relevance to and may appeal to a group of organisms that includes most (but not all) birds, bats, and a large and miscellaneous set of insects. In general, convergent evolution and the acquisition or loss of traits within any sizeable monophyletic group make it clear that no perfect coincidence between monophyletic groupings and the extension of physiologically interesting traits can be anticipated. Whether this calls for a distinct, nonphylogenetic system of classification is less clear. To pursue the example given, there is no particular reason why the student of flight should attach any particular significance to the miscellaneous group of organisms that fly.

Ecology, on the other hand, raises more difficult issues. Ecology, it may be said, is the microstructure of evolution. Nevertheless, it is not obvious that evolutionarily based taxa will be ideal or even well suited to ecological investigations. Certainly, there are categories—predator, parasite, or even flying predator—that are of central importance to ecological theory and that include phyletically very diverse organisms. There is no reason why phyletically diverse sets of organisms might not be homogeneous (for example as fully substitutable prey) from the perspective of an ecological model. On the other hand, such concepts may reasonably be treated as applying to a higher level of generality than the classification of particular organisms. At a more applied level of ecology, however, some kind of taxonomic scheme must be applied to the particular organisms in a particular ecosystem. Ecology will often be concerned with the trajectory of a population without addressing competition between different subgroups within that population. It may, that is, abstract from distinctions within a population, perhaps corresponding to distinct lineages, which could be fundamental in understanding the longer-term evolutionary trajectory of the population. Groups of sibling species may prove ecologically equivalent (or demographically exchangeable) and thus provide another example of a kind of distinction that may be phylogenetically significant, but ecologically irrelevant. On the other hand it is possible that behavioral distinctions within a phyletic taxon, perpetuated by lineages of cultural descent, might provide essential distinctions from an ecological perspective. It is at least a theoretical possibility that a group of organisms might require radically diverse classification from phyletic and ecological perspectives. Perhaps a population of rats, consisting of several related species, divide into scavengers, insectivores, herbivores, and so on in ways that do not map neatly onto the division between evolutionary lineages. Ecology may therefore, in principle at least, require either coarser or finer classifications than evolution, and it may need to appeal to classifications that crosscut phyletic taxa.[9]

This distinction leads me to the third objection to pluralism, the metatheoretical desirability of a monistic taxonomy. Here, it is relevant to distinguish two possible aspects of pluralism. One might be a taxonomic pluralist because one believes that different groups of organisms require different principles of classification, or one might be a pluralist because one thinks that the same group of organisms require classification in different ways for different purposes. Monistic objections to the first kind of pluralism seem to me to have no merit. Taking the extreme case of bacterial taxonomy, there seem to be very good reasons for doubting the possibility of a phylogenetic taxonomy. The various mechanisms of genetic transfer that occur between bacteria suggest that their phylogenetic tree should be highly reticulated, and standard concepts of monophyly have little application to such a situation. The significance of bacteria as pathogens, symbionts, or vital elements of ecosystems make the goals of classification quite clear in many cases regardless of these problems with tracing phylogenies. Of course, it is possible that new insights into bacterial evolution might nevertheless make a phylogenetic taxonomy feasible. But no vast theoretical problem would be created if bacterial taxonomy appealed to different principles from those appropriate, say, to ornithology.[10] In this sense, the assumption that there is *any* unitary answer to "the species problem" is no more than an optimistic hope. The suggestion that the use of different taxonomic principles might lead to serious confusion is absurd. It is of course possible that an ornithologist might mistakenly suppose that a bacterial species name referred to a monophyletic group of organisms, just as it is possible that a nuclear physicist might suppose that the moon was a planet. Not every possible misunderstanding can be forestalled.

The danger of confusion is a more plausible concern regarding the idea that the same organisms might be subject to different principles of classification for different biological purposes. In one sense, I am happy to agree that this type of confusion should be avoided. It would be undesirable for a particular species name, say *Mus musculus*, to be variously defined and to have varying extensions according to the taxonomic theory espoused by various authors. We should aim to agree as far as possible which organisms are house mice. In the concluding section of this paper, I explain how I think such species names should be understood. If, to recall my hypothetical example about rats, it proves useful to treat scavenging rats as a basic kind in some ecological model, it would be misguided to insist that scavenging rats constitute a species. Equally clearly, however, this concession to standardized terminology does not at all require that all species names be conceived as answering to the same criterion of what it is to be a species. The other consequence of insisting on an unambiguous interpretation of particular species names is that we cannot assume a priori that the canonical taxonomy incorporating standard species names will be suitable for all biological purposes. The question here is, again, an empirical one that depends ultimately on how orderly biological nature turns out to be. If it should prove to be disorderly in the relevant sense, then biology would prove to be a more complicated discipline than is sometimes assumed. But once again I cannot see that any unavoidable confusion need be introduced.

## CONCLUSION: A CASE FOR TAXONOMIC CONSERVATISM

Many taxonomists and almost everyone who uses the results of taxonomic work have complained about the genuine confusion caused by changes in taxonomic nomenclature. Some of these changes seem entirely gratuitous—for example, changes in the names of taxa grounded in the unearthing of obscure prior namings and in appeals to sometimes esoteric rules of priority. Other changes are more theoretically based adjustments of the extent of particular taxa. Many such theoretically motivated changes have been alluded to in this paper. BSC-committed theorists will urge that discoveries of substantial gene-flow between otherwise apparently good species should lead us to apply one species name to what were formerly considered several species. Phylogenetic taxonomists will want to amend the extensions of any higher taxa that fail their favored tests for monophyly, and strict cladists will promote the breaking up of prior "species" into various smaller units when their favored criteria for lineage splitting demand it.[11] Less theoretically committed taxonomists may promote the splitting or lumping of higher taxa on the basis of general principles about the degree of diversity appropriate to a particular rank.

There is no doubt that the taxonomic system we now possess is a highly contingent product of various historical processes. Walters (1961) gives a fascinating account of how the size of angiosperm families and genera can very largely be explained in terms of earlier biological lore available to Linnaeus. Considering the data collected by Willis (1949) in support of the idea that the large families—families, that is, with large number of genera—were those of greater evolutionary age, Walters argues compellingly that the data much more persuasively support the hypothesis that larger families are those that have been recognized for longer. Very crudely, one might explain the point by arguing that the existence of a well-recognized type provides a focus to which subsequently discovered or distinguished types can be assimilated. Thus, plants of ancient symbolic significance, such as the rose and the lily, have provided the focus for some of the largest angiosperm families, Rosaceae and Liliaceae. Walters makes the suggestive observation that even Linnaeus, recognizing the similarities between the Rosaceous fruit trees, apple, pear, quince, and medlar (*Malus, Pyrus, Cydonia,*[12] and *Mespilus*), attempted to unite them into one genus, *Pyrus*. This attempt was unsuccessful, however, presumably because of the economic significance of these plants, and modern practice has reverted to that of the seventeenth century. Walters comments: "Can we doubt that, if these Rosaceous fruit trees had been unknown in Europe until the time of Linnaeus, we would happily have accommodated them in a single genus?" A general feature of Walters's argument is that our taxonomic system is massively Eurocentric. The shape of taxonomy has been substantially determined by which groups of plants were common or economically important in Europe.

The crucial question, of course, is whether this bias is a matter for concern and a reason for expecting wholesale revision of our taxonomic practices. To answer this question, we must have a view as to what taxonomy is for, and we come back to the major division introduced at the beginning of this essay: should we see taxonomy as answering to some uniform theoretical project or more simply as providing a general reference scheme to enable biologists to organize and communicate the wealth of biological information? The central argument of this paper is that the more we have learned about the complexity of biological diversity, the clearer it has become that any one theoretically motivated criterion for taxonomic distinctness will lead to taxonomic decisions very far removed from the desiderata for a general reference scheme. Of course, the contingencies of taxonomic history will no doubt have led, in many instances, to a scheme that is less than optimal even as a mere device for organizing biological information. On the other hand, in the absence of a theoretical imperative for revision, it is essential to weigh the benefits of a more logical organization of diversity against the costs of changing the extensions of familiar terms. My intuition is that on this criterion taxonomic revisions will seldom be justified.

We might begin by recalling part of Huxley's account of the function of classification (in the epigraph to this essay): to facilitate the operations of the mind in clearly conceiving and retaining in the memory the characters of the objects in question. Plainly to the extent that taxonomic names are undergoing constant modification, what any one person "conceives and retains in the memory" will be potentially incommunicable to others, and the possibility of reliably adding further information obtained from the work of others will be constantly jeopardized. This is not to say that taxonomic revision is never justified. If a species is included in a genus in which it is highly anomalous, and if that species is much more similar to other species in some other genus, then the goals of organizing information will be better served by reassigning it. It is of course also true that monophyletic taxa will tend to be more homogeneous than polyphyletic taxa, and that in paraphyletic taxa—taxa in which some of the descendants of the common ancestors of a particular taxon are excluded—there will be often be a case, on grounds of similarity, for including the excluded parts of the lineage. My point is just that these consequences rather than monophyly itself should provide the motivation for taxonomic change, and the benefits of such change must be weighed carefully against the potential costs. In this weighing process, the presumption that taxon names retain constant extension should probably be kept as strong as possible to maximize the ability of biologists to maintain reliable and communicable information.

To take perhaps the most familiar example, it seems to me that there is no case at all for revising the class Reptilia (reptiles) to include Aves (birds). This move is required by a strict cladistic concept of monophyly because it is believed that birds are descended from ancestral reptiles. We cannot exclude these avian ancestors from the class that includes modern reptiles because crocodiles, still classed as reptiles, are believed to have diverged from the main reptilian lineage earlier than birds did. The fact remains, however, that most zoologists, I suppose, would consider crocodiles much more like other reptiles than either is like any bird. The attempt to convince the learned or the vulgar world that birds are a kind of reptile strikes me as worse than pointless. It may be said that the only important claim is that Aves should be classified as a lower-level taxon included within Reptilia, and that this classification has nothing to do with our common usage of the terms *reptile* and *bird*. Although it is certainly the case that scientific taxonomic terms frequently differ considerably from apparently related vernacular terms, this differentiation is a source of potential confusion that should not be willfully exacerbated (see Dupré 1993, ch. 1, and forthcoming). It is also unclear what advantage is to be gained from insisting on such a revision. All evolutionists, I suppose, are likely to be familiar with recent thinking on the historical relationships within the main groups of vertebrates, and if they are not, their ignorance is not likely to be relieved by terminological legislation. Similarly, experts on smaller groups of organisms will presumably be familiar with current thinking on phylogenetic relationships within those groups. To celebrate every passing consensus on these matters with a change in taxonomic nomenclature is an inexcusable imposition of a particular professional perspective on the long-suffering consumers of taxonomy outside these phylogenetic debates.

In conclusion, I am inclined to dissociate myself from the strongest reading of the taxonomic pluralism I advocated earlier (1993; see also Kitcher 1984). In view of the limited success of theoretical articulations of the species category, it would seem to me best to return to a definition of the species as the basal unit in the taxonomic hierarchy, where the taxonomic hierarchy is considered as no more than the currently best (and minimally revised) general purpose reference system for the cataloguing of biological diversity. This system should provide a lingua franca within which evolutionists, economists, morphologists, gardeners, wildflower enthusiasts, foresters, and so on can reliably communicate with one another. Where special studies, such as phylogeny, require different sets of categories, it would be best to avoid using the term *species* (the desirability of rejecting this concept is sometimes asserted by evolutionists). Of course, such specialized users will be free to advocate changes in taxonomic usage, but should do so only in extreme circumstances. Although I am inclined to doubt the desirability of a pluralism of overlapping taxonomies, a general taxonomy will evidently draw broadly and pluralistically on a variety of considerations. Perhaps the most important will be history, not an unattractive idea in a science in which evolutionary thought is so prominent: a goal of general taxonomy should be to preserve the biological knowledge accumulated in libraries and human brains as far as possible. In addition, there would be a range of the morphological, phylogenetic, and ecological considerations that have figured in various monistic attempts to define the species. The importance of these considerations may vary greatly from one class of organisms to another. My feeble monism is my recognition of the importance of such a general reference system. My recognition of the likelihood that different enquiries may need to provide their own specialized classifications and my tolerance of diverse inputs into the taxonomic process will leave serious monists in no doubt as to which side I am on.

The position I am advocating provides, incidentally, a quick and possibly amicable resolution to the species as individuals debate. Species, I propose, are units of classification and therefore certainly not individuals. Lineages, on the other hand, are very plausibly best seen as individuals. Often, it may be the case that the members of a species (or higher taxon) are identical to the constituents of a lineage, but of course this coincidence does not make the species a lineage. And it is doubtful whether all species, or certainly all higher taxa, are so commensurable with lineages.

Resistance to or even outrage at the kind of position I am advocating may derive from the feeling that I am flying in the face of Darwin. Darwin, after all, wrote a well-known book about the origin of species, and he was writing about a real biological process, not a naming convention. Of course, the problem is that Linnaeus (or for that matter Aristotle) also talked about species and had in mind kinds, not things. Arguably, the tension between these two usages is at the root of the great philosophical perplexity that the concept of species has generated in this century. In arguing for reversion to the earlier usage of the term *species*, I am at least honoring conventions of priority. What I am proposing, however, is not much like a Linnaean taxonomy either. As many have observed, Darwin forced us to give up any traditionally essentialist interpretation of taxonomic categories and even any objectively determinate taxonomy. But almost a century and a half of biological work in the Darwinian paradigm have also shown us that evolution does not reliably produce units of biological organization well-suited to serve the classificatory purposes for which the concept of species was originally introduced, so perhaps rather than a reversion to Linnaeus, it would be better to see my proposal as a quasi-Hegelian synthesis. At any rate, if I seem to have been implying that Darwin may have been responsible for introducing some confusion into biology, I am sure no one will take this as more than a peccadillo in relation to his unquestionably positive contributions.

## ACKNOWLEDGMENTS

I would like to thank Rob Wilson and Chris Horvath for valuable comments on a draft of this essay.

## NOTES

1. It is not entirely clear how to make this idea precise. Obviously, not every organism founds a lineage, unless every organism is to constitute a species. A natural idea is that every organism in any way genetically distinct from its parent should found a new lineage. Given, however, the possibility of the same point mutation occurring more than once, it could turn out that a set of genetically identical organisms might constitute two or more distinct species.

The proposal also leads to the surprising conclusion that the vast majority of species are asexual. As Hull notes, this conclusion may mitigate the well-known difficulty in explaining the origin of sex by showing that sexual reproduction is a much rarer phenomenon than is often supposed (1989, 109). I should also mention that Hull's proposal is made in connection with the thesis that species are individuals, and is thus not necessarily an explicit defense of the BSC.

2. It appears that the same is probably true for some kinds of flowering plants (see Niklas 1997, 74 ff.).

3. This claim is perhaps less true now than it was twenty years ago. An influential evolutionary classification of bacteria was proposed by Woese (1987); see also Pace (1997). On the other hand, Gyllenberg and others (1997) aim explicitly to produce a classification that is optimal from an information-theoretic perspective, a goal that there is no reason to suppose would be met by any imaginable phylogenetic scheme. See also Vandamme and others (1996) for a related proposal. It is clear, at any rate, that any possible phylogenetic classification of bacteria, if it is to be of any practical use, must define taxa with great clonal diversity. Gordon (1997), for instance, reports that the genotypic diversity of *Escherichia coli* populations in feral mice was an increasing function of the age of the mouse, indicating the development of distinct clones during the lifetime of the mouse. I assume one would not want to think of this development as speciation, but

given this clonal diversity, it is difficult to see how any useful taxonomy could avoid being arbitrary from a phylogenetic perspective. The situation is still worse in view of the partially reticulate phylogeny consequent on genetic exchange between bacteria.

4. An extreme statement of this optimistic view can be found in Ruse (1987, 237): "There are different ways of breaking organisms into groups and they coincide! The genetic species is the morphological species is the reproductively isolated species is the group with common ancestors."

5. On *R. fruticosus*, the common blackberry, Bentham and Hooker (1926, 139) wrote: "It varies considerably. The consequence has been an excessive multiplication of supposed species ... although scarcely any two writers will be found to agree on the characters and limits to be assigned to them." The same "species" is described by Schauer (1982, 346) as "aggregate, variable with very numerous microspecies". More optimistically, *The Oxford Book of Wildflowers* (Nicholson, Ary, and Gregory 1960) states that "[t]here are several hundred species and hybrids in the *Rubus* group, and only an expert can identify all of them".

6. See Sober (1992) for a very clear exposition of this distinction. Although the debate here is a fundamental one, it is not of central concern to my essay.

7. See Davis (1978, 334–338) for a discussion of some of the difficulties in subspecific classification of angiosperms.

8. For further elaboration, see Mishler and Brandon (1987). For more general arguments against any fundamental distinction between species and higher taxa, see Ereshefsky (1991 and chapter 11 in this volume) and Mishler (chpater 12 in this volume).

9. Some more realistic examples have been discussed by Kitcher (1984).

10. For references to bacterial taxonomy and brief discussion, see note 3.

11. De Queiroz and Gauthier (1990, 1994) claim that taxonomic changes they advocate will promote constancy of meaning, or definition, for taxa. Mammalia, for example, should be defined as the set of descendants of the most recent common ancestor (i.e., ancestral species) of monotremes and therians. The extension of such a term, however, will be constantly revisable in the light of changes in opinion about the details of evolutionary history. From the point of view of the consumer of taxonomy, at least, I suggest that constancy of extension is surely more valuable than constancy of definition.

12. Subsequent to Walters' paper, the quince appears to have been reconceived as *Chaenomeles* (though not unanimously according to the few sources I consulted on this matter). This reconception effects a conjunction with the ornamental flowering quinces. One might speculate that the increasing obscurity of the quince as a fruit might have exposed it to this annexation, which one doubts could have happened to the apple.

## REFERENCES


Bentham, G., and J. D. Hooker (1926). *Handbook of the British Flora*, 7th ed., revised by, A. B. Rendle. Ashford, Kent: L. Reeve and Co.

Davis, P. H. (1978). The moving staircase: A discussion on taxonomic rank and affinity. *Notes from the Royal Botanic Garden* 36, 325–340.

de Queiroz, K., and J. Gauthier (1990). Phylogeny as a central principle in taxonomy: Phylogenetic definitions of taxon names. *Systematic Zoology* 39, 307–322.

de Queiroz, K., and J. Gauthier (1994). Toward a phylogenetic system of biological nomenclature. *Trends in Ecology and Evolution* 9, 27–31.

Dupré, J. (1993). *The disorder of things: Metaphysical foundations of the disunity of science.* Cambridge, Mass.: Harvard University Press.

Dupré, J. (forthcoming). Are whales fish? In D. L. Medin and S. Atran, eds., *Folkbiology.* Cambridge, Mass.: MIT Press.

Ehrlich, P. R., and P. H. Raven (1969). Differentiation of populations. *Science* 165, 1228–1232. (Reprinted in Ereshefsky 1992.)

Ereshefsky, M. (1991). Species, higher taxa, and the units of evolution. *Philosophy of Science* 58, 84–101.

Ereshefsky, M., ed. (1992). *The units of evolution: Essays on the nature of species.* Cambridge, Mass.: MIT Press.

Floodgate, G. D. (1962). Some remarks on the theoretical aspects of bacterial taxonomy. *Bacteriology Review* 26, 277–291.

Ghiselin, M. T. (1987). Species concepts, individuality, and objectivity. *Biology and Philosophy* 2, 127–143.

Ghiselin, M. T. (1997). *Metaphysics and the origin of species.* Albany, N.Y.: SUNY Press.

Gilmour, J. S. L. (1951). The development of taxonomic theory since 1851. *Nature* 168, 400–402.

Gordon, D. M. (1997). The genetic structure of *Escherichia coli* populations in feral mice. *Microbiology* 143, 2039–2046.

Gyllenberg, H. G., M. Gyllenberg, T. Koski, T. Lund, J. Schindler, and M. Verlaan (1997). Classification of *Enterobacteriacea* by minimization of stochastic complexity. *Microbiology* 143, 721–732.

Hull, D. L. (1965). The effect of essentialism on taxonomy: Two thousand years of stasis. *British Journal for the Philosophy of Science* 15, 314–326, and 16, 1–18. (Reprinted in Ereshefsky 1992.)

Hull, D. L. (1989). *The metaphysics of evolution.* Albany, N.Y.: SUNY Press.

Kitcher, P. (1984). Species. *Philosophy of Science* 51, 308–333. (Reprinted in Ereshefsky 1992.)

Locke, J. (1689; 1975 edition). *An essay concerning human understanding,* edited by P. H. Nidditch. Oxford: Oxford University Press.

Mayr, E. (1961). Cause and effect in biology. *Science* 134, 1501–1506.

Mayr, E. (1987). The ontological status of species: Scientific progress and philosophical terminology. *Biology and Philosophy* 2, 145–166.

Mill, J. S. (1862). *A system of logic,* 5th ed. London: Parker, Son and Bourn.

Mishler, B. D., and M. J. Donoghue (1982). Species concepts: A case for pluralism. *Systematic Zoology* 31, 491–503. (Reprinted in Ereshefsky 1992.)

Mishler, B. D., and R. N. Brandon (1987). Individuality, pluralism, and the biological species concept. *Biology and Philosophy* 2, 397–414.

Nicholson, B. E., S. Ary, and M. Gregory (1960). *The Oxford book of wildflowers.* Oxford: Oxford University Press.

Niklas, K. J. (1997). *The evolutionary biology of plants.* Chicago: University of Chicago Press.

Pace, N. R. (1997). A molecular view of microbial diversity and the biosphere. *Science* 276, 734–740.

Ruse, M. (1987). Biological species: Natural kinds, individuals, or what? *British Journal for the Philosophy of Science* 38, 225–242. (Reprinted in Ereshefsky 1992.)

Schauer, T. (1982). *A field guide to the wild flowers of Britain and Europe*, translated by R. Pankhurst. London: Collins.

Sober, E. (1992). Monophyly. In E. A. Lloyd and E. F. Keller, eds., *Keywords in evolutionary biology*. Cambridge: Harvard University Press.

Sneath, P. H. A., and R. R. Sokal (1973). *Numerical taxonomy*. San Francisco: W. H. Freeman.

Stebbins, G. L. (1987). Species concepts: Semantics and actual situations. *Biology and Philosophy* 2, 198–203.

Templeton, A. R. (1989). The meaning of species and speciation: A genetic perspective. In D. Otte and J. A. Endler, eds., *Speciation and its consequences*. Sunderland, Mass.: Sinauer. (Reprinted in Ereshefsky 1992.)

Vandamme, P., B. Pot, M. Gillis, P. De Vos, K. Kersters, and J. Swings (1996). Polyphasic taxonomy, a consensus approach to bacterial taxonomy. *Microbiological Review* 60, 407–438.

Van Valen, L. (1976). Ecological species, multispecies, oaks. *Taxon* 25, 233–239. (Reprinted in Ereshefsky 1992.)

Walters, S. M. (1961). The shaping of angiosperm taxonomy. *New Phytologist* 60, 74–84.

Willis, J. C. (1949). *Birth and spread of plants*. Geneva: Conservatoire et Jardin Botanique de la Ville.

Woese, C. R. (1987). Bacterial evolution. *Microbiological Review* 51, 221–271.

# 2 On the Plurality of Species: Questioning the Party Line

David L. Hull

In the nineteenth century, one of the hot topics of debate was the plurality of worlds. Did God create a single Earth inhabited by all and only those souls that Jesus gave the opportunity to be saved from eternal damnation, or did He create millions of worlds inhabited by just as many morally responsible beings? On the first alternative, God would appear to be as profligate as the most extravagant wastrel. He created millions of nebulae, each containing just as many stars, each of which might have planets circling it in stately regularity, but on only one of these planets circling a single star did He breath soul into a single species. What a waste. But if we assume that God is the Protestant God of "waste not, want not," surely He would not have let so many opportunities slip through his fingers for creating subjects to worship him. Not only did Jesus come down to Earth to be sacrificed for our sins, but apparently he also repeated this ritual in world after world after world (for a history of this controversy, see Dick's 1982 *Plurality of Worlds*).

No less a figure than William Whewell entered into this debate—on the side of the monists. For fear of damaging his hard-earned reputation as a sober seeker after truth, Whewell anonymously published *The Plurality of Worlds* (1853). To those who complained that God and hence nature did nothing in vain, Whewell cited all the waste that was already so apparent in *this* world:

> We reply, that to work in vain, in the sense of producing means of life which are not used, embryos which are never vivified, germs which are not developed; is so far from being contrary to the usual proceedings of nature, that it is an operation which is consistently going on, in every part of nature. Of the vegetable seeds which are produced, what an infinitely small proportion ever grow into plants! Of animal ova, how exceedingly few become animals, in proportion to those that do not; and that are wasted, if this be waste! (p. 249)

A similar question might be asked of God about species in general. God created numerous different species, but did He create a *single sort* of species or *many different sorts* of species? Each and every organism belongs to one species and one species only, but are all these species of the same *sort*? Or possibly, does any one organism belong to many different sorts of species?

These questions are also central to the present controversy between monists and pluralists with respect to species. In Whewell's day, the plurality of worlds was a very open question, but today the party line on pluralism conflicts with respect to there being a party line on species. As Sterelny (chapter 5 in this volume) remarks, "Evolutionary theory has moved close to a consensus in seeing species as historical individuals (Ghiselin 1974, Hull 1978)," but how can consensus exist with respect to the ontological status of species if pluralism is the party line among philosophers of science, especially philosophers of biology? Everyone seems to feel obligated to espouse the position held by all thoughtful scholars—a nuanced pluralism, as distinct from any crude, simplistic monism.

One problem unfortunately characteristic of such contrasts as monism versus pluralism is that the apparent differences between them tend to disappear under analysis. Numerous senses of *monism* blend imperceptibly into just as many senses of *pluralism*. For example, Ereshefsky (1992, 688) concludes his discussion of "eliminative pluralism" with the observation that "Some may view eliminative pluralism as just a complicated form of monism. If that is the case, then the arguments of this paper have been successful." A clear contrast exists between more simplistic notions of monism and pluralism, but no one seems to hold any of these simplistic alternatives. When pushed, most authors retreat to some platitudinous middle ground. In this respect, the issue of pluralism mirrors the conflicts over nature nurture and genetic determinism. Does anyone think that genes are sufficient for anything? If this view is what genetic determinism entails, then genetic determinists are most noteworthy for their nonexistence.

In the first section of this chapter, I discuss some very general issues with respect to pluralism before turning to one example—biological species. I take a look at the connection between the monism versus pluralism dispute and *(a)* the contrast between realism and antirealism, *(b)* prerequisites for communication, and *(c)* reflexivity. I then turn to some illustrative examples —does HIV cause AIDS, and does smoking cause lung cancer? Philosophers of science have produced a variety of analyses of the notion of causation. Some of the cruder, simplistic analyses allow us to reject certain claims made by scientists who find the HIV hypotheses a conspiracy, not to mention scientists working for the Tobacco Institute who argue that smoking does not cause lung cancer or any other disease for that matter.

However, these scientists are able to hide behind the smoke screens generated by more sophisticated, pluralist analyses of "cause" produced by equally sophisticated philosophers. Causal situations are so various and complicated that nothing can be identified as causing anything, just as the biological world is so varied and complicated that no one analysis of "species" will do. Instead, indefinitely many species concepts are needed for indefinitely many contexts. The great danger of pluralism is "anything goes." In order to avoid this end of the slippery slope, criteria must be provided for distinguishing between legitimate and illegitimate species concepts. I end this paper by evaluating these evaluative criteria for various species definitions. My conclusion is that if we retain the traditional organizational hierarchy of genes, cells, organisms, colonies, demes, species, and so on, any and all species definitions appear inadequate. One possible way for a monist to avoid this conclusion is to abandon, not the traditional Linnaean hierarchy (Ereshefsky, chapter 11 in this volume), but the organizational hierarchy.

## REALISM AND ANTIREALISM

One reason why philosophers find the monism-pluralism debate so interesting is its apparent connection to the dispute over realism and antirealism. Of the four possible combinations of these philosophical positions, two seem quite natural: monism combined with realism, and pluralism combined with antirealism. Monists argue that scientists should strive to find the best way to divide up the world and that such a best way does exist, even though we may never know for sure (whatever that means) whether or not we have found it. Yes, scientists are fallible. Yes, conceptual revolutions do occur in science, revolutions that require us to start not all over again, but at least a few steps back. Given these assumptions, the goal of finding a single, maximally informative conceptualization of nature seems both desirable and reasonable. If one is going to be flat-footed and simpleminded with respect to one philosophical position, why not two? (Holsinger [1987] interprets Sober [1984a] as a monist-realist.)

Conversely, pluralists argue that the differences of opinion so characteristic of science will continue indefinitely into the future. The nature of these differences surely changes, but rarely does consensus ever emerge, and when it does, it is likely to be short-lived. How come? Because "the" world can be characterized in indefinitely many ways, depending on differences in perspectives, worldviews, paradigms, and what have you. Even though not all of these ways are equally plausible, acceptable, or promising, no one way is clearly preferable to all others once and for all. One must keep an open mind. If one is going to be sophisticated and nuanced with respect to one philosophical position, why not two? (Stanford [1995] portrays himself as being a pluralist-antirealist.)

The other two combinations of the two philosophical distinctions are somewhat strained. It would seem a bit strange to argue that one and only one way exists to divide up the world, but that the groups of natural phenomena produced on this conceptualization are not "real." They are as real as anything can get! Of course, *real* can be defined in such a way that nothing could possibly be real, just as philosophers have defined *know* so that no one ever knows anything and *law* so that no generalizations ever count as laws, but this game, as much fun as it is to play, seems nonproductive in the extreme. A combination of pluralism and realism seems equally peculiar (but see Dupré 1981 and Kitcher 1984a). The world can be divided up into kinds in numerous different ways, and the results are all equally real! Once again, *real* can be defined in such a way that two contradictory classifications of the same phenomena can refer to real groups of entities. Or one might get really sophisticated and claim that these two ways of viewing the world are incommensurable, but the sort of holistic semantics that generates incommensurability again seems a high price to pay. (For a discussion of realism and pluralism with respect to the units of selection controversy, see Sober 1984b; Sterelny and Kitcher 1988; Kitcher, Sterelny, and Waters 1990; Waters 1991; Sober and Wilson 1994; Shanahan 1997.)

## A RETURN TO BABEL

Kitcher (1984a, 326) suggests that one worry possibly bothering opponents of pluralism is that it might "engender a return to Babel." Do monists treat language monistically? Do they think that every utterance has one and only one meaning? Are pluralists pluralistic when it comes to language? Should I treat every utterance made by pluralists pluralistically, or is it just possible that texts do on occasion constrain interpretations? As far as I know, no one denies the existence of ambiguity and vagueness in language. In fact, in the face of imperfect knowledge, vagueness may be necessary for communication (Rosenberg 1975). Again, I find it difficult to tell, but even the most rabid deconstructivist shies away from anything—literally *anything*—goes. Even they seem to assume that they are saying something or trying to say something with varying degrees of success. Possibly, they are not saying one and only one thing with absolute clarity, but not all interpretations are equally acceptable—Emily Dickinson as a Marxist feminist. Perhaps our intended meaning is neither patent nor all there is to the story, but at the very least I have been taught that we should all aim to present our views as clearly and unambiguously as possible, and I see no reason to give up the ghost at this late date. Perhaps I can be a selective pluralist about a half-dozen concepts at a time, but I cannot treat all of language pluralistically all at once, not if I want to say something, not if I want other human beings to understand me.

## REFLEXIVITY

Early on in the prehistory of what has come to be known as the Science Wars, young sociologists who were in the midst of rediscovering epistemological relativism also stumbled upon reflexivity when an occasional opponent asked why they were gathering so much evidence to cast doubt on the efficacy of evidence. These sociologists tried a variety of ways to extricate themselves from this "tension" in their position. Some argued that sociologists, when they are acting as sociologists, must "treat the social world as real, and as something about which we can have sound data, whereas we should treat the natural world as something problematic—a social construct rather than as something real" (Collins 1981c, 217; see also Collins 1981a and 1981b). However, Collins also acknowledges that even sociologists periodically need to step back from their own work and treat it as problematic as well. The knowledge claims made by sociologists are also socially constructed.

Because sociologists are themselves scientists, reflexivity poses especially poignant problems for them, but reflexivity can also be brought to bear on contrasts between object-level and metalevel positions. Philosophers are not scientists. As philosophers, we do not *do* science. We comment on it. (Of course, someone who is officially a philosopher can join in the scientific enterprise, and vice versa.) Thus, a certain tension exists in Hempel's position that data play crucial roles in science, but no role whatsoever in his logical empiricist analysis of science. Just because scientists claim to explain phenomena in ways different from Hempel's covering-law model does not mean his covering-law model of scientific explanation is false. If no scientists ever explained anything by deducing it from laws of nature and statements of particular circumstances, Hempel's covering-law analysis of explanation still could count as a totally adequate analysis of scientific explanation!

Hempel's holding these two positions on the role of data with respect to the object level and metalevel respectively may seem to be a bit implausible, but it is not in the least contradictory. Parallel observations hold for pluralists. With respect to science—so pluralists claim—serious, respectable alternative positions always exist for every issue, but when one steps back to view philosophy of science, one and only one position is acceptable: pluralism. John Maynard Smith (1998) raises precisely this objection in his review of Sober and Wilson (1998). "A second reason why this book is confusing is that, although the authors argue for pluralism, they are not themselves pluralists: for them, the only right way to describe a model is in group selection language" (p. 640).

As in the case of Hempel's logical empiricist analysis of science, pluralists are not contradicting themselves in holding such different positions on science and philosophy, but it is a bit difficult to swallow their position. If one can be a monist with respect to—of all things—philosophical debates, one can certainly on occasion think that monism is justified in science. If one and only one position is warranted with respect to the monism-pluralism dispute, then certainly one and only one position is warranted with respect to the evolutionism-creationism dispute. Possibly one and only one position is warranted with respect to the species problem. If not, I am missing something of massive significance.

## PROFESSIONAL CONSIDERATIONS

In general, people find it much more plausible and desirable to counsel pluralism with respect to *other* people's areas of expertise than their own. As I have argued above, philosophers find pluralism extremely attractive in science, much more so than in philosophy. Scientists in turn do not find pluralism all that attractive in their own area of expertise and usually stay well clear of philosophy. At least sometimes, scientists think that they have the right answer to a particular question. What would science be like in the absence of such convictions? Picture hundreds of scientists, each being terribly considerate of each other's hypotheses:

"You think that selection occurs only at the level of the genetic material? That may be so, but in addition, individual organisms are the main target of selection."

"I agree with everything you say and want only to add that selection wanders up and down the organizational hierarchy in biology."

"To supplement the nuanced position being expressed, I submit that no such thing as selection exists. It is really two processes, not one."

"So true, and selection is of only peripheral importance in evolution."

"Speak on, oh wise one."

A preference for monism and pluralism waxes and wanes as various groups gain and lose power. Right now, advocates of developmental systems theory are trying to supplant the current gene-centered world view (Moss 1992, Griffiths and Gray 1994). None too surprisingly, these developmentalists are urging pluralism. In general, groups who hold minority opinions at a particular time find pluralism to be the correct philosophical view, whereas the groups in power are not nearly so attracted to it. During the heyday of the biological species concept, Mayr saw no reason to give ground to his opponents. He insisted that there are basic units in the evolutionary process, that these units are delineated in terms of reproductive isolation, and that making these units coincide with the basic units of classification is both possible and desirable. I predict that if and when developmentalists see their views prevailing, they will cease their pleas for pluralism and become staunch monists. Pluralism looks good to outsiders—regardless of whether they belong to different disciplines (e.g., philosophers looking at science)—or to groups currently a minority within a particular discipline (e.g., developmentalists trying to muscle evolutionary biologists out of their positions of power).

## ANALYSES OF CAUSATION

Much of the problem that I am having with the monism-pluralism issue stems from the general character of philosophical analyses as they function in philosophy of science. At times, they seem to help; at other times, they seem to obfuscate understanding. The notion of cause can serve as an illustrative example. In the early days of AIDS, scientists floundered around trying to find "the" cause of AIDS. Possibly drug addicts and sexually promiscuous homosexual men were battering their immune systems with so many different diseases that eventually their immune systems gave up and closed down. Or perhaps an inhalant termed *poppers* was the cause. Or possibly some infectious agent was responsible. Rather rapidly, the evidence indicated that the most likely explanation was also the correct explanation—a virus transmitted primarily by infected needles and sexual intercourse. At the time that scientists turned their attention to the AIDS virus in 1982 and 1983, the evidence was far from conclusive. In a very real sense of the term, they "might" have been mistaken.

However, little by little a consensus emerged that AIDS is a contagious disease, and the causative agent is a virus eventually termed the *human immunodeficiency virus* or HIV for short (Blattner, Gallo, and Temin 1988). One virologist, Peter Duesberg (1987), disagreed. He argued that AIDS is not one disease, but at least four. What is mistakenly called "the AIDS virus" has nothing to do with the disease sweeping through Europe and North America. Instead, the use of drugs, such as poppers, are the real cause. Hence, the antiviral drugs such as AZT being used to combat AIDS are doing nothing but exacerbating the problem. One important feature of science is persistence in the face of considerable opposition. On occasion, scientists who were thought to be in the lunatic fringe in their own day eventually are vindicated. Doubting the role of HIV in causing AIDS in 1987 is one thing. Continuing to doubt it today is quite another matter (Duesberg 1996).

Right now, the only explanation of AIDS in the least plausible is that it is caused by HIV. However, Duesberg (1996) uses all sorts of specious arguments about causation to salvage his position. Yes, yes, HIV *alone* can't cause AIDS. It needs hosts to invade. Without people, the various strains of virus that can infect only human beings would be in real trouble. In addition, if no one had ever isolated the blood-clotting factor in people, hemophiliacs would not have been put at greater risk than other people. If no one ever shared needles, if everyone used virus-impermeable rubbers whenever they engaged in sex, etcetera, etcetera, we would never have had an AIDS epidemic.[1] And if God were truly good, etceter, etcetera. In addition, contagious diseases are commonly defined in terms of the presence of the causative agent. In order to have tuberculosis, you must be infected with the tuberculosis bacterium. It is *necessary*—by definition. But it is not *sufficient*. In order for people infected by the tuberculosis bacterium to come down with tuberculosis, all sorts of other factors are also required. In sum, any philosopher worth his or her salt could explain to Duesberg that just because the presence of HIV is not *sufficient* for a person coming down with AIDS, he can't conclude that it is irrelevant.

The issue I wish to raise is what advice advocates of pluralism would have given scientists and funding agencies in the early 1980s, the later 1980s, and today. In the early 1980s, scientists entertained all sorts of hypotheses, but by the late 1980s they had settled on HIV as the contagious agent that causes AIDS. If the prophets of pluralism had had the power in the late 1980s, would they have required that research time and money be distributed equally across all possible causes of AIDS? Should Duesberg have been funded? If so, would these pluralists be willing to live with the disastrous results of their decision? If not, why not? How about today? Are poppers still a plausible causative agent for AIDS?

If the contrast between monism and pluralism is to be of any significance at all, advocates on both sides of this divide have to admit that at least on occasion the position they prefer might be wrong. Monists and pluralists alike have to present a list of criteria to help in deciding when one, two, three, or more possible alternatives are justified. The danger is that every situation, no matter how apparently straightforward, turns out to be hopelessly complex. It also must be noted that the decisions that we make in this connection are likely to have effects on society. People do not, as a rule, pay excessive attention to what we philosophers have to say, but sometimes the distinctions that we make and the positions that we espouse find their way into the public at large. For example, the tobacco industry for years has claimed that the scientific data are inadequate to prove that smoking causes lung cancer. After all, some people smoke three packs of cigarettes a day and die at age ninety-five in a car accident, whereas others contract lung cancer at an early age although they have never been exposed to cigarette smoke.

In most scientific contexts, the factors that scientists pick as causes are rarely necessary conditions; they are even more rarely sufficient conditions; and they are hardly ever both necessary and sufficient conditions. Perhaps finding necessary and sufficient conditions for the occurrence of natural phenomena is the ideal, but in most contexts we have to settle for much less. The issue is statistical correlations. Holding everything else constant, how much does smoking increase one's chances of contracting lung cancer—or heart disease, for that matter?

> If "Smoking causes heart disease" turns out not to be true for humankind, I take it that its truth has been established for the various populations of Western countries in which it has been systematically investigated. It is now known, for example, that smoking causes heart disease for the human population of the United States. There may be inductive hazards in the extrapolation of this result to other identifiable human populations; but it seems unnecessarily cautious to restrict the claim to an as-yet-unidentified subpopulation of inhabitants of the United States. (Dupré 1993, 200)

If we philosophers were to stop here, the effects of our analysis would be, from my perspective, decidedly beneficial. Smoking is a major cause of both lung cancer and heart disease, and anyone with a shred of intellectual integrity has to concur, pluralism be damned.

However, we do not stop here. Promiscuous pluralists (not to be confused with Dupré's promiscuous realists) feel obligated to resist the conclusion that smoking causes lung cancer, even in a statistical sense, because it is insufficiently nuanced. Even when people do smoke and do contract lung cancer, it does not follow that smoking is "the" cause of lung cancer. All sorts of alternatives and combinations of alternatives might and probably do play a role in people coming down with lung cancer. In fact, as the father of statistics (not to mention the synthetic theory of evolution), Ronald Fisher (1958, 108) mused, perhaps lung cancer is one of the causes of smoking cigarettes!

Causal situations are extremely complicated. All sorts of supplemental factors can play a role, and the reasons for selecting one factor and terming it "the" cause while demoting all the others to the position of "supplemental factors" are far from obvious. This issue arises time and again in science. Natural selection is the cause of organic complexity, while every other influence is a "constraint." Genes are the cause of various traits, whereas everything else is demoted to being part of the "background knowledge." Many evolutionary biologists claim that gene exchange is crucial in the individuation of species as units of evolution, but all other factors only "contribute" to the cohesion that is so important in species being species. On what grounds are causes assigned to these various categories? This literature throws up a picture of the empirical world so murky that even the tobacco industry can hide in the fog.

## SPECIES AND MUDDLED METAPHYSICS

At long last, after all the preceding preparatory discussion, it is time to turn to the species problem. Numerous philosophers have urged pluralism with respect to biological species. As a point of departure, I take Philip Kitcher's (1984a, 1984b, 1987, 1989) discussion of species. As is usually the case, deciding what an author's position is can pose serious problems. Too many alternatives present themselves. (In this respect, pluralists are correct more often than I would prefer.) With respect to his own general philosophical position, Kitcher (1984a, 308) is not very pluralistic. In his papers on species, he sets himself the dual tasks of explaining a position about species that he terms "pluralistic realism" and of indicating in a general way why he thinks that "this position is true." Kitcher does not say that from some perspectives and in certain circumstance, pluralism is preferable, or that from other perspectives and in other circumstances monism is the correct position to hold. If I read Kitcher correctly, he believes that no form of monism is acceptable by anyone no matter what. With respect to his own *philosophical* outlook, Kitcher is inclined to monism. So am I with respect to mine.

Kitcher then turns to the issue of the ontological character of species. Periodically, authors have tentatively suggested that species are like individual organisms more than they are like universals such as triangularity, but not until Michael Ghiselin (1974) did this position become widely discussed. According to Ghiselin and later Hull (1976), if species are to fulfill their role in the evolutionary process, they must be conceived of as spatiotemporally localized entities connected in space and time (see also Mayr 1999; Mishler and Theriot 1999: and Wiley and Mayden 1999). At any one time, species must exhibit a certain degree of cohesiveness (though the mechanisms producing this cohesiveness might vary), and through time, they must be connected.[2] They are chunks of the genealogical nexus. The term *species* can be and has been used in a variety of other senses, but when species are supposed to be the things that evolve, they fit more naturally in the category *individual* (or historical entity) than the category *class* (or *kind*).

Kitcher rejects this position in no uncertain terms. However, in the end, our differences seem to be primarily terminological. If only Ghiselin and I had explicated our view in terms of *sets*, much of the controversy would have been avoided. Instead of talking about individuals and spatiotemporally unrestricted classes, we should have distinguished "historically connected sets" from those sets that are not historically connected (Kitcher 1984a, 314; 1987, 186; and 1989). The notion of a class is a "bastard notion that deserves no place in anybody's ontology. The respectable concept is that of a set." Although I find this position too monistic, I am willing to adjust my terminology if it improves communication. However, I think that there is more to this dispute than just terminology. I am not at all sure what a historically connected sets could be. They certainly could not function in traditional extensional set theory (Sober 1984a).

Early on in this dispute, I wondered how far Kitcher was willing to extend his notion of historically connected sets. If he could interpret species as historically connected sets of organisms, I saw no reason why organisms could not be interpreted as historically connected sets of cells. If so, Kitcher was on the verge of everything being a set. Kitcher responded that, no, anything that might pass for internal structure in species is much weaker and rarer than in organisms. If you destroy the internal organization of most organisms, they die, whereas only a relatively few species would go extinct if their population structure was drastically destroyed (Kitcher 1989, 186).

Two points are at issue—internal organization and historical connection. Most organisms do exhibit more internal organization than most species, but this differences is one of degree, not kind. Most species do not exhibit the internal organization common in vertebrate organisms, but the same can be said for plants as organisms. Most plants do not exhibit the internal organization common in vertebrate organisms. They tend to be much more modular (Sterelny, chapter 5 in this volume). Holsinger (1984, 293) argues that "Although it is possible to regard a species as a set with a special internal structure, it is preferable to regard a species as an individual precisely to emphasize this internal structure" (see also Sober 1984a). What, however, are the character and extent of this structure? Even if most species do not exhibit the degree of organization that most organisms exhibit, the issue of historical connectedness remains, and it alone raises serious problems for traditional conceptions of species. Calling species *historically connected sets* implies that these "sets" can function in set theory, but to my knowledge no such version of set theory exists.

Although Boyd (chapter 6 in this volume) does not treat species as sets, he reasons along much the same lines as Kitcher. Ghiselin and I assumed the traditional distinction between classes and individuals (see de Sousa forthcoming). Classes are spatiotemporally unrestricted, whereas individuals are spatiotemporally localized and connected. Given this fairly traditional distinction, we argued that species are more like individuals than classes. Kitcher distinguishes between two sorts of sets: those that are spatiotemporally connected and organized, and those that are not. Given this distinction, just about everything becomes a set, including such paradigm individuals as organisms. For Boyd (this volume), at least three different sorts of kinds exist—traditional kinds that can be defined in terms of necessary and sufficient conditions (very rare!), kinds defined as simple cluster concepts, and homeostatic property cluster kinds. In his view, species are not individuals but homeostatic property cluster kinds. So are organisms. So is the Rock of Gibraltar! In both cases, the notions of set and kind have been so redefined that the traditional distinction between individuals and sets or kinds becomes all but obliterated. Species are not individuals, but then neither are organisms.

The key substantive issue with respect to Boyd's suggestion is, of course, the nature of the mechanisms responsible for these homeostatic property clusters, but this issue to one side, I fail to see the advantage of the preceding maneuvers. Although the distinction between spatiotemporally unrestricted kinds and spatiotemporally localized and restricted individuals may well stem from a "profoundly outdated positivist conception of kinds" (Boyd, chapter 6 in this volume), I fail to see how we can avoid such a distinction and still talk sense. Of course, if nothing of any importance rides on the apparent difference between "$f = ma$" and "Richard Nixon died before going bald," then I must revise my understanding of science from the ground up. (For an example of the impact that a biological perspective can bring to the philosophical understanding of individuality, see de Sousa forthcoming, and Wilson 1999.)

However, as Kitcher (1984b) makes clear, logical issues about set theory and mereology are not his primary concern. Instead, he objects to "genealogical imperialism" as the "monism of the moment." For my part, if I were more of a pluralist, I would object to Kitcher's set-theoretic imperialism as *his* monism of the moment and Boyd's homeostatic property cluster kinds as *his*. In any case, Ghiselin and I argued that from the perspective of the evolutionary process, species must be viewed in a particular way. Kitcher objects to that way, but even more strongly he objects to our monistic view of the evolutionary perspective. He argues at some length that the evolutionary perspective is not the *only* legitimate perspective in biology. Ghiselin and I certainly agree. Not everyone thinks that the evolutionary perspective is so basic. Certainly, present-day "idealists" think that a science of form can be developed without any reference to evolution (Webster and Goodwin 1996), and advocates of developmental systems theory would like to dethrone evolution (Moss 1992, Griffiths and Gray 1994).

However, although acknowledging other legitimate perspectives in biology, Ghiselin and I nevertheless insist that the evolutionary perspective is the most basic perspective in biology (see Eldredge 1985, 200, and Ghiselin 1989, 74). Perhaps the claim so often repeated by evolutionary biologists that nothing in biology makes sense except in the light of evolution is a bit of an exaggeration, but not much. In sum, I think that *(a)* the evolutionary perspective in biology is in a significant sense "basic" to all of biology; *(b)* from the evolutionary perspective, species must be treated as historical entities; and *(c)* Kitcher's interpretation of species as historically connected sets and Boyd's interpretation of species as homeostatic property cluster kinds are, to say the least, strained. Even if we accept the alternatives that Kitcher and Boyd suggest, the distinction between universals and particulars must be reintroduced by distinguishing between spatiotemporally restricted and connected sets or kinds and those sets or kinds that lack these restrictions. What such an exercise accomplishes, I fail to see.

## A PLETHORA OF SPECIES CONCEPTS

All of the preceding discussion has concerned very general philosophical and metaphysical issues. How about species themselves? Kitcher (1984a) distinguishes between two families of species concepts—species defined in terms of structural similarities and species defined by their phylogenetic relationships (see also Dupré, chapter 1 in this volume). He divides structural concepts into common genetic structure, common chromosomal structure, and common developmental program. He divides historical concepts according to whether continuity or division is paramount. He then subdivides each of these concepts according to the same three principles of division: reproductive isolation, ecological distinctness, and morphological distinctness. The end result is nine different species concepts, each one of which Kitcher thinks produces real groups.

Whether or not one is willing to go along with Kitcher's entire classification of species concepts, his first division cannot be ignored. No two ways of classifying the world could be more different. On the structural similarity alternative, genesis is irrelevant. If two structures are similar enough, they are the same structure. At some level, the vertebrate eye and the eye of the octopus are the same trait. On the historical alternative, genealogy can override similarity. The Eustachian tube in humans is the same character as the spiracle in sharks. They are evolutionary homologies. The contrast is between homoplasies and homologies or, if you prefer, sets that do not have to be spatiotemporally connected in any way (homoplasies) and sets that must be (homologies). The same distinction also applies to periods in human history—for example, between feudal and Feudal. As a homoplasy, "feudal" can be defined in such a way that it can occur at different times and places as Marxist theories require, or it can be defined so that it is restricted to a single period in the development of Europe. If anyone finds this distinction too difficult to understand, just compare the prices of Tiffany lamps and Tiffany-style lamps.

Both homoplasies and homologies pose serious problems. Since at least Goodman (1972), philosophers have realized that the notion of similarity so pervasive in our conceptions of the world is currently unanalyzed, if not unanalyzable. A question that must be asked of species definitions in terms of similarity is, How similar is "similar enough" and in what sense of "similar"? Can one level of similarity be specified—one level that can be applied equally across all organisms to produce even a minimally acceptable classification? The answer to this question is, thus far, no. One implication of this conclusion is that no general purpose classification of plants and animals is possible. *General purpose* remains as unanalyzed, as does *overall similarity* (but see Dupré, chapter 1 in this volume).

The historical alternative avoids the problems that plague species definitions in terms of similarity, but it has problems of its own—chiefly difficulties involved in reconstructing phylogeny, which in turn requires that homologies be distinguished from homoplasies. Organisms on Earth evolved the way that they did and no other way. Even though phylogeny includes lots of merger, there is one and only one phylogenetic "tree." But reconstructing phylogeny can be extremely difficult, if not impossible. Most of the information that we have of the biological world is of extant organisms. Our knowledge of extinct forms is even sketchier—much sketchier. The principles of cladistic analysis were devised to establish transformation series in which genuine characters (homologies) nest perfectly and hence produce perfectly monophyletic taxa. Any character that does not nest perfectly is not a genuine character. As Sterelny (chapter 5 in this volume) argues, we must view "phenomenological species—identifiable clusters of organisms—as fallible clues to the existence of evolutionarily linked metapopulations."

With respect to monism versus pluralism, systematists are put in a bind. The rules of nomenclature do not allow them to be pluralists. They have to produce one and only one classification—not nine as Kitcher would have them do. Systematic principles that take history as basic seem appealing because they can promise a single classification. It may not be equally useful for all sorts of purposes, but it is at least attainable in principle. Possessing a single classification as a reference system, no matter its faults, is better than having dozens of alternative and incommensurable classifications. The metric system of measurement is not the only possible system of measurement. God did not deliver it from on high. It is not equally good for all purposes. Given different contingencies, we might have constructed very different systems. In fact, we did. We have two systems of weight, distance, and so on—the metric and the English. I can't speak for others, but I find the presence of these two systems a persistent irritation. I would hate to think how inconvenient having a dozen such systems would be.

The same goes for naming and organizing the elements in the periodic table. All sorts of different ways of organizing the elements were proposed. The end result that anyone who ever took an introductory course in chemistry stared at for hours on end arose in a hit-or-miss, highly contingent fashion. In fact, we are only now getting around to formalizing the names of the elements with atomic numbers from 101 to 109 (see *Pure and Applied*

*Chemistry* 69, 2471–2473, 1997). But having a single faulty system that depends in large measure on the historical contingencies of its genesis is better than having no general reference system at all. Kitcher (1984a, 326–327) disagrees. He thinks that we do not need to worry about a return to Babel. Biology can and does function just fine without such a general reference system. Generations of natural historians and systematists have been misguided in trying to produce a single, coherent, and consistent classification of plants and animals. Kitcher would no doubt disagree with a recent editorial in *Nature* (1997, vol. 389, 1) that took molecular biologists to task for the "promiscuity" of the numerous conflicting names that they coin for proteins and other molecular structures.

## PRUNING THE TREE OF KNOWLEDGE: THE NEED FOR CRITERIA

For Kitcher (1984a), systems of classification are theory dependent: that is, given one theory, you might divide up your subject matter one way; given a different theory, you might divide it up differently (see Dupré, chapter 1 in this volume). I agree, but in his early work on the species concept, Kitcher made only the faintest gestures toward specifying these all-important theoretical contexts. He refers to nothing more specific than the "legitimate interests of biology," "biologically interesting relations," and so on. Sober (1984a, 334) raises just this complaint against Kitcher's pluralism. Sober distinguishes between species concepts that are currently playing a significant role in active research programs and species concepts that depend on a "mere hope" and that lack "any serious degree of theoretical articulation." Too many of the species concepts that contribute to Kitcher's pluralism are of this second sort. There may well be laws of form, and some very good biologists are currently working on the theoretical background of such a view, but it still remains largely promissory.

Kitcher is not a total pluralist. He rejects both the creationist and the phenetic species concepts, but he does not explicitly set out the criteria that he has used in making these decisions. All he says is that we are obligated to pay attention to those concepts that are biologically interesting and ignore the "suggestions of the inexpert, the inane and the insane!" (Kitcher 1987, 190). Those people who are pushing the creationist species concept are certainly inexpert and inane, but they are hardly insane. I am not sure how to decide whether or not the creationist species concept is biologically interesting. I don't think it is. Pheneticists in their turn are neither inexpert nor insane. I also do not find their work on the species problem all that inane either. Phenetic (or numerical) taxonomists are genuine scientists working on genuinely scientific issues in genuinely scientific ways. They also obtain money from the National Science Foundation and publish in scientific journals. Many biologists have also found their views biologically interesting. If their species concept is to be rejected (as I think it must), it will have to be on grounds other than those that Kitcher mentions.

What both sides of the controversy need to do is to state criteria for including or excluding particular definitions of the species category and to give reasons why these criteria are appropriate. Ereshefsky (1992), Stanford (1995), Hull (1997), and Mayden (1997) have attempted to do just that. Ereshefsky groups the myriad definitions of the species category into three basic types: interbreeding, ecological, and phylogenetic. He then evaluates each type according to two basic kinds of principles—sorting and motivating. *Sorting principles* "sort the constituents of a theory into basic units," and motivating principles "justify the use of sorting principles" (Ereshefsky 1992, 682). With respect to the species category, Ereshefsky provides a single sorting principle and four motivating principles. The sorting principle for a taxonomic approach "should produce a single internally consistent taxonomy." The *motivating principles* are empirical testability, consistency with well-established hypotheses in other scientific disciplines, as well as consistency with and derivability from the tenets of the theory for which the taxonomy is produced (Ereshefsky 1992, 682).

When Ereshefsky applies his criteria to a variety of species definitions, he concludes that the only respectable definitions are historical; none of Kitcher's structural definitions are up to snuff. The interbreeding, ecological, and phylogenetic definitions are acceptable because they are based on evolutionary theory (characterizing a process) and phylogeny (the product of this process). Ereshefsky (1992, 684) dismisses both the creationist and phenetic species concepts, as one might expect, but he also dismisses the typological species concept of the idealists because it is "incompatible with current evolutionary biology." Kitcher would surely reply that there is more to biology than evolutionary biology. Perhaps some other area of biology, such as developmental biology, which is (largely?) independent of evolutionary biology might need a nonevolutionary way of grouping organisms, call them species or not.

I feel required to add that the typological or Aristotelian species concept is not quite as dead as Ereshefsky might lead one to believe. It is still alive and well in Catholic universities around the world (even after the current pope put his imprimatur on evolution); it plays a central role in the objectivist philosophy of Ayn Rand (a.k.a., Alisa Rosenbaum); it raises its head once again in connection with pattern cladists (Platnick 1979, 1985); and Atran (1990) argues that perceiving salient species of living creatures typologically is built into our genes. We are all born typologists, like it or not. Finally, Ereshefsky's requirement that each approach must produce a *single* internally consistent classification might strike some pluralists as too monistic.[3]

Stanford (1995) reacts primarily to Kitcher's brand of pluralism, arguing that it lacks adequate criteria for distinguishing between acceptable and unacceptable species concepts. (He also argues that Kitcher's brand of realism is incompatible with his pluralism, but that is another story.) Building on Kitcher's book *The Advancement of Science* (1993), Stanford (1995) proposes to base his criteria on the progress that we make in the questions that we ask in science. "Species divisions are the handmaidens of erotetic progress: *They enable us to make the significant questions through which we extend successful schemata more tractable*" (79, emphasis in original). Thus, we can reject species definitions if they are redundant, boring, or wrongheaded. A species division is redundant if it fails to "make any significant questions more tractable." A species division is boring if it does not "help us to pursue further goals." Finally, a species division is wrongheaded if the schemata on which it is based "involve presuppositions we believe are incorrect" (Stanford 1995, 80).

In his ensuing discussion, Stanford clarifies his criteria by applying them to particular cases. Creationism is wrongheaded. The explanatory schemata of the creationists "rest upon substantially mistaken presuppositions" (Stanford 1995, 80). For example, they attempt to argue away the implications of carbon dating by postulating a directional change in the rate of radioactive decay. They explain the patterns to be found in the fossil record in terms of which organisms could climb, swim, or fly the highest during the Great Flood. I need go on no further. According to Stanford, pheneticism is boring. The "natural dependencies identified by the pheneticists' Operational Taxonomic Units are trivial and unhelpful in pursuing any practical end" (but see Dupré, chapter 1 in this volume). The pheneticists would find this objection especially damning because they take the practical usefulness of their classifications to be one of their chief virtues. In defense of the phenetic species concept, even Mayr (1981), the chief opponent of pheneticism, finds the establishment of phena an important first step in the recognition of genuine species.

In the discussion of his examples, Stanford makes it clear that applications of his criteria are historically contingent. Right now, the creationist and phenetic species concepts can be rejected, but in the past they might well have led to scientific progress. For example, the appeal of pheneticism rested on a quite common, if not universal, conviction that something out there exists that answers to the name "overall similarity." If Atran (1990) is right, this mode of perception may have a significant genetic basis. If nothing else, the pheneticists showed that, contrary to their own goals, no such thing as overall similarity exists. If such bright, hardworking, and creative scientists in a period of twenty years or so could not come up with anything even approximating overall similarity, it is very likely not to exist in the first place. In Cuvier's day, there might have something to say for structuralism, but less so today.

## HULL AND MAYDEN ON SPECIES

In a recent paper (Hull 1997) I used three philosophical criteria—universality, applicability and theoretical significance—to evaluate nine species concepts currently being entertained by professional biologists to see if any of them score more highly on these criteria than do others. The goal was to see if the very general philosophical criteria that philosophers have developed to evaluate scientific concepts can distinguish between the nine species concepts. To begin, scientists value the *universality* of their concepts (but see Boyd, chapter 6 in this volume). For example, any definition of *element* must apply to all matter, not to just a subset. Physicists would be less than pleased if their element concept applied only to metals or to nonradioactive substances. Biologists would like their preferred species definition to apply to all organisms, not just some, but fulfilling this desideratum has proven to be very difficult. For example, the biological species concept applies to only those organisms that reproduce sexually, at least on occasion.

Biologists, like all scientists, would prefer that their concepts be *applicable*. Perhaps they need not be totally applicable in all circumstances, the way that operationists propose, but the more applicable they are, the better. At the very least, defining concepts in such a way that they can never be applied runs counter to the testability criterion of science. The testing may be difficult, indirect, and fallible, but it must be possible. Philosophers are usually content once we have decided that a particular concept in science is in principle applicable, but scientists want more, much more. They want *grouping* criteria—criteria that enable them to decided whether two or more organisms belong or do not belong to a species in a significant percent of the cases. They also want *ranking* criteria—criteria that enable them to decide whether a taxon is a subspecies, species, or genus (Mishler and Brandon 1987).

Finally, philosophers of science are currently convinced that to be useful, all scientific concepts must be *theoretically significant*. They must function in a significant scientific theory. Because no scientific concept can be totally theory free, the issue becomes which theory colors which concepts (see Dupré, chapter 1, this volume). Next we must rank these theories according to how fundamental they are. To use the traditional example, Newton's theory of universal gravitation (once fixed up) is more fundamental than Kepler's laws of planetary motion (once fixed up). On this view, those concepts required by the most fundamental theories take precedence to those concepts required by less fundamental theories. And if theory reduction is possible, all of these various upper-level theories can be reduced to the lower-level theories. With respect to the connection between process theories and the patterns discernible in nature, many scientists disagree with philosophers about the primacy of theories. Perhaps philosophers have worked their way free of inductivism, but many scientists have not. They insist that all scientific investigations must begin with direct, theory-free observation and proceed as cautiously as possible, avoiding idle speculation (see the papers in Claridge, Dawah, and Wilson 1997).

In my paper, I grouped the species concepts that I evaluated into three families. The first family includes species concepts that determine species status in terms of some form of similarity—for example, traditional morphological species concepts, the phenetic species concept, as well as certain molecular concepts. Until the past few decades, all of these species concepts were typological in the sense that a single list of characters was developed and only those organisms exhibiting all of these characters are considered to belong to the same species. Now, cluster analysis is the norm for estimating similarity (see Hull 1965 and Boyd, chapter 6 in this volume). The second family of species concepts has been generated by evolutionary biologists. The intent is to discern basic units of evolution. Included in this group are the evolutionary species concept that can be traced back to Simpson (1961) and Hennig (1966), especially as developed by Wiley (1981); Mayr's (1969) biological species concept; Paterson's (1981) species mate recognition concept; and Templeton's (1989) cohesion concept.

The third family of species concepts overlaps the preceding two families of concepts because I have distinguished it by means of historical considerations. I have constructed it in this way because of the historical connection to Hennig's phylogenetic systematics.[4] Hennig (1966, 32) considered himself to be adopting a species concept common in his day—something like the concepts of Simpson and Mayr (but see Meier and Willmann 1999). When the reticulate relationships of sexual organisms are rent and splitting occurs, then one species has evolved into two species. However, Hennig's descendants have developed species concepts that, though they may be descended from his ideas, are not exactly coincident with them. The first is the monophyletic (or autapomorphic) species concept of Mishler and Donoghue (1982), Mishler and Brandon (1987), de Queiroz and Donoghue (1988, 1990), and Mishler and Theriot (1999). According to the monophyletic species concept, a species is the least inclusive monophyletic group definable by at least one autapomorphy. Hennig limited the application of the term *monophyly* to higher taxa. The advocates of the monophyletic species definition extend it to cover species as well.

The diagnostic species concept also grew out of Hennig and, in this case, Rosen (1978, 1979), but it has taken its own direction (Platnick 1977, Eldredge and Cracraft 1980, Nelson and Platnick 1981). The most influential formulation of the diagnostic species concept was first presented by Cracraft (1983) and then further developed by McKitrick and Zink (1988), Nixon and Wheeler (1990), Wheeler and Nixon (1990), and Vrana and Wheeler (1992), among others. According to this view, a species is the smallest diagnosable cluster of individual organisms within which there is a parental pattern of ancestry and descent. The monophyletic species concept emphasizes phylogeny, though not the processes that give rise to phylogeny. The diagnostic species concept acknowledges the importance of genealogy, ancestor-descendant relations among organisms, but not necessarily phylogeny. It does not depend on the evolutionary process over and above genealogy.

In the span of this essay, I cannot go through all of the particulars of the preceding species definitions and their variants. At the very least, the presence of so many species definitions taken seriously by professional biologists should warm the hearts of pluralists. As much as each of these authors is sure that he has the correct species definition, such monistic inclinations have yet to narrow this list of species definitions significantly. As far as *universality* is concerned, the phenetic species concept is the most general concept because it applies equally to all organisms. The price that one pays for universality of this sort is that males and females may end up not being classed in the same species. The females of one species may be more similar to the females in another species than they are to the males of their own species. On a strictly phenetic species concept, that eventuation would be perfectly acceptable. The biological and mate recognition concepts are the least universal because they apply only to those organisms that reproduce sexually with reasonable frequency. Where these species concepts apply, they distinguish significant evolutionary units. The trouble is that they do not apply to organisms that existed during the first half of life on Earth or to many groups of organisms today. During the first half of life on Earth, evolution occurred, but not with the aid (or effect) of species (see Ereshefsky, Sterelny and Nanney, this volume). All other species concepts are arrayed between these two extremes.

As far as *applicability* is concerned, the species definitions that I treated are arrayed much more continuously from the phenetic species at one end to the evolutionary species concept at the other end. The phenetic species concept is the most operational because that was the main reason for developing it. The diagnostic species concept is the next most applicable concept. Systematists must know who tends to mate with whom and what the results of these unions are for sexual organisms, as well as who gives rise to whom for asexual organisms. The only other bit of information that they need is character covariation. The monophyletic species concept is as operational as the methods of cladistic analysis permit. The goal of cladistic analysis is the individuation of characters so that they nest perfectly. The mate recognition, biological, and cohesion concepts are even more difficult to apply because the forces and mechanisms that they specify are more difficult to discern. Finally, the evolutionary species concept is most difficult to apply because it explicitly specifies that species are extended in time. Decisions about species status are contingent upon what will happen in the future. Species can be determined only in retrospect (see Sterelny, chapter 5 in this volume).

With respect to *theoretical significance*, only the phenetic species concept is designed to be totally theory neutral or theory free. The diagnostic species concept assumes only some very low-level, unproblematic theories, whereas all other species concepts are openly theoretical in their content. Whether or not this characteristic is a virtue or a vice varies in the systematics community. Some concepts assume knowledge of the evolutionary process. Others assume only phylogenies regardless of the processes that produced them. Some are "nondimensional"—that is, extended only minimally in space and time. Others treat species as lineages.

When I first set myself the task of evaluating representative species concepts on the basis of widely assumed philosophical criteria for good scientific classifications, I assumed that one or two would emerge as better than the others. After all, I am a monist. However, no matter how I massaged the data, I could not produce the result I had anticipated. All of the species concepts I evaluated scored about the same! One reason for this outcome is that the most easily applied concepts tend to be those with the least theoretical commitment, whereas those concepts that produce theoretically significant species tend to be the most difficult to apply. Universality, in its turn, does not covary with either theoretical significance or ease of application. Some theoretically committed species concepts, such as the monophyletic species concepts, apply to all organisms and are moderately easy to apply.

The grudging conclusion of my paper (Hull 1997) is that none of the species concepts that I evaluated are all that superior to the others—that is, if universality, applicability, and theoretical significance are weighted equally. However, in this same volume, Mayden (1997) set himself the same task, but came to quite a different conclusion. Mayden's goal was to find *the* primary species concept. The differences between our papers is instructive. First, Mayden recognizes twenty-two different species concepts, and he combines as single concepts several formulations that others take to be separate and distinct species concepts. For example, the two formulations of the phylogenetic species concept that I classified as two separate species concepts are classified by Mayden as a single species concept. Next, he evaluates these twenty-two concepts according to their "convenience, accuracy, precision, and the successful recovery of natural biological diversity" (Mayden 1997, 381).

Finally, Mayden evaluates his species concepts on several additional criteria that are a good deal more specific. To serve as the primary species concept, a concept must be theoretically significant and include sexual, asexual, and hybrid species; it must be a nonrelational lineage concept that treats species as individuals rather than as classes; and it must place no constraints on necessary attributes that a species must possess in order to be validated. As the primary species concept, it need not be operational because other, secondary species concepts provide the operational basis for this primary species concept. For the primary species concept, theoretical significance is of primary importance. Only after a species concept passes this test do the other criteria come into play. The only species concept that fulfills all of these criteria is the evolutionary species concept as reworked by Wiley (1981) and by Wiley and Mayden (1999).

The main reason why Mayden and I came to very different conclusions in evaluating various species definitions is that I combined theoretical significance and operationality. A species concept might score quite highly on theoretical significance, but if it was not very operational, it ended up with a mediocre cumulative score. Mayden, to the contrary, took theoretical significance as necessary and then ranked theoretically significant species according to his other criteria, including operationality. A second reason why Mayden's conclusion is so different from mine is that he included more substantive criteria. Yes, I think that any adequate species concept must treat species as lineages, but as a philosopher evaluating species concepts from the outside, I felt committed to being open-minded about such issues. As a scientist engaged in these disputes, Mayden felt no such compunction. As a result, he was able to arrive at a *single* primary species concept.

## CONCLUSION

Postmodernists have made "positivists" all-purpose whipping boys, usually parodying their views in the process. Other authors have also joined in these parodies. For example, just about everyone claims that attempting to demarcate science from nonscience or pseudoscience is terribly wrongheaded, but then what do we do about creation science? Some of it is very bad science; some of it is not science at all (Reisch 1998). Has philosophy really become so sophisticated and nuanced that we can't distinguish between science and creation science? My fellow philosophers are likely to respond that courts of law are not graduate seminars, and they insist on limiting themselves to graduate seminars. The rest of society is irrelevant.

In deciding which species concepts to take seriously, we seem inextricably caught up in the issue of what counts as genuine science and what not. Kitcher rejects the creationist and phenetic species concepts for philosophical reasons. The problem with pheneticism is that it comes into conflict with his philosophical monism of the moment—namely, that no such things as theory-free observations, let alone concepts, exist. Hence, any attempt to define the species category in a theoretically neutral way is impossible. Kitcher is putting his bet on this philosophical position prevailing for awhile. I share Kitcher's prediction. Not only will philosophers continue to value theoretical significance highly, but I am betting that an increasing numbers of systematists will come to share this conviction as phylogenetic cladists win out over their pattern cladist brethren.

But theoretical significance only narrows the number of philosophically acceptable species concepts. For those of us who are more monistically inclined, traditional philosophical criteria alone are not sufficient for evaluating species concepts. On this score, *there is more to science than philosophy*. Right now, an extremely powerful, well-articulated theory actually exists in biology—evolutionary theory. This theory places constraints on both species definitions and traditional philosophical desiderata. Of course, other theories are possible. Some scientists are making a little headway in articulating alternative ways of viewing the living world, but until the promise of these alternatives is realized, we cannot treat them on a par with evolutionary theory. In connection with our understanding of the evolutionary process, Sober (1984a, 335) is "guessing that [the] species-are-individuals perspective will win" (but see Wheeler and Platnick 1999). I share Sober's conviction, but someday, way down the road, this perspective on species may be overturned. The possibility of future alternatives is not, however, a sufficiently strong reason for accepting the pluralist philosophical monism of the moment.

Mayden (1997) concludes that one species concept is preferable to all others for the role of the primary species concept. None too surprisingly, the preferable species concept is the one that he prefers—the revised evolutionary species concept. Isn't Mayden simply reasoning in a circle? He started his investigations preferring the revised evolutionary species concept, and he concluded that it was preferable. However, even biased investigations can turn out differently from our expectations. I know that mine did. Mayden might have discovered to his dismay that the revised evolutionary species concept does not stand head and shoulders above its competitors. Stranger things have happened. Mayden introduces criteria for evaluating species concepts that are closely connected to the sort of science that he wants to conduct. As a scientist engaged in the process that he is investigating, he cannot play it coy. He must commit himself. These additional commitments are what allow him to select one species concept as preferable.

As far as strategies in science are concerned, sometimes scientists work themselves into a tight corner. They can see one view of the world and one view only. During such times, I would join in the cry for pluralism. We need to get flexible, proliferate alternatives even if they are not very well supported, and so on. Sometimes, however, scientists are lost in conceptual brambles. Too many alternatives present themselves, and there seems to be no way to decide among them. During these times, a strong dose of monism is called for to help prune the tree of knowledge. With respect to the species problem right now, the situation clearly seems to exemplify the second extreme. We are drowning in a sea of species concepts. Hence, scientists are justified in being more monistic than they have in the past. Perhaps more than one species concept is justified, but twenty-two?

But what do I think? Did God create a single sort of species or many different sorts? Is there a single level of organization across all organisms that is in some significant sense the "same"? If we take for granted the traditional organizational hierarchy of cells, organs, organisms, colonies, populations, and so on, the answer is clearly no. What if we take as the basic level the level at which reticulation is converted to divergence? This level is certainly significant as far as the evolutionary process is concerned, but in certain areas of the phylogenetic tree, organisms exhibit this level of organization, in other areas colonies exhibit this level of organization, in other areas species, and in still other areas higher taxa. In asexual uniparental organisms that do not exchange genetic material even parasexually, splitting occurs at the level of single organisms. Are we to call each of these organisms a separate species (see Dupré, chapter 1, Nanney, chapter 4 in this volume)? In sexual species, reticulation does not cease until speciation has occurred, but in many groups of organisms, reticulation is relatively prevalent even among organisms commonly classed as belonging to different genera. The problem seems to be the traditional organizational hierarchy. As long as it seems so right to us, any efforts to distinguish a single level of organization that counts as the same across all organisms may be very strongly counterintuitive. Perhaps we need to change our intuitions.

## ACKNOWLEDGMENTS

I wish to thank John Dupré, Marc Ereshefsky, Kim Sterelny, M. H. V. Van Regenmortal, Ed Wiley, Rob Wilson, and an anonymous referee for reading and commenting on this paper.

## NOTES

1. As Dupré remarked in response to the example given, always using virus-impenetrable rubbers would defeat the AIDS epidemic in two ways: it would prevent the transmission of the virus to new hosts via sexual intercourse, but in the process it would also preclude the birth of new human beings because such prophylactics would also be sperm impenetrable, a cure decidedly worse than the disease.

2. Kitcher (1987, 187) enlists Mishler and Donoghue (1982) as fellow pluralists, but they protest. "Kitcher's (1984a, 1984b) brand of pluralism implies that there are many possible and permissible species classifications for a given situation (say, the *Drosophila melanogaster* complex), depending on the needs and interest of particular systematists. In contrast, Mishler and Donoghue's (1982) brand of pluralism implies that a single optimal general-purpose classification exists for each particular situation, but that the criteria applied in each situation may well be different" (Mishler and Brandon 1987, 403). Nor is Templeton (1989) a Kitcher-style pluralist. He thinks that one character and one character only is relevant to species status—cohesiveness—even though different mechanisms can contribute to this cohesiveness (see also Donoghue 1985 and Ereshefsky 1992).

3. Pluralism is at bottom incompatible with Whewell's consilience of inductions. Because Michael Ruse is among the most enthusiastic supporters of this principle, it comes as a surprise to find him on Ereshefsky's list of pluralists, albeit as highly conservative pluralist. Although Ruse (1987, 238) argues that there are "different ways of breaking organisms into groups, and they *coincide*!" I happen to think that he is far too sanguine on this point.

4. I am aware that the decisions I have made at this higher level of abstraction mirror the decisions others have made at the level of species concepts. My allowing historical considerations to override similarity with respect to how I classify species concepts is likely to imply something about which considerations I find most important at the species level.

## REFERENCES

Atran, S. (1990). *Cognitive foundations of natural history: Towards an anthropology of science*. Cambridge: Cambridge University Press.

Blattner, W., R. C. Gallo, and H. M. Temin (1988). HIV causes AIDS. *Science* 241, 515.

Claridge, M. F., H. A. Dawah, and M. R. Wilson, eds. (1997). *Species: The units of biodiversity*. London: Chapman and Hall.

Collins, H. M. (1981a). Stages in the empirical programme of relativism. *Social Studies of Science* 11, 3–10.

Collins, H. M. (1981b). Son of seven sexes: The social destruction of a physical phenomenon. *Social Studies of Science* 11, 33–62.

Collins, H. M. (1981c). What is TRASP?: The radical programme as a methodological imperative. *Philosophy of the Social Sciences* 11, 215–224.

Cracraft, J. (1983). Species concepts and species analysis. In R. F. Johnson, ed., *Current Ornithology*. New York: Plenum Press.

de Queiroz, K., and M. J. Donoghue (1988). Phylogenetic systematics and the species problem. *Cladistics* 4, 317–338.

de Queiroz, K., and M. J. Donoghue (1990). Phylogenetic systematics and species revisited. *Cladistics* 6, 83–90.

de Sousa, R. (forthcoming). Biological individuality. *Biology and Philosophy.*

Dick, S. J. (1982). *Plurality of worlds: The origins of the extraterrestrial life debate from Democritus to Kant.* Cambridge: Cambridge University Press.

Donoghue, M. J. (1985). A critique of the BSC and recommendations for a phylogenetic alternative. *The Bryologist* 83, 172–181.

Duesberg, P. H. (1987). Retroviruses as carcinogens and pathogens: Expectations and reality. *Cancer Research* 47, 1199–1220.

Duesberg, P. H. (1996). *Inventing the AIDS virus.* Washington, D.C.: Regnery.

Dupré, J. (1981). Natural kinds and biological taxa. *Philosophical Review* 90, 66–90.

Dupré, J. (1993). *The disorder of things: Metaphysical foundations of the disunity of science.* Cambridge, Mass.: Harvard University Press.

Eldredge, N. (1985). *Unfinished synthesis.* Oxford: Oxford University Press.

Eldredge, N., and J. Cracraft (1980). *Phylogenetic patterns and the evolutionary process: Method and theory in comparative biology.* New York: Columbia University Press.

Ereshefsky, M. (1992). Eliminative pluralism. *Philosophy of Science* 59, 671–690.

Fisher, R. A. (1958). Lung cancer and cigarettes? *Nature* 182, 108.

Ghiselin, M. T. (1974). A radical solution to the species problem. *Systematic Zoology* 23, 536–544.

Ghiselin, M. T. (1989). Sex and the individuality of species: A reply to Mishler and Brandon. *Biology and Philosophy* 4, 73–76.

Goodman, N. (1972). *Problems and projects.* Indianapolis: Bobbs-Merrill.

Griffiths, P. E., and R. D. Gray (1994). Developmental systems and evolutionary explanations. *Journal of Philosophy* 91, 277–304.

Hennig, W. (1966). *Phylogenetic systematics.* Chicago: University of Illinois Press.

Holsinger, K. E. (1984). The nature of biological species. *Philosophy of Science* 51, 293–307.

Holsinger, K. E. (1987). Discussion: Pluralism and species concepts, or when must we agree with one another? *Philosophy of Science* 54, 480–485.

Hull, D. L. (1965). The effects of essentialism on taxonomy. *British Journal for the Philosophy of Science* 15, 314–326, and 16, 1–18.

Hull, D. L. (1976). Are species really individuals? *Systematic Zoology* 25, 174–191.

Hull, D. L. (1978). A matter of individuality. *Philosophy of Science* 45, 335–360.

Hull, D. L. (1997). The ideal species definition and why we can't get it. In M. F. Claridge, H. A. Dawah, and M. R. Wilson, eds., *Species: The units of biodiversity.* London: Chapman and Hall.

Kitcher, P. (1984a). Species. *Philosophy of Science* 51, 308–333.

Kitcher, P. (1984b). Against the monism of the moment: A reply to Elliott Sober. *Philosophy of Science* 51, 616–630.

Kitcher, P. (1987). Ghostly whispers: Mayr, Ghiselin and "the philosophers" on the ontology of species. *Biology and Philosophy* 2, 184–192.

Kitcher, P. (1989). Some puzzles about species. In M. Ruse, ed., *What philosophy of biology is not.* Kluwer: Dordrecht.

Kitcher, P. (1993). *The advancement of science.* New York: Oxford University Press.

Kitcher, P., K. Sterelny, and K. Waters (1990). The illusory riches of Sober's monism. *Journal of Philosophy* 87, 158–161.

Mayden, R. L. (1997). A hierarchy of species concepts: The denouement in the saga of the species problem. In M. F. Claridge, H. A. Dawah, and M. R. Wilson, eds., *Species: The units of biodiversity.* London: Chapman and Hall.

Maynard Smith, J. (1998). The origin of altruism. *Nature* 393, 634–640.

Mayr, E. (1969). *Principles of systematic zoology.* New York: McGraw-Hill.

Mayr, E. (1981). Biological classification: Toward a synthesis of opposing methodologies. *Science* 214, 510–516.

Mayr, E. (1999). The biological species concept. In Q. D. Wheeler and R. Meier, eds., *Species concepts and phylogenetic theory: A debate.* New York: Columbia University Press.

McKitrick, M. C., and R. M. Zink (1988). Species concepts in ornithology. *Condor* 90, 1–14.

Meier, R., and R. Willmann (1999). The Hennigian species concept. In Q. D. Wheeler and R. Meier, eds., *Species concepts and phylogenetic theory: A debate.* New York: Columbia University Press.

Mishler, B. D., and R. N. Brandon (1987). Individualism, pluralism, and the phylogenetic species concept. *Biology and Philosophy* 2, 397–414.

Mishler, B. D., and M. J. Donoghue (1982). Species concepts: A case for pluralism. *Systematic Zoology* 31, 491–503.

Mishler, B. D., and E. Theriot (1999). Monophyly, apomorphy, and phylogenetic species concepts. In Q. D. Wheeler and R. Meier, eds., *Species concepts and phylogenetic theory: A debate.* New York: Columbia University Press.

Moss, L. (1992). A kernel of truth? On the reality of the genetic program. In D. Hull, M. Forbes, and K. Okruhlic, eds., *PSA 1992,* vol. 1. East Lansing, Mich.: Philosophy of Science Association, 335–348.

Nelson, G., and N. Platnick (1981). *Systematics and biogeography: Cladistics and vicariance.* New York: Columbia University Press.

Nixon, K. C., and Q. D. Wheeler (1990). An amplification of the phylogenetic species concept. *Cladistics* 6, 211–223.

Paterson, H. E. H. (1981). The continuing search for the unknown and unknowable: A critique of contemporary ideas on speciation. *South African Journal of Science* 77, 113–119.

Platnick, N. (1977). Cladograms, phylogenetic trees, and hypothesis testing. *Systematic Zoology* 26, 438–442.

Platnick, N. (1979). Philosophy and the transformation of cladistics. *Systematic Zoology* 28, 537–546.

Platnick, N. (1985). Philosophy and the transformation of cladistics revisited. *Cladistics* 1, 87–94.

Reisch, G. A. (1998). Pluralism, logical empiricism, and the problem of pseudoscience. *Philosophy of Science.*

Rosen, D. E. (1978). Vicariant patterns and historical explanation in biogeography. *Systematic Zoology* 27, 159–188.

Rosen, D. E. (1979). Fishes from the uplands and intermontane basins of Guatemala. *Bulletin of the American Museum of Natural History* 162, 267–376.

Rosenberg, A. (1975). The virtues of vagueness in the language of science. *Dialogue* 14, 281–305.

Ruse, M. (1987). Biological species: Natural kinds, individuals, or what? *British Journal for the Philosophy of Science* 38, 225–242.

Shanahan, T. (1997). Pluralism, antirealism, and the units of selection. *Acta Biotheoretica* 45, 117–126.

Simpson, G. G. (1961). *Principles of animal taxonomy*. New York: Columbia University Press.

Sober, E. (1984a). Discussion: Sets, species, and evolution. Comments on Philip Kitcher's "Species." *Philosophy of Science* 51, 334–341.

Sober, E. (1984b). *The nature of selection*. Cambridge, Mass.: MIT Press.

Sober, E. (1993). *Philosophy of biology*. Boulder, Colo.: Westview Press.

Sober, E., and D. S. Wilson. (1994). A critical review of philosophical work on the units of selection problem. *Philosophy of Science* 61, 534–555.

Sober, E., and D. S. Wilson. (1998). *Unto others: the evolution and psychology of unselfish behaviors*. Cambridge: Harvard University Press.

Stanford, P. K. (1995). For pluralism and against monism about species. *Philosophy of Science* 62, 70–91.

Sterelny, K., and P. Kitcher (1988). The return of the gene. *Journal of Philosophy* 85, 339–361.

Templeton, A. R. (1989). The meaning of species and speciation. In D. Otte and J. A. Endler, eds., *Speciation and its consequences*. Sunderland, Mass.: Sinauer.

Vrana, P., and W. Wheeler (1992). Individual organisms as terminal entities: Laying the species problem to rest. *Cladistics* 8, 67–72.

Waters, K. (1991). Tempered realism about the force of selection. *Philosophy of Science* 58, 553–573.

Webster, G., and B. Goodwin (1996). *Form and transformation: Generative and relational principles in biology*. Cambridge: Cambridge University Press.

Wheeler, Q. D., and K. C. Nixon (1990). Another way of looking at the species problem: A reply to Queiroz and Donoghue. *Cladistics* 6, 77–81.

Wheeler, Q. D., and N. I. Platnick (1999). The phylogenetic species concept *sensu* Wheeler and Platnick. In Q. D. Wheeler and R. Meier, eds., *Species concepts and phylogenetic theory: A Debate*. New York: Columbia University Press.

Whewell, W. (1853). *Of the plurality of worlds*. London: John W. Parker.

Wiley, E. O. (1981). *Phylogenetics: The theory and practice of phylogenetic systematics*. New York: John Wiley.

Wiley, E. O., and R. L. Mayden (1999). The evolutionary species concept. In Q. D. Wheeler and R. Meier, eds., *Species concepts and phylogenetic theory: A debate*. New York: Columbia University Press.

Wilson, D. S., and E. Sober (1994). Re-introducing group selection to the human behavioral sciences. *Behavioral and Brain Sciences* 117, 585–654.

Wilson, J. (1999). *Biological individuality*. Cambridge: Cambridge University Press.

# 3 The General Lineage Concept of Species and the Defining Properties of the Species Category

Kevin de Queiroz

*There is nothing more common than that the meaning of an expression varies in such a way that a phenomenon is now considered as a symptom and now as a criterion of a state of affairs. And then for the most part in such a case the change of meaning is not noticed. In science it is usual to turn phenomena which allow exact measurements into defining criteria of an expression; and one is then inclined to think that now the genuine meaning has been found. An enormous number of confusions arise in this way.*
—Wittgenstein (1967)

Given the proliferation of species concepts in recent years, it might seem that the species problem—the difficulty of reaching agreement about the definition of the species category—is as far from being solved as it has ever been. On the contrary, the species problem has, for the most part, already been solved. Despite the considerable diversity among contemporary views on species, all are encompassed by a single, general concept that equates species with segments of population-level lineages. Because this population lineage concept underlies virtually all modern ideas about species, it bears on almost every historical and philosophical question that one would care to ask about those ideas, including the major themes of this volume. In this essay, I describe the general concept of species as segments of population lineages and show how it encompasses the diversity of modern views on species. I then discuss two assumptions that, despite widespread agreement about the general nature of species, lead to incompatible species concepts. I show how eliminating one of those assumptions, which entails reconsidering the defining properties of the species category, effectively solves the species problem. I then use this perspective to clarify several philosophical issues concerning species, including the role of the species concept in biology, the individuality of species, whether the species category is a relational concept, monistic versus pluralistic views of species, and species realism. Finally, I briefly describe the history of the lineage concept of species.

## THE GENERAL LINEAGE CONCEPT OF SPECIES

In a previous paper (de Queiroz 1998), I argued that all modern species concepts are variants of a single general concept of species. In that paper, I presented evidence that every modern species definition in a diverse sample either explicitly or implicitly equates species with segments of population lineages. I also argued that most of the differences among what have been called *species concepts* in the literature of the last thirty years involve species criteria, and I proposed a revised terminology that more clearly distinguishes between the various concepts, criteria, and definitions.[1] Rather than repeating the same arguments in the present essay, I emphasize here how the most fundamental differences among modern views on species are nonetheless compatible with the general concept of species as population lineages. First, however, I must describe the general lineage species concept itself. Because the concept of a lineage is fundamental to this concept, I start by clarifying some things about lineages.

## Lineages

I have used the term *lineage* (de Queiroz 1998; see also Simpson 1961, Hull 1980) for a series of entities forming a single line of direct ancestry and descent. For example, a lineage can be traced from a given organism backward though a parent, grandparent, great-grandparent, and so on, and forward through a child, grandchild, great-grandchild, and so on. Biological entities at several different organizational levels form lineages. Thus, biologists speak of gene lineages, organelle lineages, cell lineages, organism lineages (as described in the above example), and population lineages. Because entities that form lineages often make up, or are made up of, entities at different organizational levels, the same is also true of the lineages themselves. An organism lineage, for example, is (often) made up of multiple cell lineages, and multiple organism lineages make up a population lineage.

Lineages in the sense described above are unbranched; that is, they follow a single path or line anytime an entity in the series has more than one direct descendant (figure 3.1a). Consequently, lineages are not to be confused with clades, clans, and clones—though the terms are often used interchangeably in the literature.[2] Clades, clans, and clones include all paths or lines of descent from a given ancestor and thus are branched, which is to say that they are composed of multiple lineages (figure 3.1b). Moreover, clades, clans, and clones are monophyletic by definition; a clade, for example, is defined as a monophyletic group of species.[3] Lineages, in contrast, can be paraphyletic or even polyphyletic in terms of their lower-level components (see "Phyly"). They can even be paraphyletic in terms of their segments at the same organizational level. Thus, the later segments of a lineage commonly share more recent common ancestors with separate but recently diverged lineages than they do with earlier segments of their own lineage (figure 3.2).

## Species

Definitions that equate species with lineages refer to lineages at a level of organization commonly referred to as the *population level* (e.g., Griffiths

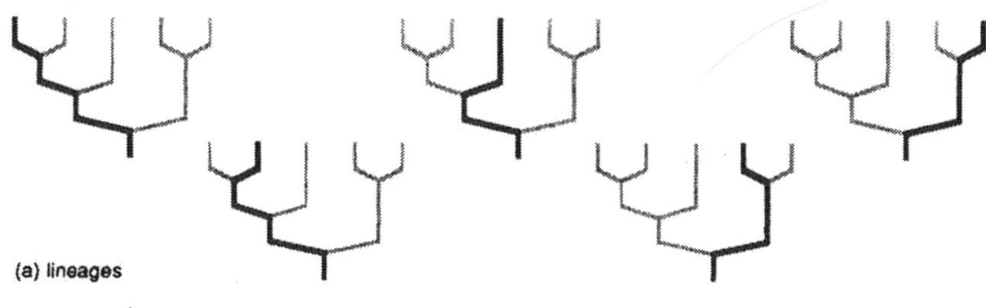


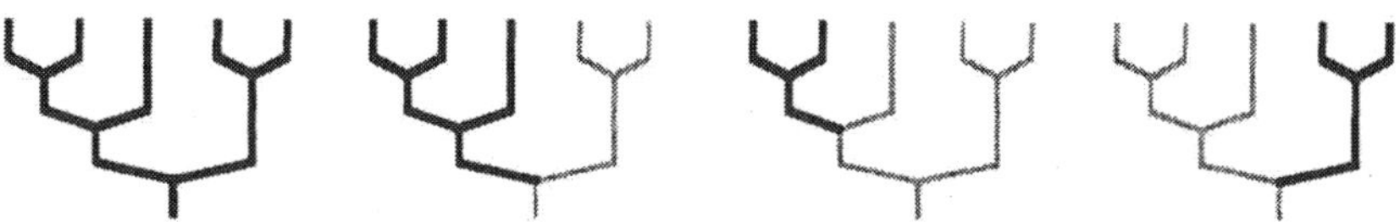


**Figure 3.1** Lineages contrasted with clades, clans, and clones (after de Queiroz 1998). All of the branching diagrams represent the same phylogeny with different lineages highlighted in (a) and different clades, clans, or clones highlighted in (b). Notice that the lineages are unbranched and partially overlapping, whereas the clades, clans, or clones are branched and either nested or mutually exclusive. Additional (partial) lineages can be recognized for paths beginning at various internal nodes.

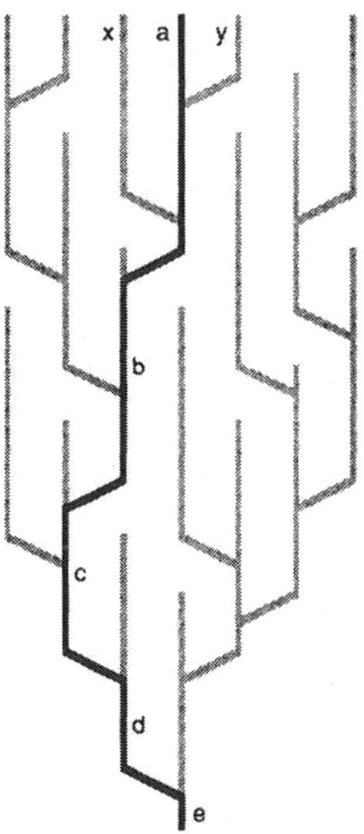


**Figure 3.2** Paraphyly of lineages. The later parts—e.g., (*a*)—of the highlighted lineage share more recent common ancestors (*b*) with separate but recently diverged lineages (*x*, *y*) than they do with earlier parts of their own lineage (*c*, *d*, *e*).

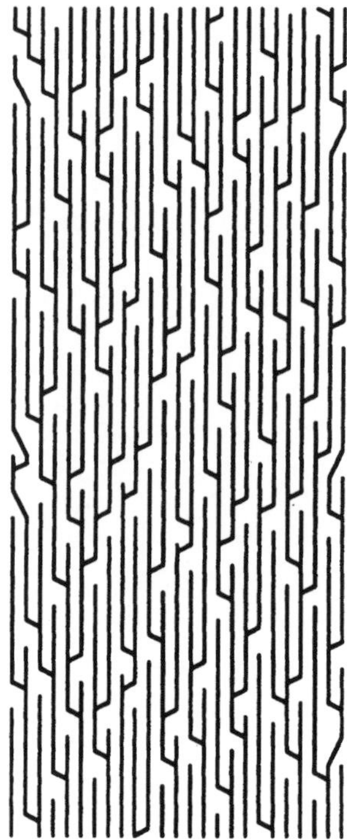


**Figure 3.3** Population lineages in sexually and asexually reproducing organisms (adapted from Brothers 1985). (a) Under sexual reproduction, organism lineages are connected through the process of reproduction itself (represented by connections [^] between vertical lines) to form a population-level lineage. (b) Under asexual reproduction, no such reproductive connections exist, but it is possible that the organism lineages are bound into a population lineage by other processes (represented by the spatial localization of the organism lineages). In both diagrams, organisms are represented by vertical lines.

1974)—that is, to groups of organism lineages that are united to form higher-level lineages. The formation of population-level lineages is most evident in the case of biparental organisms, where the process of sexual reproduction continually reconnects temporarily separated organism lineages to form a unified nexus (figure 3.3a). At least some authors, however, believe that uniparental organisms also form species (figure 3.3b). Because a general species concept (i.e., one that can encompass the diversity of modern views about species) must allow for this possibility, I use the term *population* in the general sense of an organizational level above that of the organism, rather than in the specific sense of a reproductive community of sexual organisms.

The population level is really a continuum of levels. Lineages at lower levels in this continuum (e.g., demes or deme lineages) often separate and reunite over relatively brief time intervals. Toward the other end of the continuum, lineage separation is more enduring and can even be permanent.

Thus, when I say that a lineage is unbranched, I do not mean that it can never exhibit internal branching; however, any such branching that it exhibits would have to be judged as ephemeral. In any case, most authors equate species with lineages toward the latter end of the continuum, though they differ with regard to the precise point that they consider the line of demarcation for species.

Under the lineage concept of species, species are not equivalent to entire population lineages, but rather to segments of such lineages. Just as a cell lineage is made up of a series of cells and an organism lineage of a series of organisms, a species (population) lineage is made up of a series of species. Not just any lineage segment qualifies as a species, however. Instead, a species corresponds with a lineage segment bounded by certain critical events. Authors disagree, however, about which events are critical.

In short, species are segments of population-level lineages. This definition describes a very general conceptualization of the species category in that it explains the basic nature of species without specifying either the causal processes responsible for their existence or the operational criteria used to recognize them in practice.[4] It is this deliberate agnosticism with regard to causal processes and operational criteria that allows the concept of species just described to encompass virtually all modern views on species, and for this reason, I have called it the *general lineage concept of species* (de Queiroz 1998).

## THE UNITY AND DIVERSITY OF SPECIES CONCEPTS

By identifying the unity of contemporary species concepts, the general lineage concept of species provides a context for understanding their diversity. Stated in the most general terms, that diversity results from different authors emphasizing different aspects or properties of the entities conforming to the general lineage concept. In the remainder of this section, I describe some of the major differences among contemporary ideas about species as well as the relationship of those ideas to the general lineage concept. This exercise is not intended to describe the diversity of such ideas exhaustively, but rather to illustrate that even what seem to be the most fundamental differences among contemporary views on species are compatible with the general lineage concept.

### Populations and Lineages

One of the major differences among contemporary views on species concerns the terms used to describe the entities in question and the temporal perspectives that they imply. Some authors describe species as *populations* (e.g., Wright 1940; Mayr 1942, 1963; Dobzhansky 1950, 1970; Paterson 1978; Rosen 1979; Templeton 1989), whereas others describe them as *lineages* (e.g., Simpson 1951, 1961; Van Valen 1976; Wiley 1978, 1981; Mishler

1985). These two classes of species definitions are not at odds with one another, and both are entirely consistent with the general lineage concept of species. As has been noted by several authors, a lineage (at the population level) is a population extended through time, whereas a population (in itself) is a short segment—a more or less instantaneous cross section—of a lineage (see Simpson 1951, 1961; Meglitsch 1954; George 1956; Newell 1956; Rhodes 1956; Westoll 1956).[5] Thus, definitions that equate species with populations consider the entities of interest over relatively short time intervals, whereas those definitions that equate species with lineages consider them over longer time intervals. In other words, the two categories of definitions do not describe different concepts of species; they merely describe time-limited and time-extended versions of the same species concept.

### Processes and Products

Related to the difference in the timescale within which species are considered is a difference in whether to emphasize the processes responsible for the existence of population-level lineages or the products of those processes—the lineages themselves. Because putative unifying processes, such as gene flow and natural selection, are most easily studied in the present, those processes tend to be emphasized by neontologists, particularly population biologists (e.g., Wright 1940; Dobzhansky 1950, 1970; Mayr 1942, 1963; Paterson 1985; Templeton 1989). But even species that exist in the present are not restricted to that time plane, and most of the species that have ever existed are long extinct. Because it is difficult to study processes such as gene flow and natural selection as they occurred in the past, the lineages themselves, rather than their putative unifying processes, tend to be emphasized by paleontologists (e.g., Simpson 1951, 1961; Rhodes 1956; Westoll 1956; Newell 1956; George 1956; Polly 1997). In any case, processes and their products are intimately related, so that an emphasis on one or the other does not reflect a fundamental difference regarding ideas about the nature of species.

### Relative Importance of Different Processes

Even authors who emphasize unifying processes disagree about the relative importance of different processes for the existence of species. Many have considered interbreeding—or more generally, gene flow—the most important process (e.g., Dobzhansky 1937, 1950, 1970; Mayr 1963, 1969; Grant 1963). Others have called attention to the maintenance of apparently separate species despite interbreeding between their component organisms (e.g., Simpson 1951; Van Valen 1976; Templeton 1989) and have favored natural selection as the process responsible for maintaining separation (e.g., Ehrlich and Raven 1969; Van Valen 1976; Andersson 1990). Still others have discussed common descent and the processes that underlie genetic, developmental, ecological, and historical constraints (e.g., Mishler and Donoghue 1982; Templeton 1989). To the extent that all of these proposals are theories about the process or processes responsible for unifying organism lineages to form population lineages, advocacy of any one (or more) of them is entirely compatible with the general lineage concept of species.

## Sexual and Asexual Reproduction

Related to the differences about the processes responsible for the existence of species is a difference regarding whether asexual (uniparental) organisms form species. Some authors (e.g., Dobzhansky 1937; Hull 1980) maintain that asexual organisms do not form species, whereas others (e.g., Meglitsch 1954, Templeton 1989) argue that they do.[6] Whether asexual organisms form species is more or less the same question as whether sexual reproduction (gene flow) is the only process that unites organism lineages to form populations and thus population-level lineages (figure. 3.3). Not surprisingly, those authors who believe that asexual organisms form species also tend to view processes other than gene flow as important for the existence of population-level lineages (e.g., Templeton 1989), whereas those authors who believe that only sexual (biparental) organisms form species tend to view gene flow as the most important, if not the only, process. In any case, disagreements about the existence of species in asexual organisms only reinforce the equation of species with population-level lineages in that they boil down to a disagreement about whether asexual organisms form such lineages.

## Theory and Operations

Another major difference concerning views on the species category is a preference for theoretical versus operational definitions. Theoretical definitions emphasize ideas about the underlying nature of species; operational definitions emphasize the methods and evidence used to recognize species in practice (e.g., Hull 1968, 1997). It should be clear from these descriptions that the difference between the two positions reflects a difference in emphasis on ontology versus epistemology rather than fundamentally different conceptualizations of the species category. Considering views at opposite ends of the theoretical to operational spectrum supports the basic compatibility of those views.

Ideas commonly termed *phenetic species concepts* exemplify an operational emphasis. These ideas are commonly characterized as describing an atheoretical extreme in which species are treated as if they are nothing more than groups of similar organisms—that is, without regard for the relationships of those organisms in terms of biological processes such as interbreeding and common descent (e.g., Kitcher 1984, Ridley 1993, Hull 1997). This characterization misrepresents many of the views in question. Although advocates of phenetic definitions have called attention to the reliance of theoretical definitions on phenetic criteria for practical application (Michener 1970, Sokal and Crovello 1970, Sneath and Sokal 1973), at least some of the authors in question have explicitly acknowledged the importance of theoretical considerations (e.g., Michener 1970). Other advocates of operational approaches have even attempted to incorporate theoretical considerations about interbreeding and ecology into the procedures they use to analyze species (e.g., Rogers and Appan 1969, Doyen and Slobodchikoff 1974). More recent species definitions stated in terms of diagnostic characters (e.g., Nixon and Wheeler 1990) and identifiable genotypic clusters (e.g., Mallett 1995) also tend to emphasize operational considerations, but never with total disregard for theory (cf. Nanney, chapter 4 in this volume).

At the other end of the spectrum are ideas commonly designated *evolutionary species concepts*. These ideas are sometimes characterized as representing a theoretical extreme in which operational criteria are ignored to the point that the concepts are useless in practice (e.g., Sokal and Crovello 1970, Mayr 1982). This characterization is also a misrepresentation. Far from ignoring operational criteria for recognizing species, advocates of evolutionary definitions discuss such criteria in considerable detail (e.g., Simpson 1951, Wiley 1981). For both operational and theoretical ends of the continuum, misrepresentations seem to result from considering only the explicit species definitions per se and ignoring associated discussions. Although authors often differ greatly in their emphasis on operational versus theoretical considerations, those differences exist within the context of a single general concept of species.

## Models of Speciation

Other differences among contemporary views on species involve properties related to general models of speciation. The differences in question concern the relationship between cladogenesis and speciation (e.g., Hennig 1966, Wiley 1981, Ridley 1989) and the persistence of ancestral species through speciation events (contrast the views of Hennig [1966] and Ridley [1989] with those of Bell [1979] and Wilkinson [1990]). Despite describing important conceptual differences, the general unity of these views can be seen by considering the properties in question as the basis for a classification of general models of speciation (figure 3.4; modified from Wagner and Erwin 1995, Foote 1996). The *anagenetic* or *phyletic transformation model* refers to speciation within an unbranched lineage segment (figure 3.4a). In contrast, the *cladogenetic model* equates speciation with cladogenesis or lineage splitting (figures 3.4b and 3.4c). Within the cladogenetic model, the *bifurcation model* describes situations in which ancestral species fail to persist through speciation events (figure 3.4b), whereas the *blastation model*[7] describes situations in which ancestral species persist through speciation events (figure 3.4c). The difference between the anagenetic and cladogenetic models concerns the relationship between speciation and processes that affect lineages. The ana-

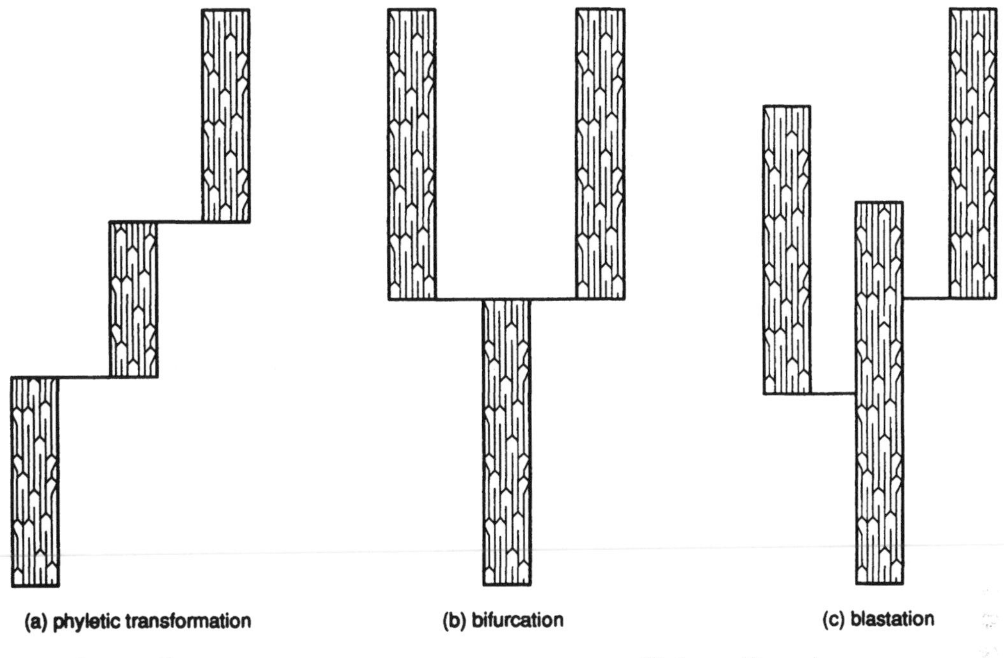


**Figure 3.4** General models of speciation (adapted from Wagner and Erwin 1995, Foote 1996, de Queiroz 1998). (a) *Phyletic transformation*, in which speciation occurs within an unbranched lineage and both the origination and the termination of species correspond with speciation events. (b) *Bifurcation*, in which speciation corresponds with lineage splitting and both the origination and the termination of species correspond with speciation events. (c) *Blastation*, in which speciation corresponds with lineage splitting and species originate in speciation events but do not terminate in such events. Species are represented by rectangles; speciation events are represented by horizontal lines.

genetic model equates speciation with lineage change, whereas the cladogenetic model equates speciation with lineage splitting. The other main difference between the models concerns how species are bounded relative to speciation events (however those events are defined). Under both the phyletic transformation and bifurcation models, species correspond precisely with the segments of lineages between speciation events (though what counts as a speciation event differs for the two models), whereas under the blastation model, species correspond with lineage segments that originate in speciation events but do not necessarily terminate in such events. The point is that all three of these models equate species with lineage segments.[8]

## Phyly

Another major difference among contemporary views on species concerns what might be termed *phyly*—that is, whether species can or must be monophyletic, paraphyletic, or polyphyletic. Different authors allow all three types

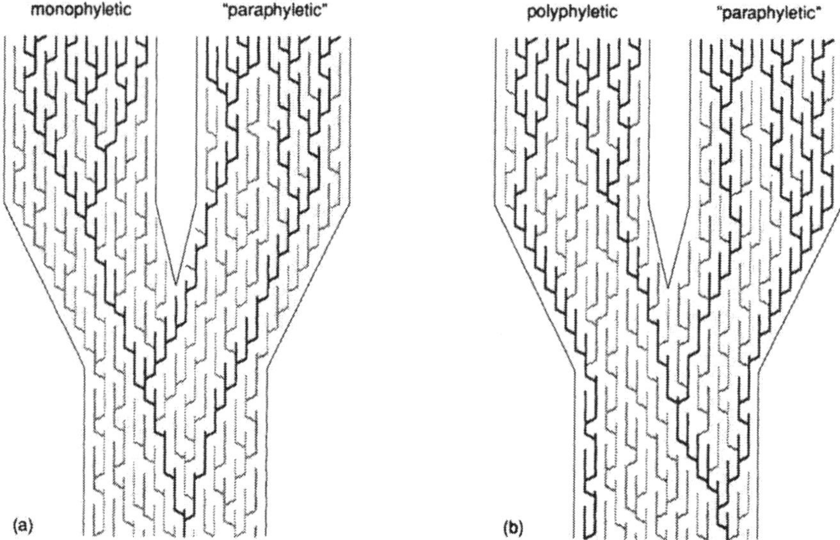


**Figure 3.5** "Paraphyly" and polyphyly of species in terms of their components genes, organelles, or organisms. (a) The species on the right side of the split is "paraphyletic" in the sense that some of its lower-level components share more recent common ancestors with the components of another species than with other components of their own species (but see note 9). (b) The species on the left side of the split is polyphyletic because some of its lower-level components are only distantly related to one another, coalescing in a remote ancestral species (not shown). In both diagrams, gene, organelle, or organism lineages that have survived to the most recent time are highlighted so that their relationships can be seen more easily.

of species (e.g., Neigel and Avise 1986); or only paraphyletic and monophyletic species (e.g., Brothers 1985, Crisp and Chandler 1996), or only monophyletic species (e.g., Rosen 1979, Mishler and Donoghue 1982). Other authors argue that the concepts of phyly do not apply to individual species but only to groups of species (e.g., Wheeler and Nixon 1990; also see note 3).

Some of the differences regarding species phyly reflect differences in the level of organization under consideration. Thus, phyly in terms of component genes or organisms, (as discussed by Neigel and Avise, 1986), should not be confused with phyly in terms of component populations (as discussed by Bremer and Wanntorp, 1979). Paraphyly[9] and polyphyly in the former sense (figure 3.5) appear to be common initial stages in the divergence of population-level lineages (Neigel and Avise 1986) and, in the case of polyphyly, when species arise as the result of hybridization. Most authors presumably would not deny that species can be either paraphyletic or polyphyletic in this sense (but see Baum and Shaw 1995). In contrast, there are probably few (if any) contemporary biologists whose concept of species includes entities that are polyphyletic in terms of their component populations—that is, who would recognize as parts of a single species two or more populations that are not particularly closely related to each other (figure 3.6a; Sosef 1997). Similarly, at least some authors (e.g., Rosen 1979, Bremer and

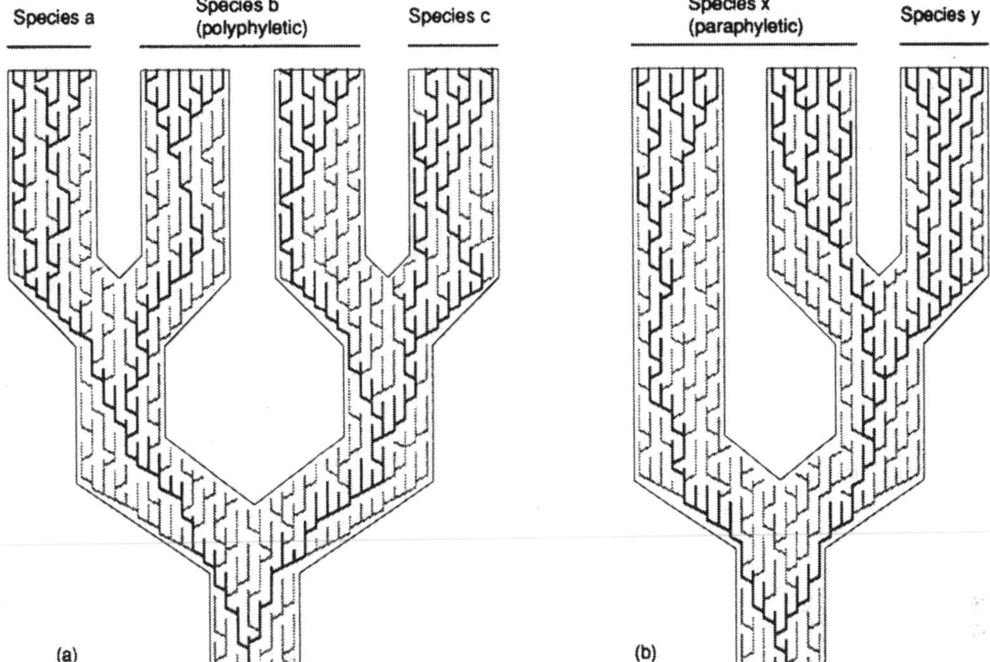


**Figure 3.6** Polyphyly and paraphyly of species in terms of their component populations. (a) Polyphyly of species b, whose two component populations both share more recent common ancestors with heterospecific populations than with one another. It is assumed that the two populations of species b are considered conspecific because of convergent rather than retained ancestral characters, so that their common ancestral population would not be considered part of species b. (b) Paraphyly of species x, one component population of which is more closely related to species y than to the other population of its own species. In both cases, phylogenies of lower-level components (e.g., genes) are shown within the population lineages, with lineages that survived to the most recent time highlighted.

Wanntorp 1979, Mishler and Donoghue 1982) do not want to recognize as a single species any assemblage of currently separate populations that is paraphyletic in terms of its component populations—in other words, if some of the populations in the assemblage share a more recent common ancestor with heterospecific populations than with conspecific ones (figure 3.6b). When these distinctions are borne in mind, the main disagreement seems to be about whether it is permissible to recognize paraphyletic groups of populations as species.[10]

The difference among contemporary views on species with regard to population-level paraphyly boils down to a question about when in the process of divergence two population lineages are to be considered distinct species. Disagreements involve cases in which characters affecting intrinsic separation (such as reproductive compatibility) diverge later than other characters, which nonetheless provide evidence of common ancestry relationships. Some authors want species to consist of mutually most closely related populations, which means avoiding paraphyly. Consequently, some of the lineages that they recognize as species will exhibit only extrinsic separation. In contrast, other authors want species to reflect intrinsic separation. Consequently, some of the lineages that they recognize as species will be demonstrably paraphyletic. Regardless of which alternative is preferred, the disagreement concerns the amount or type of differentiation considered sufficient to justify recognizing lineages as separate species; thus, both positions equate species with lineages.

### Species Criteria

The differences regarding phyly described in the previous section are related to a more general issue about species—namely, species criteria. Species criteria are standards for judging whether an entity qualifies as a species, though different interpretations of this statement are possible (de Queiroz 1998). In terms of their practical consequences, differences in species criteria are probably the most significant differences among contemporary ideas about species in that they are directly responsible for differences in the species taxa recognized by biologists.

The species criteria adopted by contemporary biologists are diverse and exhibit complex relationships to one another (i.e., they are not necessarily mutually exclusive). Some of the better-known criteria are: potential interbreeding or its converse, intrinsic reproductive isolation (e.g., Mayr 1942, 1963); common fertilization or specific mate recognition systems (e.g., Paterson 1978, 1985); occupation of a unique niche or adaptive zone (e.g., Van Valen 1976); potential for phenotypic cohesion (Templeton 1989); monophyly (e.g., Mishler and Donoghue 1982) as evidenced by fixed apomorphies (e.g., Rosen 1979) or the exclusivity of genic coalescence (e.g., Baum and Shaw 1995); and distinguishability, whether phenotypic or genotypic (e.g., Mallett 1995), qualitative (Nixon and Wheeler 1990) or quantitative (e.g., Michener 1970, Sneath and Sokal 1973). Because the entities satisfying these various criteria do not exhibit exact correspondence, authors who adopt different species criteria also recognize different species taxa.

Although different species criteria are often interpreted as the bases of fundamentally different species concepts, they all correspond with thresholds crossed by diverging lineages (de Queiroz 1998). Thus, as lineages diverge, they become distinguishable in terms of the phenotypic, genotypic, qualitative, and quantitative characters of their component organisms. At some point in the course of divergence, the lineages become mutually exclusive in terms of the common ancestry relationships among those organisms, and this result is often reflected by one or more fixed apomorphies and the exclusive coalescence of gene lineages. If divergence affects ecologically significant characters, the lineages may come to occupy distinct niches or adaptive zones. Divergence in components of the breeding system of sexual organisms leads to differences in the fertilization, mate recognition, and developmental systems that underlie intrinsic reproductive isolation. In short, the diverse species criteria adopted by contemporary biologists all correspond with properties acquired by lineages during the course of their divergence; thus, all criteria are compatible with a single general lineage concept of species.

## THE CAUSES OF THE SPECIES PROBLEM AND A SIMPLE SOLUTION

Despite nearly universal acceptance of the general lineage concept of species, at least two factors prevent a general consensus about the definition of the species category. One of these factors compromises universal acceptance of the general lineage concept itself; the other creates incompatibilities among the concept's numerous variants. Consequently, these factors are critical to solving the species problem.

### Ontological and Taxonomic Categories

The first factor concerns a basic assumption about how the species category is interpreted, which bears on acceptance of the general lineage concept itself. One interpretation is that the species category is an ontological category (see Ghiselin 1997)—that is, one of the fundamental categories of biological existence (other such categories are the cell and the organism). The other interpretation is that the species category is a taxonomic category—that is, a level or rank in the Linnean hierarchy of taxonomic categories (other such categories are the genus and the family). These alternative interpretations are not necessarily at odds with one another, but they often underlie at least partially incompatible views on species (cf. Boyd, chapter 6 in this volume).

The interpretation of the species category as an ontological category is implicit in the general lineage concept of species, which equates the species category with the ontological category whose members are the biological entities known as population lineages. On the other hand, the interpretation of the species category as a taxonomic category is implicit in its use in biological taxonomy, which equates the species category with one of the taxonomic categories in the Linnaean hierarchy. These two interpretations have several possible relationships with one another. (1) All of the taxonomic categories are artificial; none of them corresponds with an ontological category (cf. Ereshefsky, chapter 11 in this volume). (2) Each taxonomic category corresponds with a different ontological category; the species category corresponds with one ontological category, the genus with another, the family with yet another, and so on. (3) All the taxonomic categories apply to the same ontological category, the members of which form nested hierarchies; the various taxonomic categories represent different ranks or levels in those nested hierarchies. (4) The various taxonomic categories represent some combination of the first three alternatives.

Given that the general lineage concept describes an ontological category, some of the above interpretations are compatible with that concept, but others are not. The first interpretation seems to have been adopted by at least some critics of the idea that species are unified by gene flow (e.g., Ehrlich and Raven 1969). Those critics have seen little evidence of gene flow between conspecific populations and therefore consider species, like other taxa, to be groups of population lineages rather than population lineages themselves. To the extent that those groups were viewed as artificial, this position is incompatible with the general lineage concept. The second interpretation is implicit in the writings of authors who suggest that taxa assigned to different taxonomic categories originate in fundamentally different ways—that families, for example, originate by different mechanisms than genera, which in turn originate by different mechanisms than species (e.g., Jablonski and Bottjer 1991). To the extent that species (as opposed to families or genera) are equated with population lineages, those views are compatible with the general lineage concept of species. The third interpretation is implicit in the writings of authors who consider all taxa, including species, to be monophyletic entities (e.g., Mishler and Donoghue 1982, Nelson 1989). Those authors consider the species category one of the various ranks or levels to which monophyletic taxa are assigned. Thus, if, species are equated not with monophyletic population lineages, but with monophyletic groups of such lineages, then the interpretation in question is inconsistent with the general lineage concept of species. There are many possible combinations of these first three basic positions, at least two of which have been adopted commonly. (1) The species category is an ontological category; the other (higher) taxonomic categories are artificial groups of species (e.g., Dobzhansky 1937, Mayr 1969). (2) The species category is an ontological category; the other (higher) taxonomic categories refer to different levels in a nested hierarchy of entities that represent a different ontological category, usually the clade (e.g., Hennig 1966, Wiley 1981, de Queiroz 1988, Ghiselin 1997). Both of these positions are compatible with the general lineage concept of species.

Thus, many interpretations of the species category as a rank in the Linnean taxonomic hierarchy are entirely compatible with the general lineage concept of species, and even some that are potentially incompatible may not actually be. For example, it is not clear whether authors such as Ehrlich and Raven (1969) believe that species are not unified lineages or only that they are not unified by gene flow. Similarly, it is not always clear whether authors such as Mishler and Donoghue (1982) view species as monophyletic groups of population lineages as opposed to unitary population lineages that have attained monophyly in terms of their component organisms (see "Phyly"). But even authors who do not view species as unitary population lineages acknowledge the importance of such lineages; they simply equate species with groups of population lineages rather than with the lineages themselves. Therefore, all that is required to bring such views into line with the general lineage concept is a simple downward shift of the species category. Even this shift may have more to do with temporal perspective than with hierarchical level. Populations that are separated over relatively short time intervals may be a connected over longer ones. Therefore, a group of currently separate populations may be the temporarily separated parts of a single population lineage.

## The Defining Properties of the Species Category

The second of the two complicating factors that prevent consensus on species concerns the defining properties of the species category, which creates incompatibilities among the variants of the general lineage concept. The properties in question are the so-called species criteria, which form the basis of some of the most obvious differences among alternative conceptualizations of the species category. Most authors interpret those properties as defining or necessary properties of species, which is implicit both in their designation as *species criteria* and in their incorporation in explicit *species definitions*. This interpretation leads to irreconcilable concepts of the species category, each of which is based on a different defining property. Nevertheless, the properties in question are all properties of population lineages, and consequently, the alternative definitions still reflect an underlying unity with regard to a more general concept of species. In effect, the alternative species definitions are conjunctive definitions. All definitions have a common primary necessary property—being a segment of a population-level lineage—but each has a different secondary property—reproductive isolation, occupation of a distinct adaptive zone, monophyly, and so on. Under this interpretation of species criteria, reconciliation of alternative species definitions is only possible if the various secondary properties always characterize the same lineages, which they clearly do not, and thus the only potential solution to the species problem is for one of the species criteria to achieve widespread acceptance at the expense of the others.[11]

Alternatively, the various species criteria can be interpreted as contingent rather than necessary properties of species. Under this interpretation, there is only one necessary property of species—being a segment of population-level lineage.[12] Other properties, the so-called species criteria, are not necessary for a lineage to be considered a species. No one of those properties is possessed by all species, though many are acquired by numerous species during the course of their existence. Thus, some species are reproductively isolated, some are monophyletic, some occupy different adaptive zones, and many possess various combinations of these and other properties. The alternative definitions are not in conflict because they are not definitions of the species category itself but of classes of species possessing different contingent properties. Although these contingent properties are irrelevant to the definition of the species category, they are still important for assessing the separation of lineages—that is, for identifying species taxa. Furthermore, no one of these properties holds a privileged theoretical position; all of them describe potentially useful lines of evidence regarding the empirical investigation of species (de Queiroz 1998). Under this interpretation, the alternative species definitions are reconciled, and the species problem is thereby solved. In this context, the species problem is seen to result from considering descriptions of operational criteria to be descriptions of logically necessary properties. In other words, the species problem results from confusing the concept of species itself with the operations and evidence that are used to put that concept into practice.

## PHILOSOPHICAL CONSEQUENCES

In this section, I examine the implications of the perspective developed in the previous sections for various philosophical issues concerning species, including several of the major themes and topics of this volume. My purpose is to show how the general lineage concept, along with the reinterpretation of the necessary properties of the species category, either clarifies or resolves other issues about species.

### Species and the Representation of Biological Diversity

O'Hara (1993) viewed the species problem as part of the general problem of representing biological diversity (he used the term "evolutionary history") and compared it with the problem of representing the surface of the earth. Both of these endeavors, taxonomy and cartography, require decisions about which things to omit, which things to represent, and how to represent them. This perspective is very much in keeping with the views developed in the present paper and provides a useful context for illustrating those views using cartographic analogies. In this context, the species problem stems from treating the term *species* as if it is analogous to the term *city*. Determining whether a particular lineage is a species as opposed to a subspecies is much like determining whether a particular population center is a city as opposed to a town. Thus, one might choose different criteria (e.g., intrinsic reproductive isolation, distinguishability, monophyly) for deciding which lineage segments qualify as species (i.e., for representation in a taxonomy)—just as one might choose different criteria (e.g., population size, land area occupied, political status as a municipality) for deciding which population centers qualify as cities (i.e., for representation on a map). Similarly, several reproductively compatible but diagnosable allopatric populations will be represented as a single species in one taxonomy but as several in another—just as several physically contiguous but administratively separate population centers will be represented as a single city on one map, but as several on another. Because biologists adopt different species criteria, the term *species*, like the term *city*, has no universal definition.

By reinterpreting the defining properties of the species category as described in the present paper, the term *species* is no longer analogous to the term *city*. Instead, it is analogous to the general term *urban area*. That is to say, the term *species* applies to all separate population-level lineages, including demes and lineages that were formerly called subspecies, species, and superspecies—just as the term *urban area* refers to all separate population centers, including villages, towns, cities, and metropolitan areas. In this context, the problem of which lineages to recognize as species is seen as a problem about representation, rather than as a problem about the nature of species or the definition of the species category. Moreover, it is now possible to formulate a universal, if general, definition of the term *species* (see "Species" and "Species Life Cycles").

## Species and Biology

The concept of species developed in this essay plays a central role in biology. Under this concept, species are members of one of the basic categories of biological entities—in particular, one of the categories of biological entities whose members propagate themselves to form lineages. The concept of species thus has comparable importance in biology to the concepts of the gene, the cell, and the organism—ontological categories whose members are entities that form lineages at different levels of biological organization. The general lineage concept of species also plays a central role in evolutionary biology. Species are one of the kinds of entities that form lineages, and lineages are the things that evolve.[13] Furthermore, lineages form more inclusive entities of considerable evolutionary significance—namely, the historically unified collections of lineages that are termed *clades* when formed by species, and *clans* and *clones* when formed by entities at other levels in the organizational hierarchy (see note 3).

Although all modern biologists equate species with segments of population lineages, their interests are diverse. Consequently, they differ with regard to the properties of lineage segments that they consider most important, which is reflected in their preferences concerning species criteria. Not surprisingly, the properties that different biologists consider most important are related to their areas of study. Thus, ecologists tend to emphasize niches; systematists tend to emphasize distinguishability and phyly; and population geneticists tend to emphasize gene pools and the processes that affect them. Paleontologists tend to emphasize the temporal extent of species, whereas neontologists tend to emphasize the segments of species that exist in the present. Many of these differences affect which lineage segments are recognized as species taxa by different biologists, and this recognition in turn affects the study of species and speciation, as well as the use of species taxa as data in studies of diversification and extinction.

Considering alternative species definitions in the context of the role of the species concept in biology supports the idea that the so-called species criteria should not be interpreted as necessary properties of species. Because many of the commonly advocated species criteria correspond with different thresholds crossed during the process of lineage divergence (see "Species Criteria"), a consequence of the interpretation of species criteria as necessary properties is that a lineage segment is only a species if it has achieved a certain level of divergence. This consequence in turn implies that the species category designates a stage in the existence of population lineage segments (e.g., Dobzhansky 1935, 1937), which diminishes its theoretical significance (de Queiroz 1998). To use an organism-level analogy, treating one of the events that occurs during the process of population lineage divergence (e.g., diagnosability, concordant coalescence of gene trees, intrinsic reproductive incompatibility) as a necessary property of species is like treating one of the events that occurs during the process of organismal development (e.g., formation of the heart, birth or hatching, maturation of the gonads) as a necessary property of organisms. As important as those events are in the life cycles of organisms, they are not considered necessary properties of organisms. To do so would compromise the generality of the concept of the organism. For example, some of the properties just noted preclude the consideration of functionally autonomous and structurally individuated unicellular entities as organisms.[14] In addition, certain stages of the life cycle would be left in conceptual limbo. For example, if only entities that have been born are organisms, then what are earlier stages in the life cycle? For these reasons, biologists use the category *organism* to designate lineage segments that represent an entire turn of an organism-level life cycle—from initial propagation to termination.

If the concept of species is to have comparable theoretical significance, the species category must also designate lineage segments from initial propagation to termination (see "Species Life Cycles"). Rather than treating certain events in the process of lineage divergence as necessary properties of species and thus treating only some separate population lineages as "full" or "good" species (much as adults were considered "perfect" organisms by earlier workers), it would be more useful conceptually to treat all separate population lineages as species and use the various thresholds as the basis for different subcategories of a single general species category. Thus, we should talk about diagnosable, monophyletic, and reproductively isolated species just as we talk about postembryonic, sexually mature, and fully grown organisms. But organism lineage segments do not have to be born, sexually mature, or fully grown to be organisms; similarly, population lineage segments do not have to be diagnosable, monophyletic, or reproductively isolated to be species.[15] Although these conclusions are not entirely consistent with currently recognized species taxa (but see "Species and the Representation of Biological Diversity"), they grant the concept of species a more important role in biology and are logical consequences of the solution to the species problem proposed in this paper.[16]

## Species Individuality

An idea that has generated considerable discussion in the philosophically oriented literature on species is the conceptualization of species as individuals—collections of organisms united into larger wholes (e.g., Griffiths 1974; Ghiselin 1974, 1997; Hull 1976, 1978; Williams 1985). The idea is not that species are organisms or even superorganisms, but simply that they are composite wholes made up of organisms. This view is contrasted with the conceptualization of species as sets or classes—collections of organisms assigned to groups because they share certain properties (e.g., Kitcher 1984a). The general lineage concept both strengthens and clarifies the conceptualization of species as individuals (cf. Boyd and Wilson, chapters 6 and 7 in this volume).

**Species and Organisms** The individuality of species under the general lineage concept is implied by the concept of the population lineage upon which it is based. As a unified collection of organism lineages, a population lineage is a quintessential composite whole. Moreover, species, like organisms, are entities that form lineages, and organisms are paradigm individuals (Hull 1976). Thus, the analogy between organisms and species is even closer than might have been inferred from the proposition that species are individuals in a general philosophical sense. In other words, organisms and species have much more in common than merely being individuals in the sense of concrete entities or composite wholes—which is also true of individual atoms, molecules, planets, galaxies, chairs, furniture stores, corporations, cities, states, and nations. Organisms and species are not only individuals; they are very similar kinds of individuals in that both are lineage segments (see also Griffiths 1974, Hull 1976). Indeed, one could even go so far as to say that organisms and species (along with genes and cells) are members of the same general category of individuals—lineage-forming biological entities —though they obviously differ with respect to the level of organization.[17]

**Individuals and Classes** Despite its compatibility with the thesis of species individuality, the general lineage concept requires only a slight modification to accommodate the interpretation of species as sets or classes. This modification is accomplished by recognizing that the individuals in question are composite wholes and that for any composite whole, a class or set can be conceptualized whose members are the parts of that whole (de Queiroz 1992a, 1995). Therefore, a species can be conceptualized as the class or set of organisms that make up a particular population-level lineage segment. Nevertheless, several points should be kept in mind. First, this reconciliation of the individual and class/set interpretations of species in no way contradicts or compromises the proposition that the lineage segments in question are individuals; indeed, it is based on that very proposition. Second, the classes that might be equated with species are spatiotemporally restricted, which is to say that they should not be confused with the spatiotemporally *un*restricted classes that people usually have in mind when they contrast individuals with classes. Third, despite the possibility of conceptualizing species as either individuals or (spatiotemporally restricted) classes, it is important to distinguish between the two conceptualizations—that is, between the population lineage segments as wholes and the classes or sets of their organismal parts.[18] An effective way to reinforce this distinction is by using different terms for the different conceptualizations, as is already being done in some cases (e.g., "*Homo sapiens*" versus "human beings").[19]

**Processes Responsible for Unification** Although the thesis of species individuality helps to resolve some philosophical conundrums about those entities (see Ghiselin 1974, 1997), it does little by itself to clarify biological issues (de Queiroz and Donoghue 1988). The lineage concept of species answers this need by describing more precisely what kind of individuals species are, and this description in turn focuses attention on the biological phenomena responsible for their existence as wholes. Under the lineage concept, the individuality of species results from whatever processes or relationships unite organisms to form population-level lineages, and consequently, those processes or relationships are topics of central importance in biology.

The process most commonly proposed to explain the existence of species as population-level lineages is interbreeding—that is, sexual reproduction. It is probably easiest to visualize the formation of population-level lineages in organisms with sexual reproduction (see "Species"). However, to conclude that asexual organisms do not form species (e.g., Dobzhansky 1937, Hull 1980, Ghiselin 1997) is to implicitly accept the proposition that interbreeding or sexual reproduction is the *only* process that unites organism lineages to form population-level lineages. Although this proposition may turn out to be true, other processes have been proposed as important in the maintenance of population-level lineages, and at least some of them apply to asexual organisms. Templeton (1989; see also Meglitsch 1954), for example, argued that ecological factors determine the limits of populations with respect to evolutionary processes such as genetic drift and natural selection, which do not require sexual reproduction to operate. He also argued that these factors are more important than interbreeding for maintaining population-level lineages both in asexual organisms and in sexual organisms whose population lineages remain distinct despite interbreeding between them. My purpose is not to endorse these views, but only to point out that the existence of population-level lineages in organisms with different reproductive modes can potentially be investigated empirically. This issue has received surprisingly little study in view of its importance to the biology of species.

**Species Life Cycles** The realization that species and organisms are similar kinds of individuals provides insights into the life cycles of species, the existence of which is implied by the fact that species, like organisms, are lineage segments. This is not to say that species have regular and integrated ontogenies like those of many organisms, but merely that they go through cycles of genesis and termination, with other changes in between. Indeed, because different processes are responsible for the unification of organisms (e.g., cell membrane junctions, cell to cell adhesion) and that of species (e.g., interbreeding, selection), care should be taken when drawing analogies between the two kinds of individuals. On the other hand, because the implications of organismal individuality are more familiar to us, such analogies often greatly facilitate our ability to conceptualize the implications of species individuality. Thus, both the similarities and differences between organisms and species provide insight into species life cycles.

With regard to origins, an obvious analogy can be made between reproduction by fission and speciation by bifurcation (see "Models of Speciation"), where new species arise from large subdivisions of an ancestral species (reviewed by Bush 1975). In both cases, the descendants originate from major (often more or less equal) portions of their ancestors. Similarly, reproduction by budding corresponds with speciation by blastation (see "Models of Speciation"), where a species originates from a small founder population (see Bush 1975). In both cases, the descendant arises from a small portion of its ancestor. In all of these modes of genesis (fission, bifurcation; budding, blastation), the production of new organisms or species coincides with lineage splitting. If species are like organisms, then the model of speciation by phyletic transformation (see "Models of Speciation") would seem to be invalid (e.g., Hennig 1966, Wiley 1978). A single organism changes considerably during the course of its life (e.g., zygote to adult human), so the fact that a species changes during its existence does not require that it changes into a different species. Indeed, a species should be able to change indefinitely and still remain the same species, provided that the change is more or less gradual and continuous. Situations in which each organism or species in a series produces a single descendant via budding or blastation should not be confused with unbranched lineages. Although we tend to think of a such successions as linear or unbranched in the case of organisms, they are really branched if parent and offspring coexist temporally. On the other hand, if a parent dies more or less simultaneously with the propagation of a single offspring, perhaps the lineage can be considered unbranched.[20] Similarly, if an unbranched population lineage passes through a severe bottleneck, which is similar in many respects to a founder event, perhaps it is justifiable to consider the lineage segments on either side of the bottleneck as different species.

Despite the possibility of phyletic transformation, differences between the component organisms in earlier and later parts of an unbranched population lineage—even those that affect other biologically significant properties (e.g., the ability to interbreed)—are not particularly relevant to the question of whether their respective lineage segments constitute the same or different species. Undifferentiated cells in an early embryo are not particularly similar to the differentiated cells that make up the same organism later in its life, and perhaps they would not be integrated into the later organism if given a chance. Even changes in the emergent properties of lineage segments do not necessarily imply that they are different species. Some organisms change from carnivores to herbivores, or from females to males, during a single turn of an organism life cycle, so a species should be able to change from panmictic to subdivided, for example, during a single turn of a species life cycle.

Regarding termination, organisms sometimes end by ceasing to function as integrated wholes—that is, by death. Species can end in an analogous manner, normally termed *extinction*. And just as certain component cells can continue to live after their organism dies, certain component organisms can continue to live after their species becomes extinct. The most obvious example is a species composed of organisms with obligate sexual reproduction and separate sexes (and no sex-changing abilities) in which the only surviving organisms are all members of the same sex. In other cases, organisms end by separating into more than one whole, that is, by fission. The analogous situation for species is bifurcation. Because the ancestor in both cases is no longer identifiable after the lineage splits, it is considered to terminate at the splitting event.[21] This is not to say that ancestors necessarily terminate whenever lineages split. When the split is highly unequal, as in the cases of organism budding and species blastation, the ancestor can be considered to persist.

If there is a difference between organisms and species with regard to lineage splitting, perhaps it is the frequency of intermediate cases—that is, cases in which the split is only moderately unequal so that it is ambiguous as to whether the ancestor persists. Organismal reproduction appears strongly polarized into fission and budding modes, with few intermediate cases. In contrast, bifurcation and blastation modes of speciation appear to be opposite ends of a continuum in which the intermediate cases are far more common, particularly if extrinsic barriers are a common cause of speciation. In any case, considering the life cycles of species helps us to formulate a fuller description of the general lineage species concept. Species are not just any segments of population-level lineages; they are the segments of population-level lineages that correspond with a single turn of the life cycle, from genesis to termination.

An important difference between species and organisms concerns fusion. Separate organism lineages rarely fuse as wholes (but see below). Even the continual merging of sexual organism lineages usually involves only the transfer of genetic material between cells or the union of specialized cells (gametes); the organisms themselves retain their separate identities. In the case of population lineages, fusion appears to be much more common. Although certain species definitions are based on properties that would seem to be correlated with irreversible separation (e.g., intrinsic reproductive isolation), resulting in species taxa that are more like organisms regarding their likelihood of fusion, there are no guarantees. For example, premating barriers based on habitat differences can be broken down by environmental changes, and even certain postmating barriers can (in theory) be removed by selection against the genetic elements responsible for the reduced fitness of hybrids. More importantly, such definitions make species seem more like organisms than they really are. Although the separation of most pairs of population lineages probably does become irreversible eventually, in many cases that stage is reached long after the lineages have begun to function as separate entities.

Most species seem to exhibit nothing comparable to the regular and complex ontogenies of many organisms, such as the stages of the cell cycle (prophase, metaphase, anaphase, etc.) or of multicellular development (e.g., blastulation, gastrulation, neurulation, etc. of bilateral metazoans). This claim does not deny that species pass through stages; however, those stages appear far less orderly than their organismal analogs. For example, intrinsic reproductive isolation, morphological distinguishability, ecological differentiation, and genetic exclusivity can presumably be acquired in various sequences, even in sister species. Although it is at least possible that some species exhibit stages analogous to reproductive maturity and senescence, this possibility seems unlikely, particularly if speciation is commonly initiated by extrinsic factors.

Perhaps the closest organism-level analogs of species, in terms of their individuality, are certain multicellular organisms that exhibit relatively weak integration. In the aggregatory phase of cellular slime molds (Acrasiales), for example, separate cells (amoebae) aggregate to form a single mass (pseudoplasmodium), but under certain environmental conditions, this mass can fragment into smaller masses that can themselves reaggregate (Bonner 1967). Certain sponges (Porifera) can be mechanically separated into their component cells, which will then reaggregate to form several new individuals (e.g., Humphreys 1970). Other sponges (termed *multioscular*) have multiple but only partially distinct functional units united into a larger whole, and it is debated whether they should be considered individual organisms or colonies (e.g., Korotkova 1970). Such organisms appear relatively weakly integrated and thus weakly individuated, so it is not always clear whether we are dealing with one or several individual organisms. But even in more tightly integrated organisms, there can be ambiguities concerning individuality—for example, conjoined twins. In the case of species, ambiguities about individuality are common.

### Is the Species a Relational Concept?

Another philosophical controversy regarding species, although one argued primarily in the biological rather than the philosophical literature, is whether the species category is a relational concept. According to the relational view (Mayr 1957, 1963, 1988; Mayr and Ashlock 1991), the concept of species is analogous to the concept of *brother*—or more generally, *sibling*—which is to say that the term *species* describes a relationship among population lineages just as the term *sibling* describes a relationship among organisms. As Mayr (1963, 19) put it: "An individual [organism] is a brother only with respect to someone else. A population is a species only with respect to other populations." The alternative view is that the concept of species is nonrelational—that species exist not by virtue of their relation to other species but by virtue of whatever phenomena unite their component organisms to form "self-defining" composite wholes (Paterson 1985; Lambert, Michaux, and White 1987; White, Michaux, and Lambert 1990).

This debate, like several others, is tied to the question about the defining properties of the species category. The relational view is implied by accepting any property that describes a relationship between population lineages as a necessary property of species. Mayr's endorsement of the relational view can thus be seen as a logical consequence of his preferred species criterion, which treats intrinsic reproductive isolation as a necessary property of species. Under this criterion, only those population lineages that have acquired reproductive isolation are species, and reproductive isolation is a relationship between lineages (a given lineage can be reproductively isolated only in relation to another lineage). Many other species criteria also imply the relational view—including similarity, distinguishability, diagnosability, exclusivity of common ancestry, and apomorphy. Other properties—such as occupation of the same adaptive zone, having the same fertilization or specific mate recognition system, and actual or potential interbreeding—may be nonrelational when interpreted as propositions about the processes responsible for the unification of population-level lineages. However, when interpreted as necessary properties of species for delimiting species taxa, they are effectively relational (see also Templeton 1987; Coyne, Orr, and Futuyma 1988). To the extent that these properties are matters of degree rather than all-or-none phenomena, they must be assessed in terms of the relational properties of similarities and differences.

One consequence of the relational view is that it is logically impossible for a species to exist without the existence of other species (de Queiroz 1992b). This logical dependence should not to be confused with the ecological dependence of most species on other species, which makes it physically—as opposed to logically—impossible for those species to exist in isolation. According to the relational view, just as an organism cannot logically be a sibling without the existence of other offspring of the same parents, a population lineage cannot logically be a species without the existence of other separate population lineages. It follows that the first population-level lineage, the common ancestor of all species, was not itself a species—that species did not come into existence until after that lineage divided into two. Another consequence of the relational view is that the concept of species is restricted in its generality. That is to say, just as only some organisms are siblings, the relational view implies that only some separate population lineages are species.

In contrast, the nonrelational view is implied by interpreting the various relational properties as contingent rather than necessary properties of species. If the only necessary property of species is being a segment of a population lineage, then species exist not by virtue of their relationships to other species, but by virtue of whatever processes unite their component organism lineages to form population lineages. If so, then the existence of species may be physically dependent on other species, but it is not logically dependent on them. The nonrelational view allows the first population lineage (ancestor of all species) to be a species. It also grants the species category greater generality. Under this view, the species category is not analogous to relationally defined categories at the organismal level, such as brother or sibling, but to the primary ontological category at that level—that is, to the category *organism* itself.

## Monism and Pluralism

Another topic that has attracted considerable attention—in this case, mostly in the philosophical literature on species (but see Mishler and Donoghue 1982)—is the debate about monism versus pluralism with regard to species concepts (see Kitcher 1984a, 1984b; Sober 1984; Holsinger 1987; Mishler and Brandon 1987; Ereshefsky 1992, 1998, chapter 11 in this volume; Stanford 1995; Hull 1997, chapter 2 in this volume; Dupré, chapter 1 in this volume). Monists hold that there is only a single kind of species, whereas pluralists hold that there are many different kinds of species. Hull (chapter 2 in this volume) points out that there are many different forms of both monism and pluralism, so that the two categories grade into one another. For example, some forms of pluralism consider different processes important for maintaining different species, but allow a given organism to be part of only a single species taxon, thus permitting only a single species taxonomy (e.g., Mishler and Donoghue 1982, Mishler and Brandon 1987). Other forms of pluralism allow a given organism to be part of several different species taxa, one for each different species concept, thus permitting the existence of many alternative species taxonomies (e.g., Kitcher 1984a, Ereshefsky 1998).

The general lineage concept of species eliminates the conflict between monism and pluralism by encompassing both the unity and the diversity of ideas about species (see also Mayden 1997). Monism accounts for the common theme underlying all concepts of species—that is, the general lineage concept itself; it reflects the unity of ideas about species. Pluralism accounts for the numerous variations on that common theme; it reflects the diversity of ideas about species. There is no conflict between monism and pluralism because the single general concept subsumes—rather than serving as an alternative to—its many variants.[22] But the conflict between monism and pluralism arose within a context in which the unity of species concepts was not fully appreciated. Consequently, the debate has centered around the variants of the general lineage concept. Monists have granted primacy to just one of the many variants, whereas pluralists have granted all of the variants, or at least several of them, equal standing. This conflict stems once again from interpreting certain contingent properties of lineages as necessary properties of species. And once again, it can be resolved by reinterpreting the significance of the properties in question and thus also the definition of the species category.

If properties such as intrinsic reproductive isolation, ecological distinctiveness, and monophyly, are regarded as contingent rather than necessary properties of species, then none of those properties define the species category. Consequently, they cannot define fundamentally different kinds (i.e., concepts) of species. Instead, the properties in question define subcategories of a single general species category, which is to say that they merely describe differences among species of the same basic kind. In this context, terms such as *biological species, ecological species, phylogenetic species,* and so on are misleading in that they seem to imply fundamentally different kinds of species. It would be better to replace them with the terms *reproductively isolated species, ecologically distinct species, monophyletic species,* and so on—terms that more accurately describe the relevant differences, while at the same time acknowledging the fundamental unity of contemporary ideas about species (de Queiroz 1998). In any case, the terms describe different classes of entities conforming to the same basic species concept rather than fundamentally different concepts of the species category. They are comparable to terms that describe different classes of entities conforming to the same basic concept of the organism, such as "gonadally mature," "socially mature," and "fully grown organism." In this context, any perceived conflict between monism and pluralism stems from confusing different senses of the term *different kind.* Although there are many "different kinds" of species in the sense that different species possess different contingent properties, there are not "different kinds" of species in the sense that different species represent different ontological categories.[23]

## Realism and Antirealism

Another philosophical debate about species concerns positions known as realism and antirealism. Species realism is the position that species exist independently of human perceptions. Species antirealism rejects the mind-independent existence of species. Hull (chapter 2 in this volume) discusses connections between the debate about monism versus pluralism, on the one hand, and the debate about realism versus antirealism, on the other. Several authors argue that species pluralism implies antirealism (e.g., Stanford 1995; Ereshefsky 1998, chapter 11 in this volume). If diverse species definitions are legitimate and describe species taxa with noncorresponding boundaries—that is, different sets of species taxa—then species must not be real. Some authors take this statement to mean that the existence of species taxa is not independent of the theoretical interests of biologists (e.g., Stanford 1995). Others take it to mean that there is no common and unique identifying property of the species category (e.g., Ereshefsky 1998, chapter 11 in this volume).

The second form of antirealism is directly contradicted by the general lineage concept of species, which is based on the identification of a common and unique property of species taxa. All species are segments of population-level evolutionary lineages. This position is consistent with Ereshefsky's (1998) view that species are genealogical entities, but Ereshefsky argues that being a genealogical entity does not suffice as a unifying feature of species because it also applies to genera, families, and so on—that is, to higher taxa. The apparent problem is readily solved in the context of the general lineage concept by recognizing a distinction between two different kinds of genealogical entities: lineages (as defined in this essay; see "Lineages") and clades. Species differ from higher taxa in that species are lineages (or more properly, lineage segments), whereas higher taxa are clades (i.e., groups of species sharing an exclusive common ancestry). The same conclusion holds if (some) higher taxa are allowed to be paraphyletic grades.

Both forms of antirealism rest on a form of species pluralism that views alternative descriptions of the species category as irreconcilable definitions —a position that in turn rests on the interpretation of certain contingent properties of lineages as necessary properties of species. This position is what allows antirealists to conclude that a single organism can belong simultaneously to different types of species and thus to different species taxa (e.g., Ereshefsky 1998, chapter 11 in this volume). If, for example, intrinsic reproductive isolation is interpreted as a necessary property of species, it will lead to the delimitation of one set of species taxa, and that set of species taxa will likely differ (in terms of both the number of species and the assignment of organisms to species taxa) from the set of species taxa delimited under a species definition that adopts a different property—diagnosability, for example—as a necessary property of species.

I have already shown how reinterpreting certain properties as contingent rather than necessary properties of species resolves the conflict between species monism and species pluralism. Because the antirealism argument rests on species pluralism (or more accurately, antimonism), it is not surprising that reinterpreting the significance of those properties also nullifies the argument against species realism. If properties such as distinguishability, ecological distinctiveness, and reproductive isolation (to mention only a few) are contingent rather than necessary properties of species, then they imply neither alternative sets of species taxa nor the existence of fundamentally different kinds (ontological categories) of species. Instead, they merely imply that a single species can belong simultaneously to several subcategories of the general category species. For example, a species can simultaneously be phenetically distinguishable, ecologically distinct, and extrinsically isolated from other species. This is analogous to saying that an organism can simultaneously be fully grown, socially dominant, and reproductively active—which no one counts as evidence against the independent existence of organisms. Species may be less tightly integrated and sharply bounded than organisms, but they are no less real than organisms. Both species and organisms exist independent of human perceptions.

## History of the General Lineage Concept

An early version, or at least a precursor, of the general lineage species concept can be found in Darwin's (1859) *Origin of Species*. In the only illustration in that book, Darwin represented species as dashed and dotted lines, or collections of such lines, forming the branches of what would now be called a phylogenetic tree. In the accompanying text, he used the term *species* more or less interchangeably with the term *lines of descent*. On the other hand, he adopted degree of difference as his species criterion (e.g., p. 120), which led him to conclude that species were not qualitatively different from varieties or genera—all of which were either lineages or collections of lineages.[24] Consequently, Darwin's species category remained firmly embedded in the Linnean hierarchy of taxonomic categories, which is to say that it remained a rank in a hierarchy of categories applied to entities of the same kind.

The general lineage concept was adopted to one degree or another by various workers in the late nineteenth and early twentieth centuries (e.g., Poulton 1903, Jordan 1905; see also Mayr 1955, Grant 1994). Its impact, however, was felt most strongly during the Modern Synthesis (Huxley 1942, Mayr and Provine 1980), in the writings of authors such as Dobzhansky (1935, 1937), Huxley (1940, 1942), Wright (1940), Mayr (1942, 1963), Stebbins (1950), Simpson (1951, 1961), and Grant (1963). An important difference between ideas about species that emerged during the Modern Synthesis and Darwin's ideas was that in at least some of the more recent ideas species were equated with inclusive population lineages themselves rather than with groups of such lineages. As a consequence, the species category was effectively decoupled from the Linnean hierarchy (de Queiroz 1997). That is to say, the species category was no longer viewed as a mere rank in the hierarchy of Linnean taxonomic categories, but as a primary ontological category. This position was manifested in the view that the species category was more objective and less arbitrary than the higher taxonomic categories (e.g., Dobzhansky 1937, Mayr 1969).

Several authors from the period of the Modern Synthesis formulated explicit definitions of the species category, among which Mayr's (1942, 1963) and Simpson's (1951, 1961) have been the most influential. Interestingly, those definitions were not originally proposed as descriptions of novel and incompatible species concepts, although they later came to be viewed as such. Mayr (1942, 1957, 1963), for example, distinguished fairly clearly between a general *biological species concept* and his explicit *biological species definition*, using those very terms to express the distinction. He used the term *biological species concept* to contrast species concepts that applied uniquely to biological entities with concepts that could be applied to both biological entities and nonbiological objects.[25] As Mayr (1969, 26) put it: "This species concept is called biological not because it deals with biological taxa, but because ... [i]t utilizes criteria that are meaningless as far as the inanimate world is concerned." Used in this sense, the general lineage concept is a quintessential biological species concept: inanimate objects don't form lineages. On the other hand, Mayr used the term *biological species definition* for his explicit definition of the species category, which incorporated potential interbreeding and reproductive isolation as its species criterion. Later, however, the term *biological species concept* came to be associated with this particular species definition rather than the more general concept.

Although Simpson (1951, 1961) originally proposed his explicit species definition as an alternative to "genetical" species definitions, such as Mayr's, he proposed it not as the description of an alternative species concept but as a more accurate description of the same species concept, which was already adopted widely by biologists.[26] In particular, Simpson (1951) called attention to the fact that Mayr's "genetical" definition did not deal adequately with the extension of populations in space and time, and that its criterion—potential interbreeding—was at odds with situations in which "quite extensive interbreeding may occur between adjacent populations which nevertheless retain their own individualities, morphologically and genetically, so clearly that any consensus of modern systematists would call them different species" (p. 289). In a passage very much in keeping with the thesis of the present paper, Simpson noted that "Most of the vagueness and differences of opinion involved in use of the genetical definition are clarified ... by taking the genetical criterion, or interbreeding, not as definitive in itself but as evidence on whether the evolutionary definition is fulfilled" (p. 289). Moreover, although Simpson (1951) called his species definition "evolutionary" (p. 289), he referred to the general concept that it describes as the "genetical-evolutionary concept" (p. 292) or simply "the species concept" (p. 285), implying that there was no fundamental conflict between his and Mayr's concepts (as opposed to their definitions) of species. Only later did Simpson's species definition come to be known as the *evolutionary species concept* and viewed as an alternative to Mayr's biological species concept (e.g., Wiley 1978, 1981).

Mayr and Simpson encapsulated their views on species as succinct and explicit definitions, which seems to have invited criticism. Mayr's definition became both the most popular and the most criticized. Pheneticists criticized it for the difficulties of applying it in practice (e.g., Sokal and Crovello 1970), paleontologists for its failure to incorporate temporal considerations (e.g., Simpson 1951), phylogenetic systematists for the fact that it sometimes resulted in paraphyletic species (e.g., Rosen 1979), selectionists for its failure to consider the role of natural selection in determining lineage boundaries (e.g., Van Valen 1976), recognitionists for its association with the view that reproductive isolation is an adaptation rather than an incidental by-product of divergence (Paterson 1985), and speciation biologists for its association with allopatric models of speciation (Mallett 1995). Simpson's definition, on the other hand, was criticized for its failure to specify an operational criterion or a causal process (Sokal and Crovello 1970, Mayr 1982, Haffer 1986, Templeton 1989, Ridley 1993). Many of these critics proposed their own species definitions based on alternative species criteria: phenetic gaps, unique adaptive zones, monophyly (as evidenced by apomorphies or the exclusive coalescence of gene lineages), unique combinations of characters, common fertilization or specific mate recognition systems, the potential for phenotypic cohesion, and the formation of genotypic clusters.

Because these species criteria were treated as defining or necessary properties of species (but see Simpson 1951, Hennig 1966, Wiley 1978, Ridley 1989), the definitions based on them came to be viewed as descriptions of fundamentally different concepts of the species category, which was (and continues to be) reflected in their common designation as *species concepts*. Thus, we have (references are for the terms rather than the definitions) the *biological species concept* (e.g., Mayr 1969), the *phenetic species concept* (e.g., Sokal and Crovello 1970), the *ecological species concept* (Van Valen 1976), the *evolutionary species concept* (e.g., Wiley 1978, 1981), the *phylogenetic species concept* (a term used in at least three different senses—e.g., Cracraft 1983, Donoghue 1985, Panchen 1992; see de Queiroz 1998), the *isolation species concept* (Paterson 1985), the *recognition species concept* (Paterson 1985), the *cohesion species concept* (Templeton 1989), the *cladistic species concept* (Ridley 1989), the *autapomorphic species concept* (Nixon and Wheeler 1990), the *monophyletic species concept* (Smith 1994), the *Hennigian species concept* (Nixon and Wheeler 1990), and the *genealogical species concept* (Baum and Shaw 1995). In a recent review, Mayden (1997) listed more than twenty named species concepts.

At the present time, each of these alternative definitions of the species category is being promoted by a different group of biologists. The campaigns to promote these alternative definitions have resulted in a tremendous proliferation of theoretical papers on species in recent years—each extolling one definition or another, criticizing competing alternatives, and presenting the differences as fundamental. The hope among biologists seems to be that one of these definitions—or perhaps one yet to be formulated—will win over the majority of biologists in the long run, solving the species problem by consensus. The problem is that different biologists have very different ideas about which definition it will be. Philosophers, in contrast, seem to revel in the disagreements among biologists, using those disagreements to support their own ideas about pluralism and antirealism, and seeming to imply that the species problem is unresolvable.

## CONCLUSION

Fortunately, the situation is not as hopeless as it may appear. By losing sight of the common thread running through virtually all modern views on species, both biologists and philosophers have overlooked a relatively simple solution to the so-called species problem. Virtually all modern biologists have the same general concept of species. Most of their disagreements stem from interpreting certain contingent properties of lineages as necessary properties of species (i.e., species criteria), which leads to species definitions that are incompatible both in theory (because they are based on different necessary properties) and in practice (because they result in the recognition of different species taxa). This situation fosters competition among alternative species criteria and their associated species definitions, with each one vying for status as *the* defining property of the species category. As a consequence, the common theme underlying all of the alternative views tends to be obscured, and the perception of a major, unresolved problem concerning the nature of species persists.

Recognizing the common thread manifested in what I have called the general lineage concept of species reveals a simple and straightforward solution to the species problem. All that is required is to drop the interpretation of certain contingent properties of lineages as necessary properties of species, and the species problem will vanish. By reinterpreting what have been called species criteria as contingent rather than necessary properties of species, or simply as different lines of evidence concerning the separation of lineages, the conflicts among species definitions are removed. The definitions in question are not alternative definitions of the species category at all, but merely descriptions of the diverse contingent properties of species. Consequently, there is no longer any major unresolved problem regarding the nature of species or the definition of the species category.

The problem is that despite the existence of a perfectly adequate concept and definition of species, most species are more like slime molds and sponges than like highly organized and tightly integrated multicellular organisms—at least in terms of their individuality. Not only can almost any part of a species give rise to a new lineage, but those new lineages also commonly reunite after separating. Consequently, there will be many cases in which it will be difficult to determine the precise number and boundaries of species—just as it is difficult to determine the precise number and boundaries of organisms in a fragmenting acrasialian pseudoplasmodium or a multioscular sponge. But such observations have not led to the conclusion that there is a major unresolved problem concerning the concept of the organism, and similarly, they do not imply a major unresolved problem concerning the concept of the species category; instead, they merely imply a practical problem about establishing the limits of species taxa in practice. Taxonomic traditions notwithstanding, everything we know about species tells us that they are inherently difficult to circumscribe, particularly in the early stages of divergence;

that they are not always sharply distinct, easily recognized entities; and that unambiguous assignment of all organisms to species taxa will be difficult, if not impossible. Attempting to solve this problem by treating operational criteria as defining properties only aggravates the situation because it confuses a purely practical problem with a theoretical one. The appropriate solution to the practical problem is simply to accept the inherent ambiguities of species boundaries (O'Hara 1993). In any case, recognizing the conceptual unity among modern views on species allows us to transcend their differences. It helps us identify both the cause of and the solution to the species problem, which clarifies a great deal concerning the concept of species itself as well as its history and its significance for both biology and philosophy.

## ACKNOWLEDGMENTS

I thank David Wake, Michael Donoghue, Jacques Gauthier, Alan de Queiroz, Bob O'Hara, Allan Larson, Michael Ghiselin, David Hull, David Good, Molly Morris, David Baum, Kerry Shaw, Jim Mallet, Marc Ereshefsky, Stewart Berlocher, Darrel Frost, Ed Wiley, Ron Nussbaum, Linda Allison, and Maureen Kearney for discussions (some many years ago) relevant to this essay. I also thank Rob Wilson for the invitation to contribute an essay to this volume and for his comments on an earlier draft.

## NOTES

1. According to the proposed terminology, a *species concept* is an idea about the nature of the entities that make up the species category; a *species criterion* is a standard for judging whether a particular entity qualifies as a member of the species category, and a *species definition* is a statement specifying the meaning of the term *species* and thus describing a species concept, usually in terms of necessary and sufficient properties.

2. Wilson (1995), for example, developed a view that equates species with what he called *lineages*, but he used that term in a sense that includes clades, clans, and clones.

3. The terminology for these entities has not been developed adequately. De Queiroz and Donoghue (1988, 1990) used the term *monophyletic* to describe the general class of entities each of whose members consists of an ancestor and its descendants, regardless of organizational level. They noted that the term *clade* had generally been used for monophyletic entities composed of species, and *clone* for comparable entities at lower levels of organization. This terminology, however, does not distinguish between monophyletic entities at several different organizational levels below that of species, nor does it take into consideration the distinction between diverging and reticulating patterns of descent and the most common use of the term *clone* for cases involving asexual (nonreticulating) reproduction. O'Hara (1993) proposed using the term *clan* for monophyletic entities at the organismal level, regardless of reproductive mode, but terms for other levels are currently lacking. Some authors (e.g., Wheeler and Nixon 1990) object to using *monophyletic* to describe entities below the species level, based on Hennig's (1966) distinction between *phylogeny* and *tokogeny*—the former describing the descent of species, the latter the descent of organisms. The term *phylogeny*, however, is commonly used in a more general sense to describe descent at various organizational levels (e.g., "gene phylogeny").

The following terminology makes most of the distinctions that previous authors have considered important, while minimizing discrepancies with previous usage. *Phylogeny* (the genesis of tribes) is used for (predominantly) branching patterns of descent, *nexogeny* (the genesis of bonds) for (predominantly) reticulating patterns of descent. Both are general terms that can be used to describe descent at various organizational levels, though each can be modified to specify the level of organization (e.g., gene phylogeny, organism nexogeny). *Ramogeny* (the genesis of branches) is used for the descent of populations (from demes to species), and *tokogeny* (the genesis of offspring) for the descent of organisms. Corresponding terms for other levels are not proposed here. The term *phyly* (*-phyletic*) can be used in association with the prefixes *mono, para,* and *poly* to describe different patterns of descent (see Hennig 1966) regardless of organizational level; the terms *ramy* (*-rametic*) and *toky* (*-toketic*) can be used for specific organizational levels (e.g., monorametic, polytoketic). The general term *entogeny* (the genesis of things that exist), and the related term *enty* (*-entetic*), can be used to encompass different modes (branching and reticlate) and levels (species, organism, etc.) of descent. Thus, *monoentetic* would used for a single ancestor and its descendants, regardless of whether that group is mutually exclusive or partially overlapping with other such groups, *monophyletic* if the group is mutually exclusive (e.g., clades, clans/clones of uniparental organisms), and *mononexetic* if it is partially overlapping (e.g., clans within a biparental species). *Clade* is used for monophyletic (and monorametic) groups of populations (from demes to species). *Clan* is used for monoentetic groups of organisms, regardless of reproductive mode—recognizing that clans of uniparental organisms will be monophyletic, whereas clans of biparental organisms will be mononexetic. *Clone* is used for monophyletic groups of asexually reproducing entities at or below the organismal level (e.g., gene clones, organelle clones, cell clones, although cell clones in unicellular organisms are also clans).

4. Two or more causal processes are implied: first, the process of descent, which is inherent in the concept of a lineage (at any level), and second, whatever process or processes unite organism lineages to form population lineages.

5. The concept of a population is not atemporal (truly instantaneous) in that the processes viewed as determining the limits of populations are temporal phenomena. For example, the process of interbreeding is commonly viewed as important in determining the limits of populations, but as pointed out by O'Hara (1993), no population is composed of organisms that are all interbreeding at any given instant.

6. Several authors (e.g., Brothers 1985, Templeton 1989) have emphasized that the exchange of genetic material among organism lineages is not neatly dichotomized into asexual and sexual reproductive modes, but instead forms a continuum.

7. From the Greek *blastos*, meaning bud, sprout, shoot, or germ. The term is proposed to distinguish the species-level process from the analogous organism-level process termed *budding*, which has also been used to designate this model of speciation (e.g., Foote 1996, de Queiroz 1998).

8. It should be noted that although these general models of speciation are logical consequences of certain views on the properties of species (those properties used to define the models in "Models of Speciation"), the properties are most commonly stated without explicit reference to the models of speciation that they imply. Other times, the models are implied by properties that are a step further removed. For example, the view that every diagnosable lineage segment represents a different species (e.g., Nixon and Wheeler 1992) implies that speciation occurs in unbranched lineages, and this consequence in turn implies an anagenetic model of speciation.

9. Neigel and Avise (1986) used the term *paraphyly* for cases in which certain gene or organism lineages within a species share more recent common ancestors with heterospecific than with conspecific gene or organism lineages (figure 3.5). However, at least some of the species that fit this description are not paraphyletic in the sense of a group including an ancestor and some, but not all, of its descendants (e.g., Hennig 1966, Wiley 1981), because the most recent common

ancestor of the lineages in question is not part of the species identified as paraphyletic, but of a more distant ancestral species. Species of this kind are polyphyletic rather than paraphyletic in terms of their component genes or organisms; they differ from the species that Neigel and Avise considered polyphyletic only in the relative depth of coalescence of their component gene or organism lineages.

10. A similar situation involving polyphyly exists when hybridization between members of separate biparental species occurs multiple times to produce separate uniparental clones, the component organisms of which are similar in most biologically significant respects. *Cnemidophorus tesselatus* (reviewed by Wright 1993) is commonly cited as an example (e.g., Kitcher 1984a, Holsinger 1987, Wilson 1995). If interbreeding is the only process that unites organism lineages to form species, then neither the individual clones nor the collection of them are species. However, if processes other than interbreeding unite organism lineages to form species, then it might be argued either that the individual clones are species or that the collection is a species, and the collection is polyphyletic in terms of its component clones. If those clones represent separate populations (e.g., if they are allopatric), then the species is polyphyletic in terms of its component populations. This case is similar to the case of species paraphyly in that the issue is whether to recognize a single species for the entire set of populations (clones) as opposed to recognizing each individual population (clone) as a species.

11. Even if the secondary properties always characterize the same lineages, the alternative species definitions based on them might not be considered reconciled in that the entities described by those definitions are still conceptually, if not physically, distinct.

12. The situation is not quite as simple as stated in that not just any segment of a population lineage qualifies as a species (see "Species Life Cycles").

13. It is often said that populations, not organisms, are the entities that evolve (e.g., Futuyma 1986), a view reflected in the common definition of *evolution* as changes in allele frequencies in populations (e.g., Wilson and Bossert 1971, Hartl 1981). The evolution of populations, however, is not the result of their organizational level, but rather of their temporal extent. Over short time intervals (i.e., less than one generation), populations do not evolve any more than organisms do. Furthermore, organism lineages (as opposed to individual organisms) do evolve in the sense that they exhibit heritable change through descent. Thus, lineages at all levels are the things that evolve (Hull 1980), and a more accurate general definition of *evolution* is heritable changes in lineages.

14. This conclusion is analogous to the proposition that asexual (reproductively autonomous) organisms do not form species. Considering the term *species* as analogous with the term *organism* implies that the situation should be described differently. Because we talk about unicellular organisms rather than saying that unicellular entities do not form organisms, it is more appropriate to talk about uniorganismal species (provided that unisexual organisms do not form population-level lineages) than to say that unisexual organisms do not form species (see Hull 1980). An incidental benefit of this terminology is reconciliation (in theory, if not in practice) of the proposition that asexual organisms do not form population-level lineages with the taxonomic tradition that requires all organisms to be members (parts) of species.

15. Species possessing different contingent properties are useful for different kinds of studies. Thus, just as one might examine only sexually mature organisms in a study of mating behavior, one might examine only reproductively isolated species (specifically, those isolated by premating barriers) in a study of reinforcement.

16. Several authors (e.g., Chandler and Gromko 1989, Mallett 1995) have argued against species definitions that treat putative speciation mechanisms or unifying processes as necessary properties of species. They argue that such definitions tend not only to restrict the generality of the species concept, but also to confuse theories about the origin and maintenance of species with

the concept of species itself. These arguments are, in effect, arguments for a general species concept and thus are very much in keeping with the reinterpretation of the defining properties of the species category advocated in this essay.

17. Even these levels of organization do not differ absolutely, as is revealed by the existence of unicellular organisms and the possibility of uniorganismal species (see "Species and Biology").

18. One important reason for making this distinction clear is that the whole may be more than the sum of its parts.

19. Ghiselin (1997; see also Frost and Kluge 1994) considered this resolution of the individual and class/set interpretations of species "semantic trickery" because it supposedly confuses different levels in the hierarchy of biological organization. On the contrary, the resolution in question requires an explicit distinction between different levels of biological organization. If any position is to be characterized as semantic trickery, it is Ghiselin's own position that "the names of taxa remain names of the taxa themselves ... they are terms like 'Mammalia' or '*Homo sapiens*,' not 'mammal' or 'human being'" (1997, 69). This position begs the question by assuming use of the term *species* to designate the population-level wholes rather than the sets of their organismal parts. Although my own terminological preference is identical, there is nothing about the idea that population lineage segments are individuals that requires using the term *species* to designate the lineage segments themselves rather than the sets of their component organisms.

20. Most organisms, of course, do not produce single offspring, and offspring that die young or fail to reproduce still count as branching despite the early termination of their lineages.

21. If species extinction is analogous to organismal death, then termination by bifurcation should not be called extinction. The following terminology makes the relevant distinctions. At the organismal level, the process of origination is called *reproduction* (~birth), which is termed *fission* if the division is more or less equal and *budding* if it is highly unequal. At the species level, the process of origination is called *speciation*, which is termed *bifurcation* if the division is more or less equal and *blastation* if it is highly unequal. The termination of organisms is called *defunction* (death) when the lineage itself terminates; it is called *disjunction* when associated with fission. The termination of species is called *extinction* when the lineage itself terminates; it is called *distinction* when associated with fission.

22. The pluralist position is sometimes considered to include species concepts that do not conform to the general lineage concept discussed in this essay—for example, views in which species are conceptualized as sets defined solely on the basis of organismal similarity (e.g., Kitcher 1984a; Dupré 1993, chapter 1 in this volume). Although use of the term *species* to designate such groups cannot be dismissed on logical grounds (because the issue is a semantic one), it is doubtful that any contemporary biologists actually conceptualize species in this way (see "Theory and Operations").

23. This position does not deny certain conceptual differences—for example, those regarding the processes that unite organism lineages to form species. Those differences do not, however, reflect different concepts of species. Instead, they reflect different hypotheses about the processes and thus the kinds of organisms (e.g., sexual vs. asexual) that form entities fitting the general lineage concept.

24. Darwin emphasized divergence in this discussion, never mentioning the possibility that even the most recently diverged lines of descent might reunite. Thus, it is not clear whether he viewed those lines as being unified by something other than their recent common ancestry.

Beatty (1985) argued that Darwin adopted a minimalist definition of species in which species were simply those taxa recognized as species by his fellow naturalists and used it to argue that species evolved. Darwin's concept of species as lineages was, therefore, a theory to explain the existence of the entities that his fellow naturalists recognized as species rather than a prescriptive definition.

25. Linnaeus (1766–68), for example, recognized species not only of animals and plants, but also of rocks and minerals.

26. Simpson's (1951) species definition may be the earliest one that explicitly equates species with lineages. Moreover, Simpson's (1951, 1961) definition (see also Wiley 1978, 1981) is perhaps the best description of the species concept that emerges from taking the elements common to all modern species definitions (the general lineage concept) and reinterpreting the so-called species criteria as contingent rather than necessary properties of species (compare Mayden 1997). Ironically, the strengths of this definition are the very things that have been criticized by advocates of alternative definitions—namely, that it "fails" to include explicit descriptions of operational criteria and causal mechanisms. By omitting such statements, Simpson's definition avoids confusing the general *concept* of species with operational criteria for recognizing species taxa or with theories about causal mechanisms.

## REFERENCES


Andersson, L. (1990). The driving force: Species concepts and ecology. *Taxon* 39, 375–382.

Baum, D. A., and K. L. Shaw (1995). Genealogical perspectives on the species problem. In P. C. Hoch and A. G. Stephenson, eds., *Experimental and molecular approaches to plant biosystematics.* St. Louis: Missouri Botanical Garden.

Beatty, J. (1985). Speaking of species: Darwin's strategy. In D. Kohn, ed., *The Darwinian heritage.* Princeton, N.J.: Princeton University Press.

Bell, M. A. (1979). Persistence of ancestral-sister species. *Systematic Zoology* 28, 85–88.

Bonner, J. T. (1967). *The cellular slime molds.* Princeton, N.J.: Princeton University Press.

Bremer, K., and H.-E. Wanntorp (1979). Geographic populations or biological species in phylogeny reconstruction. *Systematic Zoolology* 28, 220–224.

Brothers, D. J. (1985). Species concepts, speciation, and higher taxa. In E. S. Vrba, ed., *Species and speciation.* Pretoria, South Africa: Transvaal Museum.

Bush, G. L. (1975). Modes of animal speciation. *Annual Review of Ecology and Systematics* 6, 339–364.

Chandler, C. R., and M. H. Gromko (1989). On the relationship between species concepts and speciation processes. *Systematic Zoology* 38, 116–125.

Coyne, J. A., H. A. Orr, and D. J. Futuyma (1988). Do we need a new species concept? *Systematic Zoology* 37, 190–200.

Cracraft, J. (1983). Species concepts and speciation analysis. *Current Ornithology* 1, 159–187.

Crisp, M. D., and G. T. Chandler (1996). Paraphyletic species. *Telopea* 6, 813–844.

Darwin, C. (1859). *On the origin of species by means of natural selection.* London: John Murray.

de Queiroz, K. (1988). Systematics and the Darwinian revolution. *Philosophy of Science* 55, 238–259.

de Queiroz, K. (1992a). Phylogenetic definitions and taxonomic philosophy. *Biology and Philosophy* 7, 295–313.

de Queiroz, K. (1992b). Review of *Principles of systematic zoology,* 2d edition. *Systematic Biology* 41, 264–266.

de Queiroz, K. (1995). The definitions of species and clade names: A reply to Ghiselin. *Biology and Philosophy* 10, 223–228.

de Queiroz, K. (1997). The Linnaean hierarchy and the evolutionization of taxonomy, with emphasis on the problem of nomenclature. *Aliso* 15, 125–144.

de Queiroz, K. (1998). The general lineage concept of species, species criteria, and the process of speciation: A conceptual unification and terminological recommendations. In D. J. Howard and S. H. Berlocher, eds., *Endless forms: Species and speciation*. Oxford: Oxford University Press.

de Queiroz, K., and M. J. Donoghue (1988). Phylogenetic systematics and the species problem. *Cladistics* 4, 317–338.

de Queiroz, K., and M. J. Donoghue (1990). Phylogenetic systematics and species revisited. *Cladistics* 6, 83–90.

Dobzhansky, T. (1935.) A critique of the species concept in biology. *Philosophy of Science* 2, 344–355.

Dobzhansky, T. (1937). *Genetics and the origin of species*. New York: Columbia University Press.

Dobzhansky, T. (1950). Mendelian populations and their evolution. *American Naturalist* 84, 401–418.

Dobzhansky, T. (1970). *Genetics of the evolutionary process*. New York: Columbia University Press.

Donoghue, M. J. (1985). A critique of the biological species concept and recommendations for a phylogenetic alternative. *The Bryologist* 88, 172–181.

Doyen, J. T., and C. N. Slobodchikoff (1974). An operational approach to species classification. *Systematic Zoology* 23, 239–247.

Dupré, J. (1993). *The disorder of things: Metaphysical foundations of the disunity of science*. Cambridge, Mass.: Harvard University Press.

Ehrlich, P., and P. H. Raven (1969). Differentiation of populations. *Science* 165, 1228–1232.

Ereshefsky, M. (1992). Eliminative pluralism. *Philosophy of Science* 59, 671–690.

Ereshefsky, M. (1998). Species pluralism and anti-realism. *Philosophy of Science* 65, 103–120.

Foote, M. (1996). On the probability of ancestors in the fossil record. *Paleobiology* 22, 141–151.

Frost, D. R., and A. G. Kluge (1994). A consideration of epistemology in systematic biology, with special reference to species. *Cladistics* 10, 259–294.

Futuyma, D. J. (1986). *Evolutionary biology*. Sunderland, Mass.: Sinauer.

George, T. N. (1956). Biospecies, chronospecies and morphospecies. In P. C. Sylvester-Bradley, ed., *The species concept in palaeontology*. London: Systematics Association.

Ghiselin, M. T. (1974). A radical solution to the species problem. *Systematic Zoology* 23, 536–544.

Ghiselin, M. T. (1997). *Metaphysics and the origin of species*. Albany, N.Y.: State University of New York Press.

Grant, V. (1963). *The origin of adaptations*. New York: Columbia University Press.

Grant, V. (1994). Evolution of the species concept. *Biologisches Zentralblatt* 113, 401–415.

Griffiths, G. C. D. (1974). On the foundations of biological systematics. *Acta Biotheoretica* 23, 85–131.

Haffer, J. (1986). Superspecies and species limits in vertebrates. *Zeitschrift für zoologisches Systematik und Evolutionsforschung* 24, 169–190.

Hartl, D. L. (1981). *A primer of population genetics*. Sunderland, Mass.: Sinauer.

Hennig, W. (1966). *Phylogenetic systematics.* Urbana: University of Illinois Press.

Holsinger, K. E. (1987). Pluralism and species concepts, or when must we agree with one another? *Philosophy of Science* 54, 480–485.

Hull, D. L. (1968). The operational imperative: Sense and nonsense in operationism. *Systematic Zoology* 17, 438–457.

Hull, D. L. (1976). Are species really individuals? *Systematic Zoology* 25, 174–191.

Hull, D. L. (1978). A matter of individuality. *Philosophy of Science* 45, 335–360.

Hull, D. L. (1980). Individuality and selection. *Annual Review of Ecology and Systematics* 11, 311–332.

Hull, D. L. (1997). The ideal species concept—and why we can't get it. In M. F. Claridge, H. A. Dawah, and M. R. Wilson, eds., *Species: The units of biodiversity.* London: Chapman and Hall.

Humphreys, T. (1970). Biochemical analysis of sponge cell aggregation. In W. G. Fry, ed., *The biology of the Porifera.* London: Academic Press.

Huxley, J. (1942). *Evolution, the modern synthesis.* London: Allen and Unwin.

Huxley, J., ed. (1940). *The new systematics.* Oxford: Clarendon Press.

Jablonski, D., and D. Bottjer (1991). Environmental patterns in the origins of higher taxa: The post paleozoic fossil record. *Science* 252, 1831–1833.

Jordan, K. (1905). Der Gegensatz zwischen geographischer und nichtgeographischer Variation. *Zeitschrift für wissenschaftliche Zoologie* 83, 151–210.

Kitcher, P. (1984a). Species. *Philosophy of Science* 51, 308–333.

Kitcher, P. (1984b). Against monism of the moment: A reply to Elliott Sober. *Philosophy of Science* 51, 616–630.

Korotkova, G. P. (1970). Regeneration and somatic embryogenesis in sponges. In W. G. Fry, ed., *The biology of the Porifera.* London: Academic Press.

Lambert, D. M., B. Michaux, and C. S. White (1987). Are species self-defining? *Systematic Zoology* 36, 196–205.

Linnaeus, C. (1766–68). *Systema naturae per regna tria naturae, secundum classes, ordines, genera, species, cum characteribus, differentiis, synonymis, locis,* 12th ed. Stockholm: Laurentii Salvii.

Mallet, J. (1995). A species definition for the modern synthesis. *Trends in Ecology and Evolution* 10, 294–299.

Mayden, R. L. (1997). A hierarchy of species concepts: The denouement in the saga of the species problem. In M. F. Claridge, H. A. Dawah, and M. R. Wilson, eds., *Species: The units of biodiversity.* London: Chapman and Hall.

Mayr, E. (1942). *Systematics and the origin of species.* New York: Columbia University Press.

Mayr, E. (1955). Karl Jordan's contribution to current concepts in systematics and evolution. *Transactions of the Royal Entomological Society of London* 107, 45–66.

Mayr, E. (1957). *The species problem.* Washington, D.C.: American Association for the Advancement of Science.

Mayr, E. (1963). *Animal species and evolution.* Cambridge, Mass.: Harvard University Press.

Mayr, E. (1969). *Principles of systematic zoology.* New York: McGraw-Hill.

Mayr, E. (1982). *The growth of biological thought. Diversity, evolution, and inheritance.* Cambridge, Mass.: Belknap Press.

Mayr, E. (1988). The why and how of species. *Biology and Philosophy* 3, 431–441.

Mayr, E., and P. D. Ashlock (1991). *Principles of systematic zoology*. New York: McGraw-Hill.

Mayr, E., and W. B. Provine, eds. (1980). *The evolutionary synthesis: Perspectives on the unification of biology*. Cambridge, Mass.: Harvard University Press.

Meglitsch, P. A. (1954). On the nature of species. *Systematic Zoology* 3, 49–68.

Michener, C. D. (1970). Diverse approaches to systematics. *Evolutionary Biology* 4, 1–38.

Mishler, B. D. (1985). The morphological, developmental, and phylogenetic basis of species concepts in bryophytes. *The Bryologist* 88, 207–214.

Mishler, B. D., and R. N. Brandon (1987). Individuality, pluralism, and the phylogenetic species concept. *Biology and Philosophy* 2, 397–414.

Mishler, B. D., and M. J. Donoghue (1982). Species concepts: A case for pluralism. *Systematic Zoology* 31, 491–503.

Neigel, J. E., and J. C. Avise (1986). Phylogenetic relationships of mitochondrial DNA under various demographic models of speciation. In E. Nevo and S. Karlin, eds., *Evolutionary processes and theory*. London: Academic Press.

Nelson, G. (1989). Species and taxa: Systematics and evolution. In D. Otte and J. A. Endler, eds., *Speciation and its consequences*. Sunderland, Mass.: Sinauer.

Newell, N. D. (1956). Fossil populations. In P. C. Sylvester-Bradley, ed., *The species concept in palaeontology*. London: Systematics Association.

Nixon, K. C., and Q. D. Wheeler (1990). An amplification of the phylogenetic species concept. *Cladistics* 6, 211–223.

Nixon, K. C., and Q. D. Wheeler (1992). Extinction and the origin of species. In M. J. Novacek and Q. D. Wheeler, eds., *Extinction and phylogeny*. New York: Columbia University Press.

O'Hara, R. J. (1993). Systematic generalization, historical fate, and the species problem. *Systematic Biology* 42, 231–246.

Panchen, A. L. (1992). *Classification, evolution, and the nature of biology*. Cambridge: Cambridge University Press.

Paterson, H. E. H. (1978). More evidence against speciation by reinforcement. *South African Journal of Science* 74, 369–371.

Paterson, H. E. H. (1985). The recognition concept of species. In E. S. Vrba, ed., *Species and speciation*. Pretoria: Transvaal Museum.

Polly, P. D. (1997). Ancestry and species definition in paleontology: A stratocladistic analysis of Paleocene-Eocene Viverravidae (Mammalia, Carnivora) from Wyoming. *Contributions from the Museum of Paleontology. The University of Michigan* 30, 1–53.

Poulton, E. B. (1903). What is a species? *Proceedings of the Entomological Society of London*, lxxvii–cxvi.

Rhodes, F. H. T. (1956). The time factor in taxonomy. In P. C. Sylvester-Bradley, ed., *The species concept in palaeontology*. London: Systematics Association.

Ridley, M. (1989). The cladistic solution to the species problem. *Biology and Philosophy* 4, 1–16.

Ridley, M. (1993). *Evolution*. Cambridge, Mass.: Blackwell Science.

Rogers, D. J., and S. G. Appan (1969). Taximetric methods for delimiting biological species. *Taxon* 18, 609–752.

Rosen, D. E. (1979). Fishes from the uplands and intermontane basins of Guatemala: Revisionary studies and comparative geography. *Bulletin of the American Museum of Natural History* 162, 267–376.

Simpson, G. G. (1951). The species concept. *Evolution* 5, 285–298.

Simpson, G. G. (1961). *Principles of animal taxonomy*. New York: Columbia University Press.

Smith, A. B. (1994). *Systematics and the fossil record. Documenting evolutionary patterns.* Oxford: Blackwell Scientific Publications.

Sneath, P. H. A., and R. R. Sokal (1973). *Numerical taxonomy: The principles and practice of numerical classification*. San Francisco: W. H. Freeman.

Sober, E. (1984). Sets, species, and evolution: Comments on Philip Kitcher's "Species." *Philosophy of Science* 51, 334–341.

Sokal, R. R., and T. J. Crovello (1970). The biological species concept: A critical evaluation. *American Naturalist* 104, 127–153.

Sosef, M. S. M. (1997). Hierarchical models, reticulate evolution and the inevitability of paraphyletic supraspecific taxa. *Taxon* 46, 75–85.

Stanford, P. K. (1995). For pluralism and against realism about species. *Philosophy of Science* 62, 70–91.

Stebbins, G. L., Jr. (1950). *Variation and evolution in plants*. New York: Columbia University Press.

Templeton, A. R. (1987). Species and speciation. *Evolution* 41, 233–235.

Templeton, A. R. (1989). The meaning of species and speciation: A genetic perspective. In D. Otte and J. A. Endler, eds., *Speciation and its consequences*. Sunderland, Mass.: Sinauer.

Van Valen, L. (1976). Ecological species, multispecies, and oaks. *Taxon* 25, 233–239.

Wagner, P. J., and D. H. Erwin (1995). Phylogenetic patterns as tests of speciation models. In D. H. Erwin and R. L. Anstey, eds., *New approaches to studying speciation in the fossil record*. New York: Columbia University Press.

Westoll, T. S. (1956). The nature of fossil species. In P. C. Sylvester-Bradley, ed., *The species concept in palaeontology*. London: Systematics Association.

Wheeler, Q. D., and K. C. Nixon (1990). Another way of *looking at* the species problem: A reply to de Queiroz and Donoghue. *Cladistics* 6, 77–81.

White, C. S., B. Michaux, and D. M. Lambert (1990). Species and neo-Darwinism. *Systematic Zoology* 39, 399–413.

Wiley, E. O. (1978). The evolutionary species concept reconsidered. *Systematic Zoology* 27, 17–26.

Wiley, E. O. (1981). *Phylogenetics: The theory and practice of phylogenetic systematics*. New York: John Wiley and Sons.

Wilkinson, M. (1990). A commentary on Ridley's cladistic solution to the species problem. *Biology and Philosophy* 5, 433–446.

Williams, M. B. (1985). Species are individuals: Theoretical foundations for the claim. *Philosophy of Science* 52, 578–590.

Wilson, B. E. (1995). A (not-so-radical) solution to the species problem. *Biology and Philosophy* 10, 339–356.

Wilson, E. O., and W. H. Bossert (1971). *A primer of population biology*. Stamford, Conn.: Sinauer.

Wittgenstein, L. (1967). *Zettel.* Oxford: Basil Blackwell.

Wright, J. W. (1993). Evolution of the lizards of the genus *Cnemidophorus.* In J. W. Wright and L. J. Vitt, eds., *Biology of whiptail lizards (genus* Cnemidophorus). Norman, Okla.: Oklahoma Museum of Natural History.

Wright, S. (1940). The statistical consequences of Mendelian heredity in relation to speciation. In J. Huxley, ed., *The new systematics.* London: Oxford University Press.

# II Species and Life's Complications

# 4 When Is a Rose?: The Kinds of *Tetrahymena*

David L. Nanney

## THE CILIATE POINT OF VIEW

### Areas of Dissent

Beginning biologists are sometimes told they have to get inside the skins of the organisms they study, to think like a bird or a beetle. More experienced biologists clearly have their perceptions shaped by the organisms they study. Some biologists indeed come to look like fruit flies or mice. In this essay, as a ciliatologist, I look at biological species from the perspective of the ciliated protozoa. A careful look at any particular organism might be illuminating to philosophers of science, a useful corrective to airy speculations and premature generalizations, but the ciliated protozoa have a considerable history of providing not just a "real" perspective, but a "really" different point of view (Nanney 1983).

In a thoughtful paper, Jan Sapp (1991) develops this thesis of the eccentricity of the ciliates and ciliatologists by considering episodes of historical turbulence generated by work on ciliates. His paper includes the subject of his earlier book (Sapp 1987), *Beyond the Gene: Cytoplasmic Inheritance and the Struggle for Authority in Genetics,* which describes the studies on cytoplasmic inheritance in the ciliated protozoa that challenged the information hegemony of the nucleus.

The impact of the ciliate studies in heredity is illustrated by a comment made by John Maynard Smith (1983):

> There are a few well-established exceptions [to nuclear control] of which the phenomenon of 'cortical inheritance' in ciliates is perhaps the most important. Neo-Darwinists should not be allowed to forget these cases, because they constitute the only significant threat to our views. However, the overwhelming majority of inherited differences are caused by differences between chromosomal genes. (p. 39)

Sapp also discusses some whispers of heresy generated by studies of morphogenesis in ciliated protozoa. This material has been superbly summarized by Joseph Frankel (1989) in his book *Pattern Formation: Ciliate Studies and Models.* Ciliate developmental studies raise vexing doubts about the adequacy of the current tentative resolution of the long-standing "developmental enigma." This resolution postulates that differential gene action in isolated cellular compartments is an adequate explanation for developmental differentiation in higher organisms. Ciliated protozoa, however, manifest essentially the same "field phenomena" characteristic of multicellular embryos and in the same scale of organismic design as embryos, but without cellular compartments.

Additional tensions exist in other areas, particularly with respect to the meaning of life histories, the naming of biological species, and the mechanisms of evolution (Schloegel forthcoming). Among these issues is clonal aging in ciliates, described first by Emile Maupas in the nineteenth century (1889) and widely studied by many protozoologists in the context of a "genetic program." According to Graham Bell (1988) in his book *Sex and Death in Protozoa: The History of an Obsession,* the phenomenon of clonal aging doesn't exist in ciliates, except as a purely stochastic process.

The aging controversy is related to the rationalization of genetic economies in ciliates (Sonneborn 1957); controlling the length of the clonal life span and its components seems to require the agency of group selection, but group selection is currently a disfavored if not heretical notion. Natural selection is thought to be efficacious only downward, reductionistically from the level of the individual to the level of the molecule. (See Griesemer [1996] for an alternative interpretation of reduction "upward" toward a more fundamental simplicity.)

Marginally relevant to such issues is the present description of the evolutionary relationships among a group of small freshwater ciliates; the primary focus is on the significance of the sibling species discovered in ciliates some six decades ago.

## The Domain of the Ciliated Protozoa

I probably need, however, to introduce the reader more generally to the ciliated protozoa before focusing on the tetrahymenas (see Gall [1992] and Hausmann and Bradbury [1996] for recent summaries of major research areas, especially Lynn [1996] for current treatment of ciliate systematics). In their most important features, ciliates are like everybody else. All living organisms fabricate proteins on ribosomes; these machines translate biological information from the same coded sequences of nucleic acids into ordered sequences of the same amino acids in polypeptides, using by and large the same codon dictionary. The respiratory physiology and intermediary metabolism of sugars and fats, purines and pyrimidines, are not fundamentally different in ciliates from these processes in any other form of life on Earth.

Ciliates (Hausmann and Bradbury 1996) stand, however, clearly on the eukaryotic side of the great divide in life forms—the divide between the *prokaryotes* (the archaebacteria and the eubacteria) and the *eukaryotes* ("cellular" organisms) that appeared suddenly in the geological record about halfway through the history of life. The eukaryotes share an astonishing set of complex adaptations. Unlike any prokaryotes, they all possess similar electrically excitable membranes that facilitate communication and interaction between cells and with their environment. All eukaryotes exhibit homologous fiber systems that shape their bodies and move efficient motor organelles. All eukaryotes package their DNA on nucleosomic histone spools, arranged in conventional chromosomes that undergo meiosis and mitosis at appropriate times.

Ciliates differ from higher eukaryotes, (as well as from algae and fungi), however, in the organization of their germplasm and soma. Like higher organisms, they have compounded their basic equipment to provide a larger organismic entity than can be achieved within the domain of a single genome, but they have done this without aggregating and differentiating populations of discrete cells. They have pooled many identical genomes—from dozens to hundreds—within the bounds of a single amitotic macronucleus, and they have pooled their cytoplasmic components into a communal cytoplasm with an elaborately differentiated cortex. They reserve one or a few diploid micronuclei as germinal reservoirs that dance the mitotic and meiotic pavannes, but that are incapable of ordinary transcriptional activities until the next round of sexual activity. The germinal information is periodically summoned into action in sexual episodes, after which a new somatic genetic system is developed to replace the old one.

How, or even if, these organizational features account for the discomfort generated by ciliatologists' observations and conclusions is uncertain. That ciliatologists often find themselves in possession of "exceptional" observations and "exceptionable" interpretations is, however, undeniable.

## THE TETRAHYMENINES

### The Simple Life

My principal focus is a particular kind of ciliated protozoan, an organism (or group, set, assembly, individual, category—see Ghiselin 1997) first dubbed *Tetrahymena* by Waldo Furgason in 1940. They are small, deceptively simple ciliated protozoa that have been seen and studied seriously only since the perfection of the optical microscope in the last quarter of the nineteenth century. Certain features of their organization were unknown, however, until other technologies emerged more recently for the description of submicroscopic anatomy. These technologies particularly include the silver-staining procedures developed for visualizing the detailed architecture of the cell surface (Klein 1926, Chatton and Lwoff 1936) and the transmission electron microscope that revealed the complex patterns of underlying granules, fibers, and membranes constituting the architectural superstructures. Silver staining,

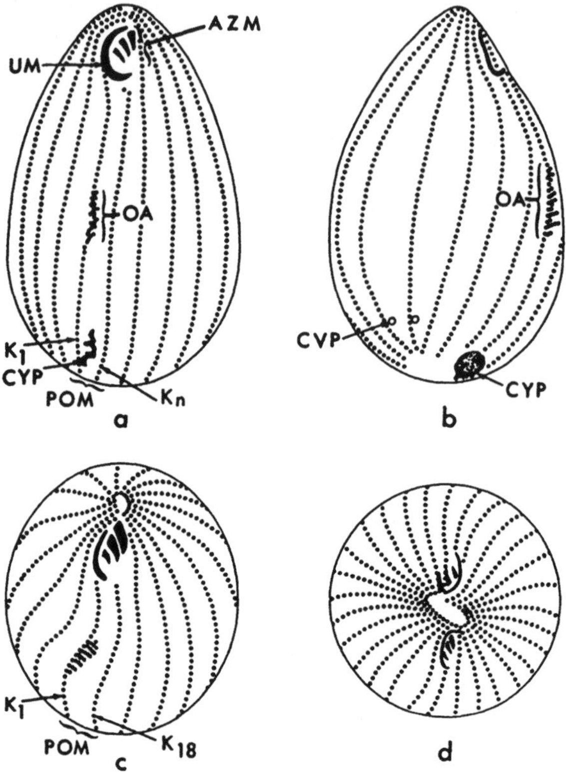


**Figure 4.1** Cortical features of a tetrahymenid ciliate revealed by Chatton-Lwoff silver stains. Each dot is a kinetosome—the base of a typical 9 + 2 cilium or flagellum. Each longitudinal row of kinetosomes is referred to as a *ciliary row* or *kinety*. The tetrahymenal oral apparatus consists of an undullating membrane (UM) and three other membranelles composing an adoral zone of membranelles (AZM); they stain by virtue of their packed ranks of kinetosomes. The ciliary row extending back from the oral apparatus is designated as kinety 1 (K1), or the stomatogenic kinety, and the new oral anlagen (OA) begins to the left of this row as a proliferating field of kinetosomes. The ciliary rows are numbered to the cell's right around the circumference of the cell so that the row to the cell's left of row 1 (or the postoral median—POM) is designated as kinety n (Kn), usually a number from 16 to 20. Also associated with kinety 1 at the posterior end of the cell is the irregularly staining outlet of the cytoproct (CYP). Approximately one-fourth of the distance around the cell to its right are the openings of the water-balance organelles—the contractile vacuole pores (CVP). (From Nanney 1966)

for example, first allowed Furgason to describe the four membranelles characterizing the remarkably complex buccal apparatus (the tetrahymenal feeding device).

The tetrahymenines (figure 4.1) are smaller than the more familiar large ciliates, such as *Paramecium* or *Stentor*. They are shaped like a U.S. football, for the same reason. They reach about 60 microns in length, but their size varies during the vegetative cell cycle when they double in volume and during starvation when they may be drastically reduced in size. The cell cycle can be completed in a couple of hours under optimal conditions. Tetrahymenas feed primarily on the bacterial lawns of freshwater habitats the earth over. Their niche is that of the generic bacterial lawn mower.

The basic ecological role of tetrahymenas was probably achieved very early during the explosive eukaryotic radiation. The elementary eukaryotic structural innovations (nucleosomic chromosomes, electrically excitable membranes, and complex fiber systems) were compounded into an elaborate multigenomic machine capable of exploiting the prokaryotic bounty.

The compounding process was different from the process later employed in the emergence of the metazoa and metaphyta—organisms with differentiated multicellularity. Tetrahymena genomes are pooled into a single large vegetative macronucleus containing initially about forty copies of the elementary chromosome set; a separate diploid micronucleus with five pairs of chromosomes is set aside as a nontranscriptional reserve.

The large cytoplasmic mass potentiated by the multigenomic nucleus is sculpted into an unexpectedly complex edifice whose major structural components are ciliary units arranged in coordinated linear rows referred to as *kineties* (figure 4.1). The tetrahymenas share a sophisticated, yet unexplained cytogeometric design worthy of a Gothic cathedral. The larger patterns are achieved by the deployment of repeated elements within global fields; they are decorated with bizarre common motifs, many of which are thus far beyond functional explanation. One is inevitably reminded of the spandrels of San Marcos (Gould and Lewontin 1979).

The tetrahymenas achieved scientific attention much later than their larger relatives, but they seem to have been preadapted to a significant artificial niche—the research laboratory (later in this essay, I consider briefly the nature of their preadaptation to this unnatural niche)—as well as to the original ecological niche in freshwater streams and lakes. In the lab they can be fed on pure strains of bacteria, but more significantly, as Andre Lwoff (1923) first showed, they grow quite well on relatively simple axenic media (i.e., with no foreign organisms). Isolates were made from several natural sources in the 1930s and 1940s, and they were welcomed in dozens of laboratories around the world. An enormous literature developed on their physiological responses, nutritional requirements, and structural details (Hill 1972).

In the late 1940s, John Corliss, then a young investigator at Yale, undertook a careful systematic evaluation of the strains that had found this new niche. Corliss was destined to become the leading expert on the taxonomy of *Tetrahymena* and indeed on the ciliated protozoa in general. Corliss's first systematic study (1952, 1953) concluded that most of the many strains of *Tetrahymena* are indistinguishable and that they constitute a remarkably uniform assembly representing a ubiquitous and essential component of freshwater habitats the world over. He assigned them to a single species, *Tetrahymena pyriformis*.

### Sex Complicates Things

At about the same time, interest was rising in *tetrahymena*'s genetic system, and crosses were attempted. These attempts failed with the available laboratory strains, which—it was belatedly discovered—were all amicronucleate and asexual. Renewed isolations from nature, however, quickly provided large numbers of strains with micronuclei, and they were fully capable of undergoing conjugation under appropriate circumstances (Elliott and Nanney 1952, Elliott and Hayes 1953). The major condition required for controlled matings was the presence of organisms of complementary mating types.

Earlier, Tracy Sonneborn (1937) had discovered mating types in another ciliate, *Paramecium aurelia*. The first mating strains of *tetrahymena* had a more complex mating system: whereas only two mating types constitute a mating system in most paramecia, the first *tetrahymena* system studied had seven mating types (Nanney and Caughey 1953). Conjugation occurred when any two mating types were brought together. Mating systems of great diversity abound in the ciliated protozoa, and the genetic economies they delineate confound easy summary. Some genetic species have only two mating types; others have multiple mating-type systems that may facilitate mating with "strangers" in an outbreeding economy. Other genetic species have characteristics of inbreeders; they have short life cycles—terminated by autogamy (self-fertilization) or death in a few weeks if a suitable mate isn't found. Others have no autogamy, but life cycles that last for years. The differences among the species may be interpreted as methods of coping with different kinds of environmental challenges by means of the differential utilization of physiological plasticity and mutational variety. None of these differences challenges the Modern Synthesis or violates the essential characteristics of the biological species concept, though they raise questions about the units of natural selection. The reader is directed to informed discussion of these breeding systems by Sonneborn (1957), Nanney (1980, chap. 10), Nyberg (1988), and Dini and Nyberg (1993). The major studies in *Tetrahymena* population genetics are those of Doerder et al. (1995).

*Tetrahymena* studies soon revealed another feature previously reported in *Paramecium* (Sonneborn 1939, 1947); the morphologically indistinguishable strains are actually divided into several genetically isolated systems (Gruchy 1955, Elliott 1973). *Tetrahymena pyriformis,* like *Paramecium aurelia,* consists of a set of noninterbreeding cryptic biological species. Because at first the biological species could only be identified by the use of living tester strains, they were referred to by numbers instead of by names, with Sonneborn's neologism *syngen* used to designate cryptic biological species identifiable only with live reference strains. Sonneborn's refusal to assign Latin binomials to sibling species earned him the scorn of Ernst Mayr.

*T. pyriformis* was clearly not a single biological species in the sense defined in the lexicon of the Modern Synthesis, but rather a collection of such species. Yet *T. pyriformis* continued to be a useful collective term for awhile, as did *P. aurelia* for the cryptic species of that complex. Individuals collected from nature can be quickly assigned to the collective set by any competent protozoologist, but special laboratory resources and reference strains are necessary for assignment to a particular biological species. The species name *T. pyriformis* was eventually reassigned (Nanney and McCoy 1976) to a collection of amicronucleate clones judged to be closely related by the near identity of their molecules. Now, the name *T. pyriformis* does not even refer to a true "biological species." Moreover, no term is available to encompass the assemblage originally represented by the term *T. pyriformis*. The revised assignment was probably a tactical mistake, though it seemed taxonomically appropriate. Some investigators (mainly biochemists) continue to use the name in its obsolete sense.

Though protozoologists generally understand and accept the biological species as the basic evolutionary unit (Corliss and Daggett 1983) some naturalists (Finlay et al. 1996; Finlay, Esteban, and Feuchel 1996) have more recently suggested that the morphospecies should be reinstated for use in faunal surveys. Separating the genetic species is just too costly and time-consuming, even when the necessary background research has been done. To ignore the biological species, however, would certainly lead to serious underestimations of species diversity. The earlier refusal of protozoologists to designate cryptic species by Latin binomials led some evolutionists to suppose that protozoologists "do not understand" modern evolutionary theory. Protozoologists, on the other hand, suspect that evolutionary geneticists do not always understand that taxonomy serves clients other than evolutionists. Different terms may be needed in different contexts.

### Will the Real Strain GL Please Stand Up?

Eventually, the means were found to identify tetrahymena strains on the basis of something besides their ability to recognize potential mates. At this time, protozoologists were happy to assign Latin binomials to cryptic species (Sonneborn 1975, Nanney and McCoy 1976).

The descriptions of strains in terms of their molecules allow them to be assigned to particular sibling species without mating tests—even strains such as the amicronucleate laboratory strains, but only with considerable laboratory effort. The first success was with isozyme analysis—a technique based on the differential electrophoretic mobilities of proteins with particular enzymatic activity (Borden et al. and 1977, Meyer and Nanney 1987). An unknown tetrahymena strain can usually be identified as a member of one of the small subsets of cryptic species by means of a few diagnostic isozyme systems.

The isozyme tests were also applied to the long-maintained laboratory amicronucleates. The most famous of these laboratory tetrahymenas was GL, presumably the one isolated by Lwoff (1923) and maintained in pure clonal culture in laboratories for tens of thousands of cell generations. Other strains of more recent origin were referred to as E, H, S, W, and so no. When evaluated in starch or polyacrilamide gels, the molecular diversity among the amicronucleate strains was nearly as great as the diversity among the mating strains. However, the pattern of distribution of the differences in these strains led to embarrassing conclusions (Borden, Whitt, and Nanney, 1973).

The deceptive similarity of the strains, both in their structural details and in their laboratory performance, seems to have led investigators into regrettable lapses in their stock-keeping methods. With unwarranted faith in the equivalence of these bioreagents, some physiologists and biochemists allowed their labels to wander freely among test-tubes and flasks. The "GL" strains re-collected from different research laboratories often manifested totally different isozyme patterns—patterns characteristic of several different cryptic species. Similar discordance was observed in all the named strains, which fell into the same limited number of isozyme patterns. Each strain belonged to one of four or five "zymotypes," which had little relevance to the assigned strain designation. In 1999, we can no longer know with certainty which of these zymotypes should characterize the strain isolated by Lwoff in 1923.

Because these amicronucleate strains (and other amicronucleates collected from nature) can be differentiated with isozyme techniques, many can also be associated with particular cryptic breeding species. Strictly speaking, however, they cannot be assigned to those breeding species, because they do not share those gene pools prospectively. Amicronucleate strains are associated with biological species because they are probably derived recently (in an evolutionary time scale) from those particular species and thus share the gene pools retrospectively.

According to a strict application of the biological species concept, the cell that loses its micronucleus instantly loses its species label, just as a human red blood cell loses its human status during the developmental loss of its nucleus. Though such rigorous application of theoretical concepts is undoubtedly worthy of praise, it violates common sense, and protozoologists may perhaps be forgiven for associating clonal lineages with named species. A different problem occurs when a micronucleate strain is collected without a suitable mate or without an understanding of the necessary mating conditions.

## Nested Boxes

Once appropriate molecular technologies are applied, individual clones can usually be assigned to biological species without using mating tests, at least when a corresponding biological species has been identified by mating tests and has been characterized molecularly. When the differences among the species are appropriately quantified, the evolutionary distances among them can be measured and their evolutionary tree constructed. When this procedure was carried out with the tetrahymena species (figure 4.2), a previously

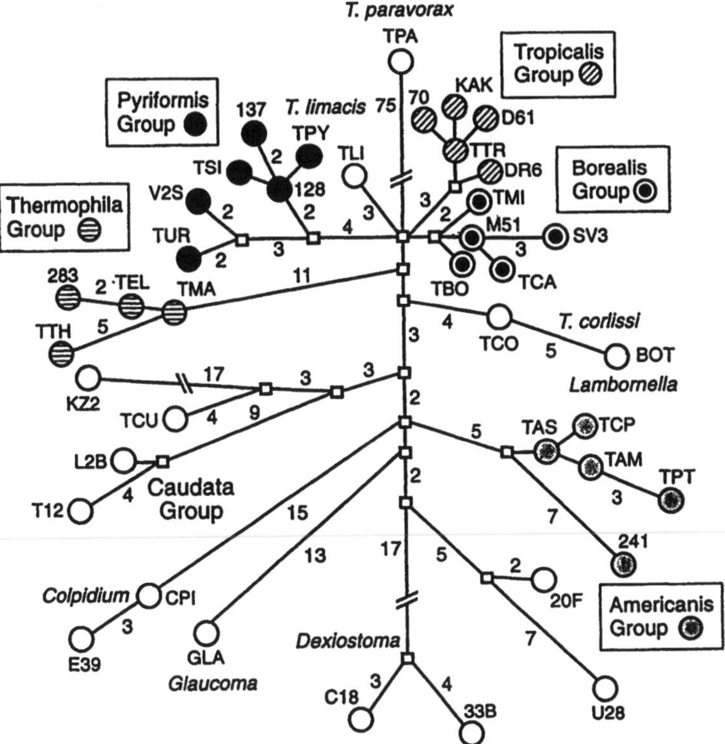


**Figure 4.2** Evolutionary tree of the tetrahymenine ciliates based on string analysis (Sankoff analysis) of 190 bases of the D2 domain of the large (23S) ribosomal RNA molecule. Although the cells are for the most part morphologically indistinguishable, at least five major clusters are apparent. The largest cluster is the Americanis Group, with eight genetically isolated sibling species sharing the same D2 sequence as TAM (for *T. americanis*). The species used most frequently in breeding studies is *T. thermophila* in the Thermophila Group. The distances among the groups are of the same order as the distances to recognizably different species assigned to other genera, such as *Glaucoma* (GLA) and *Colpidium* (CPI). (From Nanney et al. 1998, which includes the complete list of species names corresponding to the code terms appearing in this figure.)

unsuspected kind of evolutionary cluster became apparent (Preparata et al. 1989, Nanney et al. 1998).

The results of molecular analysis of tetrahymena strains do not depend significantly upon which molecules are compared or upon which tree-building program is utilized, but the most comprehensive data set is the set based on the D2 domain of the 23S rRNA, evaluated with the PHYLOGEN program. (A brief explanation of the methodologies used here can be found in the appendix to this chapter.)

Whether based on the D2 domain of the 23S rRNA (Preparata et al. 1992, Nanney et al. 1998) or on some other nucleic sequence (Sadler and Brunk 1990, 1992; Lynn and Sogin 1988), the evolutionary trees of the tetrahymenas are in almost complete agreement, except that D2 data are available for a much larger number of strains. Most of the 1950 "*T. pyriformis*" strains and those strains with equivalent morphological characteristics collected since 1950 fall into a few widely separated clusters (figure 4.2). Here, I focus attention on three clusters that contain most of the well-characterized mating systems: the *americanis* cluster, the *thermophila* cluster, and the *tropicalis-borealis* cluster.

The *americanis* cluster is the largest and most compact. It contains eight cryptic biological species with identical D2 sequences, and two others that differ by only one substitution. These species can be separated somewhat more satisfactorily by use of a more labile evolutionary standard, such as the intervening sequences of the histone cluster (Sadler and Brunk 1992). Separated from the *americanis* cluster by some twenty changes in the D2 domain is the *thermophila* cluster containing only three cryptic species that differ from each other by one to six changes. The *tropicalis-borealis* cluster also contains three cryptic species—*T. canadensis* (TCA), *T. borealis* (TBO), and *T. tropicalis* (TTR)—differing from each other by three-changes and is separated from both of the other clusters by about fifteen changes. Together, the three clusters include the twenty species that have been defined solely on the basis of mutually incompatible mating systems.

The other named species have been characterized on the basis of D2 sequence differences or by some combination of D2 sequences, isozymes, and minor morphological variations. The current bearer of the now restricted title *T. pyriformis* is a case in point. The name was awarded to a set of amicronucleate strains—labeled with titles GL, E, S, or W—that had a distinctive D2 sequence and similar zymotypes. Subsequent D2 sequencing studies placed them and some other amicronucleates collected directly from nature in a cluster well separated from the other three clusters mentioned, and that cluster was referred to appropriately as the *pyriformis* cluster. These strains seemed to have the highest probability of including the original Lwoff GL.

At least one other strain isolated from nature has this D2 sequence and very similar isozymes. A distinctive feature of this strain is an elongated caudal cilium. It is amicronucleate, but McCoy (1975) characterized several micronucleate strains with candal cilia as *T. setosa*. The *pyriformis* cluster is an anomaly in that it is a cluster identified by molecular similarities that contains no described "biological species."

Another similar "species," also named on the basis of three of the "classical" amics—labeled GL or E—is *T. elliotti* (TEL). This set of half micronucleate and half amicronucleate strains is found in the *thermophila* cluster, one base substitution from the well-characterized breeding species *T. malaccensis* (TMA) collected in Malaysia. Thus, classical amicronucleate strains labeled GL and E are found to be evolutionarily very distant from other classical strains also labeled GL and E, about twenty genetic changes apart in a system in which full reproductive isolation is observed (in the *T. americanis* (TAM) cluster) without any sequence changes.

Lest one suppose that the enormous evolutionary distances suggested here by the D2 region are an indication of an anomalous metric, I need to reiterate that all comparative analyses of particular molecules support the same general conclusions, whether the comparisons are annealing DNA molecules (Allen and Li 1974), enzyme molecules (Meyer and Nanney 1987), ciliary proteins (Seyfert and Willis 1981), cortical proteins (Williams 1986), or other rRNA molecules (Baroin et al. 1988; Brunk, Kahn, and Sadler 1990; Lynn and Sogin 1988; Sadler and Brunk 1990, 1992; Sogin, Elwood, and Gunderson 1986).

The detailed similarity among the tetrahymenas with respect to morphology (and most other phenetic characteristics), combined with evidence of substantial molecular diversification, poses an interesting evolutionary problem. It also identifies the need for a name for an evolutionary entity (a *phenospecies?*) that is not well satisfied by existing terms.

A conventional biological species has a clear operational meaning, though its application may entail lots of work. We have operationally included asexual or clonal derivatives within those species they resemble in their molecular characteristics. What kind of name is appropriate for the *thermophila* cluster and for the *borealis* cluster? Or for asexual organisms with a ribotype very distinctive from that of any major cluster?

Field biologists attempting to survey a natural habitat (Finlay et al. 1996; Finlay, Esteban, and Fenchel 1996) have appealed for a name of a thing that they see swimming in a petri dish or at least a thing that can be identified with appropriate microscopic stains. To apply a "species" name to such a thing may seem to be a minor sin, yielding to the utility of a morphospecies. The danger of allowing the camel's head inside the tent, however, is apparent and indeed appalling. It leads to gross underestimates of the number of biological species in a habitat and in the biome and of the complexity of the ecosystem. It relinquishes our understanding of diversity and evolution, hard earned over half a century.

I have not attempted to survey systematically the phenomenon of cryptic species in other ciliates, let alone in other eukaryotic protists. I would like to note, however, that Beam and colleagues (1993) found the dinoflagellate morphospecies *Crypthecodinium cohni* to consist of at least nineteen morphologically indistinguishable but genetically isolated biological species, tightly clustered on the LS-D2 map. John and Maggs (1997) give a broader perspective on the eukaryotic algae. Brasier (1997) discusses the difficulties in assorting and naming the pleomorphic fungus species. The essays in the book edited by Claridge, Dawah, and Wilson (1997) are among the few to address the species situation in protists. Authors addressing theoretical rather than practical questions (Ghiselin 1997, Lambert and Spencer 1995) characteristically pay protists little attention.

These taxonomic issues seem not to be issues of substance, but of tactics. We (think we) grasp the essential biological realities, however difficult it may be to answer the important questions. We cannot accept identifications if they mislead understanding, but we also cannot ignore the diverse legitimate needs of practicing biologists.

## Gene/Phene Complementarity in *Tetrahymena*

Having described the various evolutionary categories indicated by studies of tetrahymenid ciliates, I should now consider the mechanisms most likely to account for the observations and any insights into general evolutionary mechanisms provided by the unique perspectives of ciliates.

The phenomenon requiring most attention is the remarkable similarity of organisms separated by enormous evolutionary distances and constructed of molecules that appear superficially to be entirely different (Nanney 1977, 1982; Williams 1986).

Lacking the leisure in this section to argue each aspect of the phenomenon, I will describe instead a hypothetical basis of the phenomenon, a kind of "just so" story, and allow the reader to consider its suitability and coherence. I reject at the outset the possibility of evolutionary convergence as the basis of the overall similarity, though convergence plays an important subsidiary role. I assert that the tetrahymenal design is an ancient design that emerged and was perfected near the time of origin of the eukaryotes. Its original niche was that of bacteria feeder (unfortunately, the term *bacteriophage* is preempted), potentiated by the possibilities of scale in the compounding of multiple genomes and soma in an integrated cell. The ciliate mode of increasing size and motility was eventually found limiting when confronting the even larger-scale elaboration possible with multicellularity. But for a long evolutionary season, the ciliate design was a notable success, though obliterated from evolutionary memory by the impermanence of ciliate remains.

The original niche of bacteria feeder persisted, little changed, after the emergence of multicellular forms, and the primary adaptations to that niche were timeless. I attribute the similarities of evolutionarily dispersed forms to the persisting constraints of the original niche and the investment in complex multimolecular constructs. The ciliate design persists for the same reasons that the ribosomal design and the genetic dictionary have persisted in all life forms. The ciliate design has persisted just as the tetradic nucleosomal design and the 9 + 2 ciliary (flagellar) design have persisted in all eukaryotes. The phenomenon is described in various metaphors, but the most familiar is perhaps an adaptive peak in a complex evolutionary landscape. The phenomenon is also associated with the concept of developmental constraint.

The limits imposed by the complex adaptations are, of course, imposed not only on the larger design features (the number of histones in a nucleosome, the number of fibers in a cilium, the number of kinetosomes in a kinety, the number of membranelles in a buccal cavity), but also upon the design of the compositional molecules. The lengths of the polypeptide chains, the distribution of charges along their lengths, the particular amino acid sequences, and the folding patterns of the proteins strongly affect their interactions with other molecules and their functional efficiency. Some minor changes in the molecules—such as substituting amino acids with similar charges—do not serious modify the molecule's functions, however, and may be tolerated.

The consequence of these forces and potentialities is the persistence of the complex designs while the nucleic and amino acid sequences are slowly scrambled by "neutral" molecular changes. The genetic code no longer evolves, but the ribosomal sequences of bacteria gradually separate from those of fungi. Eventually, the histone sequences of *Sequoia* are distinguishable from those of *Chlamydomonas*, though the nucleosomic designs are the same, and the histones of one may be substituted in vitro for the other. The slow, steady modifications of the molecules represent the inevitable escapees from stabilizing selection and the basis for "neutral" evolutionary change; their steadiness provides the foundation of molecular chronometry.

Molecular modification thus continues even in the face of certain invariance. But evolutionary modifications of more substantial (adaptational) kinds also continued to occur in the tetrahymenids. The ciliates that are the results of these modifications inhabit lakes and streams. Inhabitants of freshwater must respond to enormous variations of temperature, ionicity, mineral composition, and numbers and kinds of food resources. Although remaining apparently unchanged in basic design, the tetrahymenas, acquired the capacity to respond quickly and effectively to these regular changes in the environment by changes in gene expression.

When cells are exposed to higher or lower temperatures or vicissitudes of salinity, they transform after a short interval by substituting new surface proteins more suitable for the changed environment. When the food supply becomes limited, the cells acquire a streamlined body suitable for rapid escape to another locale perhaps provided with food reserves. In the complete absence of food supplies, the cells gradually digest their own substance, acquiring a simpler version of the basic design while maintaining the energetic capacity for utilizing nutrients when they eventually become available.

Tetrahymenas also evolved the capacity to harvest their own kind when more suitable nutrients are not at hand. Several tetrahymenas respond to starvation by transforming the oral apparatus, retaining the tetrahymenostomal structure, even the same molecular constituents, but transforming them into a much larger "macrostomal" net for capturing other tetrahymenas and, of course, other eukaryotic protists. Some tetrahymenas thus evolved a facultative capacity to transform from the microbial equivalent of a browser to the equivalent of a carnivore (or, in dire need, of a cannibal).

As the biological community became more complex, populations of tetrahymenas became correspondingly specialized in their exploitation of resources, but they usually did this by evolving new species instead of new faculties. Whereas some remained generalized browsers, others shifted to the macrostome state at the slightest opportunity. Some evolved the ability to encyst as a means of surviving the evaporation of shallow pools and of dispersing to scattered habitats. Some populations acquired the capacity to sense and move toward particular food sources, such as dead or damaged larvae of aquatic insects. Having developed these specialized skills, some gained the ability to penetrate particular metazoan hosts and exploit them for food and transportation.

These originally facultative capabilities in isolated populations thus evolved in some cases to full commitment to alternative lifestyles. Macrostome specialists such as *T. vorax* and *T. paravorax* emerged independently from several evolutionary clusters. Some tetrahymenas, such as *T. chironomus* focus on dipteran larvae; *T. rostrata* may have a special affinity for molluscs and rotifers. Some organisms have become so specialized that their generic tetrahymenid design is ignored when in the dispersal mode, and these species are assigned to new genera. *Ichthyopteris multifiliis*, for example, is the name for a well-known fish parasite, responsible for what aquarium fanciers call "ick." Another curious specialist (Edgeter, Anderson, and Washburn 1986) exploits the larvae of mosquitoes that live in hollow tree stumps; but *Lambornella clarki*, in its dispersive phase is morphologically indistinguishable from *T. pyriformis*.

The curious fact is that these evolutionary specializations are not notably clustered, but bud sporadically over the evolutionary tree. A persisting question concerns the explanation for the existence of several centers for the basic tetrahymenid design. The different tetrahymena groups may reflect an early adventitious geographic separation of the primitive tetrahymena population associated with the dispersion of freshwater habitats. Biogeographical information is thus far still too fragmentary to provide much insight, however.

### The Genetic Community

One additional topic needs to be dealt with quickly before considering some of the larger issues at stake. I mentioned that tetrahymenas seemed to have been preadapted for laboratory cultivation. Lwoff (1923) had little difficulty isolating the first tetrahymena into continuous culture, and hundreds of isolations have been made from nature since that time. Once within the laboratory, tetrahymenas appeared fully competent to flourish indefinitely in axenic culture. Indeed, for many years the accepted procedure for establishing new cultures has been to add antibiotics to pond water and extract the things that grew well.

The important observation not emphasized is that many tetrahymenas fail to grow well in powerful antibiotics, and few other ciliates have been taught to live at all in axenic culture. Many ciliates harbor bacterial commensals that may be necessary for continued growth in culture or that—like the famous "killer" bacteria in *Paramecium tetraurelia*—enable the host to decimate its competitors (Quackenbush 1988). Other ciliates—such as the green paramecium, *P. bursaria*—harbor symbiotic algae, which are not essential for life, but are clearly beneficial to the hosts.

With some other kinds of eukaryotic protists, the mutual dependence of diverse lineages is more obvious—as, for example, in the case of the fungus/alga unions in lichens (Purvis 1997). Not so obvious are the multiple lineages that commonly populate the cytoplasmic communities of sarcodinids. One reason that we have so few of these protists under full domestication is that many of them are dependent on associates we feel we have to get rid of. Careful study has been given to some of the ciliate symbionts (Görtz 1988, Quackenbush 1988), mainly because of theoretical interest in the "killer" character, which was associated with a classical case of cytoplasmic inheritance.

Thanks to studies in molecular phylogenetics and to the achievements of Lynn Margulis (1981) as an advocate, we now recognize the reality of an ancient genetic conjunction at the base of the eukaryotes that converted bacteria into mitochondria and that may have converted a group of prokaryotes into the first eukaryote. Similarly, we accept the algal origin of plastids in green plants. We also have come to realize that the associations are guaranteed in part by the transfer of essential genetic elements from the peripheral to the central genetic depot.

Perhaps because we are fixated on closed gene pools, we have assumed, however, that the horizonal transfer of genetic elements from separate phylogenetic systems occurred only rarely and in an earlier geological age. Jan Sapp (1994) has discussed these issues at some length. The tight little gene pools assumed in the Modern Synthesis probably allowed much more leakage in the protists via commensal transfers and even in higher organisms and in modern times more than is often assumed.

The "cataclysmic" origin of new plant species by allotetraploid fusion of separated genealogies was accepted early in the Modern Synthesis as an "exceptional" departure from the isolation of evolutionary lineages, but we now know from DNA measurements that several rounds of polyploidization occurred within the ancestry of metazoan lineages. Did these chromosomal doublings always involve closely related lineages, or did some of these significant saltatory episodes involve hybridization of truly distinctive evolutionary stems?

## DISCUSSION

### Historical Perspectives

This essay was introduced with the observation that the ciliated protozoa often fail to conform to the generalizations that apply to more familiar laboratory denizens, and it has thus far summarized some of the major conclusions concerning the evolutionary history and genetic economies of the tetrahymenid ciliates. These organisms have been examined not because they seemed to be promising objects for evolutionary studies, but primarily because their utility as a genetic organismic technology made it possible to collect some relevant information about their more general biological characteristics.

One should not, of course, be fearful of encountering deviations among the ciliates. Indeed, the point of comparative studies is to find the contradictions that will "prove" and improve the generalizations. In most comparisons between ciliates and more familiar organisms, the phenomena described and the mechanisms discovered only support the generalizations developed elsewhere. Only recently has the ciliate work in many disciplines finally moved beyond the "me too" phase of exploration. The study of ciliates is now responsible for the first descriptions of fundamental phenomena not previously understood with more conventional organisms. I mention in this connection the discovery of *ribozymes* by Tom Cech and his coworkers (1986), a discovery made possible in part by the extensive DNA processing that occurs in the macronuclei of ciliates. I note Elizabeth Blackburn's (1986) pioneering characterization of *telomeres*, the important end structures of chromosomes in all eukaryotes, again made possible in part because of the enormous numbers of "ends" on the small macronuclear chromosomes of *Tetrahymena*. Indeed, many of the recent generalized advances in our understanding of chromosomal organization (Gorovsky 1986) have come from studies of *Tetrahymena* and its evolution.

Even so, our contrarian gustations would be disappointed if we failed to find some dissonance between standard doctrine and ciliate testimony, even in the field of evolution. Most of the dissonance that I am aware of comes as a result of the attempt to assign arbitrarily the doctrines of the Modern Synthesis to protists and to relatively remote evolutionary times.

*Evolutionary Synthesis* (Provine 1986) is the term applied to a sprawling multidisciplinary consensus developed over a short period of time in the first half of this century. It is interpreted as the reconciliation and integration of Mendel's (1866) understanding of the mechanisms of genetic transmission with Darwin's (1859) interpretation of the forces of evolutionary change.

Hull's (1988) dissection of the roles of the several major participants in the Modern Synthesis gives it a story line that I missed in my disciplinary indoctrination. The ferment following the rediscovery of Mendel in 1900 was loosely consolidated by Dobzhansky in his 1937 *Genetics and the origin of species*. The idealized "Mendelian" population studied by Hardy, Weinberg, and Castle was subjected to selective "Darwinian" forces in the quantitative analyses by Fisher (1930), Wright (1931), and Haldane (1932). Richard Goldschmidt (1940) played devil's advocate and in doing so evoked a firmer and more complete statement of the consolidation (as well as its name) by Julian Huxley (1942). Speaking from different disciplinary perches, other biologists, Mayr (1942) and Simpson (1944) particularly, endorsed and codified the Modern Synthesis. Ernst Mayr, the youngest and most autocratic of the apostles, undertook the task of elaborating and enforcing the orthodox conclusions through his long and productive career (Mayr, 1998), even after the full range of evolutionary process had been opened up through the application of molecular technology, and after the inadeguate scope of the Modern Synthesis had been made manifest (Woese, 1998).

The Modern Synthesis continued to absorb new information and points of view concerning the tempos and modes of evolution in higher organisms, and negotiated an uncomfortable truce with respect to molecular chronometry (Dietrich 1998), but the discussions did not include molecular characterization of species. Important recognized but unresolved issues were the hierarchy of life forms and the integration of ecology and genetics (Eldridge 1985, Hull 1980; cf. Sterelny, chapter 5 in this volume).

The most critical area of neglect was that of microbial and molecular genetics. The genetics of the Modern Synthesis was the genetics of Mendel, Morgan, and Muller—of corn and peas and fruit flies. The evolutionary area at issue was the last 250 million years in a span of life now recognized as extending nearly four billion years. The development of microbial genetics was precisely contemporary with the Modern Synthesis, but almost wholly independent. It started feebly with the study of the genetics of algae in the late 1920s, then went on to the study of controlled matings with *Sacharomyces, Neurospora*, and *Paramecium* in the 1930s and into the mainstream of modern genetics with the study of bacteria and viruses in the 1940s.

In 1943, Luria and Delbruck established the beginnings of microbial mutation research. Lederberg demonstrated sexuality in *Escherichia coli* in 1947. Important new understanding quickly emerged about the chemical composition of genes from work on *Pneumococcus* and tobacco mosaic viruses. We were first taught how genes act by the fungus *Neurospora* (Beadle 1945) and the bacterium *E. coli*. The central genetic issue of self-replication was resolved in principle by the dramatic announcement of the double helix (Watson and Crick 1953), followed in due time by formal analysis of the genetic code.

The synthesizers, however, were content to accept the chromosomal location of Mendel's genes and to consider their chemical composition and their mode of action to be of little relevance. The domain of the Modern Synthesis was Mendelism, but it was overtaken by a far more dynamic and better funded research agenda. The synthesizers insisted that the evolutionary modalities that were revealed in higher organisms be imposed upon microbes without significant modifications. The "biological species," the closed gene pool, was the doctrinaire universal unit of evolutionary diversification (cf. Dupré, chapter 1 in this volume).

The ruling regarding this universal unit was quickly found to be impractical in the practice of microbiology—even if it might have been correct in principle. As soon as genetic studies were undertaken on eukaryotic protists, the plethora of sibling species made impossible the task of identifying all the biological species and assigning Latin binomials to each. I have summarized the effort required to sort out the biological species in one set of organisms confidently assigned to a single Linnaean ciliate species in 1950—probably the most extensively researched "species" among the 7,200 such species, as estimated by Corliss (1979).

The multiplicity of sibling species does not contradict the postulate that the "biological species" is the unit of evolution in eukaryotic organisms. It simply underscores the impracticality of assigning names to eukaryotic protists only and always when they have been confirmed to be biological species.

### Prokaryotic Species

The situation with respect to prokaryotic species is far more serious. Microbiologists have essentially bypassed the Modern Synthesis, considering it irrelevant within their territories. Modern microbiology textbooks abound with Latin binomials, but the "species" associated with the Linnaean terms make no claims of association with closed gene pools. The bacterial species is essentially a set of organisms with an arbitrarily defined degree of molecular similarity and hence a group of presumed "recent" common origin (Goodfellow, Manfio, and Chun 1997; cf. Mishler, chapter 12 in this volume). The mechanisms of genetic exchange and recombination in bacteria are not Mendelian and no Mendelian population may exist (Maynard Smith 1995) in them. Conjugation—the mechanism most similar to synkaryon formation in eukaryotes—only rarely involves whole genomes. A large but uncertain portion of genetic exchanges in natural populations is mediated by bacteriophages or plasmids, which insert genetic elements and sometimes themselves (as in lysogenic phages) into the genomes of the organisms invaded. When released, the progeny may be inserted (via transduction) into another host with another load of genetic baggage. The early studies of host range (*h*) mutants demonstrated genetic control of the limits of transmission of viral genetic elements, but little evidence has been reported that the gene pools have any particular association with named species. Indeed, the rapid spread of antibiotic resistance among pathogens in clinical settings appears to be a consequence of low host specificity of resistance (*R*) plasmids among even the most diverse microbes. What goes on in nature is of course much more obscure than what happens in the research lab (Embley and Stackebrandt 1997).

The concept of the "selfish gene" (Dawkins 1976) is right at home among the prokaryotes, where individual genes appear capable of moving easily across highly diverse genomes under appropriate selective conditions. The modern genetic fluidity among the bacteria may be a relic of primordially "open" genetic systems. The coadapted gene complexes of the Modern Synthesis and the isolating devices required to maintain their integrity may have evolved gradually though gradients of accessibility. In such an evolutionary scenario, the "species"—i.e., the closed gene pool—might not have its "origin" until halfway through the history of life. As Doolittle (1998) recently observed, the evidence continues to accumulate that eukaryotes are organisms constructed by a committee and carry within their genomes significant memories of several diverse genetic adventures.

## Rediscovering Time

Perhaps the theory of the Modern Synthesis, despite its acknowledged success, has been relegated to a marginal position in modern biology because its custodians failed to learn well the first Darwinian lesson. In their book *Discovery of Time* (1965), Toulmin and Goodfield propose that the recognition of the indefinite extensibility of time was the most fundamental feature of the so-called Darwinian revolution. Like all intellectual discontinuities, the scaling of temporal reality continues to have aftershocks. Neither the age of the earth nor the time course of life has permeated fully the thinking of biologists. We still focus on the last chapter of the book and ignore the long stretches back to a beginning nearly four billions of years ago.

The classification scheme we employ for living creatures, with its "higher categories," was invented before the Darwinian realization of the depth of time and was modeled on higher plants and animals, which are a fairly recent phenomenon on the geological playing field (cf. Ereshefsky, chapter 11 in this volume). Moreover, the playing field considered by the synthesizers was refereed by umpires of Lyellian persuasion, and Cuvier was dismissed by whiggish Darwinians. We now accept that life is long and well-punctuated, though we dispute the agencies of discontinuity. In any case, the number of punctuation marks in an organism's evolutionary history probably reflects the length of time lapsed since the origin of its "kind"; the "natural" hierarchies may be explained as consequences of evolutionary pulses. We can hardly impose on all forms of life the same set of nested boxes by which we rationalize the evolutionary history of organisms only a few million years old—the species, genera, families, classes, phyla, and kingdoms of the metazoa. The number of nested boxes in the metazoan domain is insufficient to order the ciliates, so protozoologists have rushed to elevate the ciliates to a phylum. This move is interpreted, probably correctly, as a political move instead of a scientific judgment, but clearly the ciliates need some new boxes for the products of old genetic ferments.

A goal of molecular chronometricians is to formulate appropriate standards for dating the evolutionary branches generated by cladistic analysis. Though skirmishes over choices of molecular chronometers and modes of analysis are fierce, a faithful cladist does not doubt the eventual resolution of the details of evolutionary history. We shall soon date the evolutionary junctions, and those junctions will define the problems to be resolved in a new synthesis.

## APPENDIX

A few words of explanation are necessary about the methodologies used in the molecular analysis of tetrahymena strains because they are not very familiar even to many molecular evolutionists. The methods share the basic foundation of all molecular tree constructions (Zuckerkandl and Pauling 1963, Fitch and Margoliash 1967), but they differ in specific applications. The biological history of an organism is embedded in the sequences of the four DNA bases that encode the genetic instructions passed from generation to generation. (It is embedded also in a derived form in the sequences of the twenty amino acids in the proteins of an organism, but we will not be concerned with proteins here.) Organisms with identical DNA scripts must have a recent common ancestor; organisms with a few base differences are likely to be several generations away from their common ancestor; and organisms with many base differences must be separated from their common ancestor by many generations.

The pattern of distances among a set of organisms is expressed in a *difference matrix,* in which are recorded each of the paired comparisons possible. The distance matrix is converted into a "tree" by connecting first the sequences with few differences and then gradually connecting these to sets somewhat more distant until all the sequences are joined into a single tree. The information existing initially as simple numbers differentiating linear sequences is thus converted into a planar diagram, representing an evolutionary tree in three dimensions (or polynomial space).

This procedure, simple in conception, becomes complicated in practice, and the practice is what might be considered a discipline of low consensus. For starters, one must choose which parts of the genetic library are to be compared, and then one must prescribe the particular method of transforming a matrix into a tree. A major determinant in choosing which genetic transcript to employ in constructing a tree is the *heterochronicity* of the nucleic information. Some portions of the DNA are remarkably conserved, whereas others undergo changes much more rapidly—by orders of magnitude. The relations among recently derived organisms are best assessed through comparisons of reasonably labile sequences. Ancient junctions can be studied with only the most conserved genetic regions.

Many molecular evolutionists are now convinced that the nucleic information underlying the ribosomal RNA (that constitutes the core of all organisms' translation machine) represents the most generally suitable basis for broader tree constructions. The rationale here is that once an essential but complex biological structure is pushed to an "adaptive peak," its main features are "frozen," except for trivial modifications. At this time, its genetic base begins to function as a chronometer. The ribosome is the most ancient of such complex structures, common to all life forms on the earth, and the most stable chronometer. Different portions of the ribosomal apparatus are, however, under greater or lesser constraint, and their chronometers are correspondingly slower or faster. For example, because nearly all mammals have exactly the same sequence of bases in their 5S rRNA (a small highly constrained component of their translation machine), one cannot make a meaningful mammalian tree using 5S rRNA. This small molecule may, however, provide invaluable information about relationships among invertebrate animals and among protists.

The most popular of the rRNAs among molecular evolutionists is the 16S rRNA or small subunit SSrRNA. This molecule is, however, a very long molecule (1200 + bases), and it has to be sequenced in pieces. It also has to be edited to remove the more variable regions between conserved domains. The sequence we have used in the studies now at issue is a domain of yet another rRNA, the 23S or large subunit rRNA. This domain of about 190 bases is relatively unstable compared with the 5S or 16S rRNA molecules, so the number of changes within it is nearly as large as the number in the entire 16S molecule.

The last distinctive feature of our study is the tree-building program used. Here, we have to discuss briefly the assumptions made about the modes of evolutionary change. One major reason for using the ribosomal apparatus as the basis for evolutionary interpretations is its conservatism, not only with respect to the frequency of genetic change, but also with respect to the mode of change. The molecules are highly constrained in length, and most of the differences can be interpreted in terms of simple substitution mutations; a cytosine may be replaced by a guanine or a thymine or an arginine, for example. However, when several changes are observed between sequences, though the net result can be explained in terms of substitutions, the actual mechanisms of change might have been other kinds of mutations—particularly inversions and transpositions, and compensating insertions and deletions. The explanation of the differences in terms of the entire array of molecular changes is often simpler (more parsimonious) than the explanation entirely in terms of substitutions.

The analysis of complicated sequence differences, allowing for all kinds of molecular mutations, involves much more sophisticated calculations and is a far more laborious procedure. The theory is described in a book by Sankoff and Kruskal (1983) and is carefully ignored by most molecular evolutionists. The reader is cautioned that the findings reported in the present essay are based on an unusual molecular segment and involve the PHYLOGEN string analysis program. Where comparisons are possible with other methodologies and other sequences, the merits of this approach can be supported, but that issue is not the focus of this discussion. I simply warn the reader that the results may be contested and that the contestation will be based on important theoretical and analytical issues.

## REFERENCES


Allen, S. L., and C. I. Li (1974). Nucleotide sequence divergence among DNA fractions of different syngens of *Tetrahymena pyriformis*. *Biochemical Genetics* 12, 213–233.

Baroin, A., R. Perasso, L. Q. Qu, G. Brugerolle, J. P. Bachellerie, and A. Adoutte. (1988). Partial phylogeny of the unicellular eukaryotes based on rapid sequencing of a portion of the 16S ribosomal RNA. *Proceedings of the National Academy of Science* (U.S.) 85, 3474–3478.

Beadle, G. W. (1945). Genetics and metabolism in *Neurospora*. *Physiological Reviews* 25, 643–663.

Beam, C. A., R-M. Preparata, M. Himes, and D. L. Nanney (1993). Ribosomal RNA sequences of members of the *Crypthecodinium cohni (Dinophyceae)* species complex: Comparisons with soluble enzyme studies. *Journal of Eukaryotic Microbiology* 40, 660–667.

Bell, G. (1988). *Sex and death in protozoa: The history of an obsession*. Cambridge: Cambridge University Press.

Blackburn, E. H. (1986). Telomeres. In J. G. Gall, ed., *The molecular biology of ciliated protozoa*. New York: Academic Press.

Borden, D., G. S. Whitt, and D. L. Nanney (1973). Isozymic heterogeneity in *Tetrahymena* strains. *Science* 181, 279–280.

Borden, D., E. T. Miller, G. S. Whitt, and D. L. Nanney (1977). Electrophoretic analysis of evolutionary relationships in *Tetrahymena*. *Evolution* 31, 91–102.

Brasier, C. M. (1997). Fungal species in practice: Identifying species units in fungi. In M. F. Claridge, H. A. Dawah, and M. R. Wilson, eds., *Species: The units of biodiversity*. London: Chapman and Hall.

Brunk, C. F., R. W. Kahn, and L. A. Sadler (1990). Phylogenetic relationships among *Tetrahymena* species determined using the polymerase chain reaction. *Journal of Molecular Evolution* 30, 290–297.

Cech, T. R. (1986). Ribosomal RNA gene expression in *Tetrahymena:* Transcription and RNA splicing. In J. G. Gall, ed., *The molecular biology of ciliated protozoa*. New York: Academic Press.

Chatton, E. and A. Lwoff (1936). Techniques pour l'étude des protozoaires, spécialement de leur structures superficielles (cinétome et argyrome). *Bulletin de la Société de Microbiologique de France* 5, 25–39.

Claridge, M. F., H. A. Dawah, and M. R. Wilson (1997). Practical approaches to species concepts for living organisms. In M. F. Claridge, H. A. Dawah, and M. R. Wilson, eds., *Species: The units of biodiversity*. London: Chapman and Hall.

Corliss, J. O. (1952). Systematic status of the pure culture ciliate known as *"Tetrahymena geleii"* and *"Glaucoma piriformis."* *Science* 116, 188–191.

Corliss, J. O. (1953). Comparative studies on holotrichous ciliates in the *Colpidium-Glaucoma-Leucophrys-Tetrahymena* group. II. Morphology, life cycles and systematic status of strains in pure culture. *Parasitology* 43, 49–87.

Corliss, J. O. (1979). *The ciliated protozoa: Characterization, classification and guide to the literature*. New York: Pergamon Press.

Corliss, J. O., and P-M. Daggett (1983). *"Paramecium aurelia"* and *"Tetrahymena pyriformis"*: Current status of the taxonomy of these popularly known and widely used ciliates. *Protistologica* 19, 307–322.

Darwin, C. (1859). *On the origin of species*, facsimile of the first edition (1859) with an introduction by Ernst Mayr (1966). Cambridge: Harvard University Press.

Dawkins, R. (1976). *The selfish gene*. Oxford: Oxford University Press.

Dietrich, M. R. (1998). Paradox and persuasion: Negotiating the place of molecular evolution within evolutionary biology. *Journal of the History of Biology* 31, 85–111.

Dini, F., and D. Nyberg (1993). Sex in ciliates. *Advances in Microbial Ecology* 13, 85–153.

Dobzhansky, T. (1937). *Genetics and the origin of species*. New York: Columbia University Press.

Doerder. F. P., M. A. Gates, F. P. Eberhardt, and M. Arslanyola (1995). High frequency of sex and equal frequency of mating types in natural populations of the ciliate *Tetrahymena thermophila. Proceedings of the National Academy of Science* 92, 8715–8718.

Doolittle, W. F. (1998). A paradigm gets shifty. *Nature* 392, 15–16.

Edgeter, D. E., J. R. Anderson, and J. O. Washburn (1986). Dispersal of the parasitic ciliate *Lambornella clarki*: Implications for ciliates in the biological control of mosquitoes. *Proceedings of the National Academy of Science* (U.S.) 83, 1735–1739.

Eldridge, N. (1985). *Unfinished synthesis: Biological hierarchies and modern evolutionary thought.* New York: Oxford University Press.

Elliott, A. M. (1973). Life cycle and distribution of *Tetrahymena*. In A. M. Elliott, ed., *Biology of* Tetrahymena. Stroudsburg, Pa.: Dowden, Hutchinson and Ross.

Elliott, A. M., and R. E. Hayes (1953). Mating types in *Tetrahymena. Biology Bulletin.* 105, 269–284.

Elliott, A. M., and D. L. Nanney (1952). Conjugation in *Tetrahymena. Science* 116, 23–34.

Embley, T. M., and E. Stackebrandt (1997). Species in practice: Exploring uncultured prokaryotes diversity in natural samples. In M. F. Claridge, H. A. Dawah, and M. R. Wilson, eds., *Species: The units of biodiversity*. London: Chapman and Hall.

Finlay, B. J. M., G. F. Esteban, and T. Fenchel (1996). Global diversity and body size. *Nature* 383, 132–133.

Finlay, B. J., J. O. Corliss, G. Esteban, and T. Fenchel (1996). Biodiversity at the microbial level: The number of free-living ciliates in the biosphere. *Quarterly Revicew of Biology* 72, 221–237.

Fisher, R. A. (1930). *The genetical theory of natural selection*. Oxford: Clarendon Press.

Fitch, W. M., and E. Margoliash (1967). The construction of phylogenetic trees. *Science* 155, 279–284.

Frankel, J. (1989). *Pattern formation: Ciliate studies and models*. New York: Oxford University Press.

Furgason, W. H. (1940). The significant cytostomal pattern of the "*Glaucoma-Colpidium* group" and a proposed new genus and species, *Tetrahymena gelei. Archiv für Protistenkunde* 94, 224–266.

Gall, J. G., ed. (1992). *The molecular biology of ciliated protozoa*. New York: Academic Press.

Ghiselin, M. T. (1997). *Metaphysics and the origin of species*. New York: SUNY Press.

Goldschmidt, R. (1940). *The material basis of evolution*. New Haven: Yale University Press.

Goodfellow, M., G. P. Manfio, and J. Chun (1997). Toward a practical species concept for cultivable bacteria. In M. F. Claridge, H. A. Dawah, and M. R. Wilson, eds., *Species: The units of biodiversity*. London: Chapman and Hall, London.

Gorovsky, M. A. (1986). Ciliate chromatin and histones. In J. G. Gall, ed., *The molecular biology of ciliated protozoa*. New York: Academic Press.

Gortz, H-D. (1988). Endocytobiosis. In H-D. Görtz, ed., *Paramecium*. Berlin Springer.

Gould, S. J., and R. C. Lewontin (1979). The spandrels of San Marco and the Panglossian paradigm: A critique of the adaptationist programme. *Proceedings of the Royal Society of London B* 204, 581–589.

Griesemer, J. R. (1999). Reproduction and the Reduction of Genetics. In H-J. Rheinberger, P. Beurton, and R. Falk eds., *The gene concept in development and evolution*. Cambridge: Cambridge University Press.

Gruchy, D. G. (1955). The breeding system and distribution of *Tetrahymena pyriformis*. *Journal of Protozoology* 2, 175–185.

Haldane, J. B. S. (1932). *The causes of evolution*. London and New York: Harper.

Hausmann, K., and P. C. Bradbury, eds. (1996). *Ciliates: Cells as organisms*. Stuttgart: Gustav Fischer.

Hill, D. L. (1972). *The biochemistry and physiology of* Tetrahymena. New York: Academic Press.

Hull, D. L. (1980). Individuality and selection. *Annual Review of Ecological Systematics* 11, 311–332.

Hull, D. L. (1988). *Science as a process: An evolutionary account of the social and conceptual development of science*. Chicago: University of Chicago Press.

Huxley, J. (1942). *Evolution: The modern synthesis*. London: Allen and Unwin.

John, D. M., and C. A. Maggs (1997). Species problems in eukaryotic algae: A modern perspective. In M. F. Claridge, H. A. Dawah, and M. R. Wilson, eds., *Species: The units of biodiversity*. London: Chapman and Hall.

Klein, B. M. (1926). Über eine neue Eigentumlichkeit der Pellicula von *Chilodon uncinatus* Ehrbg. *Zoologischer Anzeiger* 67, 160–162.

Lambert, D. M., and H. G. Spencer, eds. (1995). *Speciation and the recognition concept*. Baltimore and London: Johns Hopkins Press.

Lederberg, J. (1947). Gene recombination and linked segregations in *Escherichia coli*. *Genetics* 32, 505–525.

Luria, S. E., and M. Delbruck (1943). Mutations of bacteria from virus sensitivity to virus resistance. *Genetics* 28, 491–511.

Lwoff, A. (1923). Sur la nutrition des infusoires. *Comptes Rendus de L' Academie des Sciences* 176, 928–930.

Lynn, D. H. (1996). Systematics of ciliates. In K. Hausmann and P. C. Bradbury, eds., *Ciliates as organisms*. Berlin: Gustav-Fischer.

Lynn, D. H., and M. L. Sogin (1988). Assessment of the phylogenetic relationships among ciliated protists using partial ribosomal RNA sequences derived from reverse transcripts. *Biosystems* 21, 249–254.

Margulis, L. (1981). *Symbiosis in cell evolution*. New York: W. H. Freeman.

Maupas, E. (1889). La rajeunissement karyogamique chez le cilies. *Archives de Zoologie Experimentale et Générale* (ser. 2) 7, 149–517.

Maynard Smith, J. (1983). Evolution and development. In (B. C. Goodwin, N. Holder, and C. C. Wylie, eds.), *Development and evolution* Symposium of the British Society of Developmental Biology 6, 33–46. Cambridge: Cambridge University Press.

Maynard Smith, J. (1995). Do bacteria have population genetics? In S. Baumberg, J. P. W. Young, E. M. H. Wellington, and J. R. Saunders, eds., *Population genetics of bacteria*. Cambridge: Cambridge University Press.

Mayr, E. (1942). *Systematics and the origin of species*. New York: Columbia University Press.

Mayr, E. (1998). Two empires or three? Proceedings of the National Academy of Science, USA 95, 9720–9723.

McCoy, J. W. (1975). Updating the tetrahymenids: III. Natural variation in *Tetrahymena setosa Nov. Comb*. *Acta Protozoologica* 14, 253–262.

Mendel, G. (1866). Versuche über PflanzenHybriden. *Verhandlungen des naturforschenclen Vereines, Brunn* 4, 3–47.

Meyer, E. B., and D. L. Nanney (1987). Isozymes in the ciliated protozoan *Tetrahymena*. In M. C. Rattazzi, J. G. Scandalios, and G. S. Whitt, eds., *Isozymes: Current topics in biological and medical research*. New York: A. R. Liss.

Nanney, D. L. (1966). Corticotypes in *Tetrahymena pyriformis. American Naturalist* 100, 303–318.

Nanney, D. L. (1977). Molecules and morphologies: The perpetuation of pattern in ciliated protozoa. *Journal of Protozoology* 24, 27–35.

Nanney, D. L. (1980). *Experimental ciliatology*. New York: John Wiley and Sons.

Nanney, D. L. (1982). Genes and phenes in *Tetrahymena. Bioscience* 32, 783–788.

Nanney, D. L. (1983). The ciliates and the cytoplasm. *Journal of Heredity* 74, 163–170.

Nanney, D. L. (1996). Review of *Sex and death in protozoa* by Graham Bell. *Journal of Eukaryotic Microbiology* 43, 159–160.

Nanney, D. L., and P. A. Caughey (1953). Mating type determination in *Tetrahymena pyriformis. Proceedings of the National Academy of Science* (U.S.) 39, 1057–1063.

Nanney, D. L., and J. W. McCoy (1976). Characterization of the species of the *Tetrahymena pyriformis* complex. *Transactions of the American Microscopical Society* 95, 664–682.

Nanney, D. L., C. Park, R-M. Preparata, and E. M. Simon (1998). Comparison of sequence differences in a variable 23S rRNA domain among sets of cryptic species of ciliated protozoa. *Journal of Eukaryotic Microbiology* 45, 91–100.

Nyberg, D. (1988). The species concept and breeding systems. In H-D. Görtz, ed., *Paramecium*. Berlin: Springer.

Preparata, R-M., C. A. Beam, M. Himes, D. L. Nanney, E. B. Meyer, and E. M. Simon (1992). *Crypthecodinium* and *Tetrahymena*: An exercise in comparative evolution. *Journal of Molecular Evolution* 34, 209–218.

Preparata, R-M., E. B. Meyer, F. P. Preparata, E. M. Simon, C. Vossbrinck, and D. L. Nanney (1989). Ciliate evolution: The ribosomal phylogenies of the tetrahymenine ciliates. *Journal of Molecular Evolution* 28, 427–441.

Provine, W. B. (1986). *Sewall Wright and evolutionary biology*. Chicago: University of Chicago Press.

Purvis, O. W. (1997). The species concept in lichens. In M. F. Claridge, H. A. Dawah, and M. R. Wilson, eds., *Species: The units of biodiversity*. London: Chapman and Hall.

Quackenbush, R. L. (1988). Endosymbionts in killer paramecia. In H-D. Görtz, ed., *Paramecium*. Berlin: Springer.

Sadler, L. A., and C. F. Brunk (1990). Phylogenetic relations among *Tetrahymena* species determined by DNA analysis. In M. Y. Clegg, and S. J. Obrien, eds., *Molecular Evolution. UCLA Symposium on Molecular and Cellular Biology* 120, 45–252.

Sadler, L. A., and C. F. Brunk (1992). Phylogenetic relationships and unusual diversity in histone H4 proteins within the *Tetrahymena pyriformis* complex. *Molecular Biology and Evolution* 9, 70–84.

Sankoff, D., and J. B. Kruskal, eds. (1983). *Time warps, string edits and macromolecules: The theory and practice of sequence comparisons*. Reading, Mass.: Addison-Wesley.

Sapp, J. (1987). *Beyond the gene: Cytoplasmic inheritance and the struggle for authority in genetics*. New York: Oxford University Press.

Sapp, J. (1991). Concepts of organization: The leverage of ciliated protozoa. In S. Gilbert, ed., *A conceptual history of modern embryology*. Baltimore: Johns Hopkins Press.

Sapp, J. (1994). *Evolution by association: A history of symbiosis*. New York: Oxford University Press.

Schloegel, J. J. (forthcoming). Anomaly, unification and the irony of personal knowledge: Tracy Sonneborn and the species problem in protozoa, 1954–1957.

Seyfert, H-M., and J. H. Willis (1981). Molecular polymorphisms of ciliary proteins from different species of the ciliate. *Tetrahymena. Biochemical Genetics* 19, 385–396.

Simpson, G. G. (1944). *Tempo and mode in evolution*. New York: Columbia University Press.

Sogin, M. L., H. J. Elwood, and J. H. Gunderson (1986). Evolutionary diversity of eukaryotic small-subunit rRNA genes. *Proceedings of the National Academy of Science* (U.S.) 83, 1383–1387.

Sonneborn, T. M. (1937). Sex, sex inheritance and sex determination in *Paramecium aurelia. Proceedings of the National Academy of Science* 23, 378–385.

Sonneborn, T. M. (1939). *Paramecium aurelia:* Mating types and groups. Lethal interactions, determination and inheritance. *American Naturalist* 73, 390–413.

Sonneborn, T. M. (1947). Recent advances in the genetics of *Paramecium* and *Euplotes. Advances in Genetics* 1, 264–358.

Sonneborn, T. M. (1957). Breeding systems, reproductive methods and species problems in protozoa. In E. Mayr, ed., *The species problem*. Washington D.C.: American Association for the Advancement of Science.

Sonneborn, T. M. (1975). The *Paramecium aurelia* complex of fourteen sibling species. *Transactions of the American Microscopical Society* 94, 155–178.

Toulmin, S., and J. Goodfield (1965). *The discovery of time*. New York: Harper and Row.

Watson, J. D., and F. H. C. Crick (1953). Molecular structure of nucleic acids: A structure for deoxribose nucleic acid. *Nature* 171, 737–738.

Williams, N. E. (1986). An apparent disjunction between the evolution of form and substance: *Tetrahymena. Evolution* 38, 25–33.

Woese, C. (1998). Default taxonomy: Ernst Mayr's view of the microbial world. *Proceedings of the National Academy of Science, USA* 95, 11043–11046.

Wright, S. (1931). Evolution in Mendelian populations. *Genetics* 16, 97–159.

Zuckerkandl, E., and L. Pauling (1963). Molecules as documents of evolutionary history. *Journal of Theoretical Biology* 8, 357–366.

# 5 Species as Ecological Mosaics

Kim Sterelny

## SPECIES AS A GRADE

The mechanisms of evolution have produced on Earth an astounding variety of life forms. Together with adaptive design, the evolution of that diversity is the central explanatory target of evolutionary biology. Though great, however, the diversity of life on Earth is limited in important ways. Diversity is bunched or clumped. A walk in Australia's eucalypt woodlands reveals many parrots: eastern and crimson rosellas; red-rumped parrots; rainbow, scaly, and musk lorikeets; galahs, gang-gangs, and an assortment of other cockatoos. These birds are all readily recognizable as having some overall similarity of form and behavior; their parrotness is apparent. But the divisions within the group are apparent, too. There is no spectrum of birds from, say, typical galahs to typical sulphur-crested white cockatoos. Equally, parrots do not "fade into" the equally recognizable and distinctive pigeons. We do not find a spectrum of birds from typical parrots through to typical pigeons.

Life's mechanisms have produced *phenomenological species*: recognizable, reidentifiable clusters of organisms. This fact makes possible the production of bird and butterfly field guides, identification keys for invertebrates and regional floras, and the like. However, the significance of this clustering is far from uncontroversial. The species concept is one of the key concepts of "folk biology"—our set of commonsense concepts for dealing with the living world (Atran 1990, and chapter 9 in this volume; Berlin 1992). The extent to which theoretical biology vindicates folk biology's categories is an open question. We are large mammals, with a lifespan that is long by the standards of animal ecology, but not by the standards of evolutionary change, and we have a distinctive set of perceptual mechanisms. We have no guarantee that the distinctions that strike us as salient coincide with the natural kinds of evolutionary biology. For example, one natural way to read Dawkins's book *The Extended Phenotype* (1982) is to see it as arguing that folk biology's concept of an organism does not correspond to any natural kind of evolutionary biology. Whether we accept Dawkins's case or not, we cannot just assume that our ordinary concept of organism identifies a kind of evolutionary biology. Equally, it has to be shown, not assumed, that phenomenological species constitute a biological kind. Minimalists about species do not expect theoretical biology to vindicate the species category: they think it is merely a phenomenological kind. We can clump organisms into species, but the biology explaining the clumping pattern is so diverse that *species* does not name a biological kind.

In this essay, I mount a limited defence of the idea that species are objective features of the world, existing independently of our perception and categorization, and that they are not just epiphenomenal products of ecological and evolutionary processes. I argue that some species are evolutionarily linked metapopulations. Such species are ecological mosaics, for the separate populations that compose them are scattered through quite different ecological communities. When this occurs, the relationship between evolutionary unit and ecological forces is complex and complex in a way that limits the possibility of evolutionary change, so species that are ecological mosaics in this sense are distinctive evolutionary units. Not all organisms are organized into evolutionary species, so not all phenomenological species are evolutionary species. But I argue that when this mosaic organization does evolve, it profoundly influences the future evolutionary possibilities of the species.

One traditional reason for skepticism about the significance of phenomenological species is the idea that our capacity to identify them is an artifact of our limited temporal and spatial perspective. Adelaide and crimson rosellas are now readily distinguishable, but if we trace these populations back in time, they begin to merge. Equally, if we could trace them forward in time future "Adelaide rosellas" may look increasingly unlike their contemporaries. If a smooth continuum of change links these birds to the earlier parrots from which they evolved, then we might suppose that there can be no fundamental difference between the Adelaide rosella and its ancestor. Our recognition of those species depends on our temporal standpoint. If we were able to track the slowly changing lineage backward in time, we would come to organisms different enough from today's rosellas to distinguish them. But from a future vantage point, as the lineage continues to change, today's rosellas would be the intermediate gradations between two other parrot species—Adelaide rosellas as they have become and Adelaide rosellas as they have just been. Hence, our species identifications are not the recognition of objective units in nature, for if lineages change smoothly and seamlessly, they can be segmented into species in many ways. The choice depends on the baseline and the amount of change that counts as enough for us to recognize a new species. Furthermore, evolutionary change is often seamless. Though plant species are often formed rapidly through hybridization, jumps will be rare in animal lineages.

There is a second reason for skepticism about the significance of phenomenological species. These clusters of morphologically similar organisms are produced by very different evolutionary mechanisms, and the clusters themselves have very different evolutionary consequences. This thought is most obviously explored through the biological species concept, for there is a standard conception of how morphological clumping arises through adaptation to local environments. Diversity is differentiation, as natural selection enables populations to track different environments. One heron population becomes white and long-legged; another becomes streaky brown and stumpy as it adapts to life in sedges and reed beds. But the adaptations of our cryptically coloured herons require some isolating mechanism. Only isolation allows an incipient species, a small population in a new selective regime, to preserve in its gene pool the evolutionary innovations that develop within it. An unprotected gene pool will be diluted by migration in and out. It will be homogenized. The special genetic information for making streaky brown plumage will disappear if there is substantial flow between it and the parent population. There can be no special suite of adaptations without some form of isolation, no protection of that suite of adaptations without entrenching that isolation through a process of speciation that erects permanent barriers between old and descendant populations (Mayr 1942, Mayr 1976, Mayr 1988). So as these heron populations become increasingly distinct, they become less likely to treat one another as potential mates. Moreover, when mating does occur, it will often be without issue or yield less fit offspring. They may be sterile, nearly sterile, or hybrids well suited to neither environment. Selection will therefore start to favor, in both populations, any trait that makes its bearers less likely to accept a mate from the other population. Reproductive isolation evolves and entrenches the differences between the two populations. Diversity will become permanently clumped. The existence of a reproductive community is central to the role of species as an evolutionary unit. To put the point conversely, *Mayr's Brake* prevents populations within a species from establishing and entrenching adaptation to local environment. Gene flow from the rest of the population damps down evolutionary change, and even when gene flow is too limited to prevent local divergence, such divergence is permanently fragile without isolating mechanisms.

Powerful though this picture is, it applies to a minority of phenomenological species. In bacteria, the exchange of genes is occasional, divorced from reproduction, and not confined within the boundaries of named species.[1] Many plant species reproduce sexually, so gene exchange is not divorced from reproduction, but gene flow between distinct phenomenological species is common. Plants hybridize much more readily than multicelled animals. American oaks are a well-known example (Van Valen 1976). Evolutionary botanists recognize a taxonomic rank called a *section*, a group of closely related species within a genus. Members of different species within a section can often produce fertile hybrids between them, so genes often flow between species within a section, though the flow is obviously not so great that the species disappear as phenomenologically distinct groups. Gene flow is possible sometimes even between less closely related species[2] (Niklas 1997, 75–81). Finally, this picture fails to apply at all to asexually reproducing phenomenological species.

Attempts have been made to repair this problem of generality by switching from seeing species as reproductive communities to seeing them using some version or other of a phylogenetic species concept, a concept that defines species as a segment of a phylogenetic tree between a speciation event and an extinction event, or between two speciation events (Wiley 1978, Cracraft 1987). However, the generality problem tends to resurface as defenders of phylogenetic concepts develop a view of speciation—of lineage splits.[3] The biological species definition treats gene exchange within a population as the *sole factor* that unifies that population, which may well be too restrictive a conception, for nongenetic developmental resources may play a similar role to the one played by genes in evolution (Griffiths and Gray 1994, Sterelny et al. 1996). If that turns out to be right, asexual populations may be linked through the exchange of nongenetic developmental resources. Less controversially, Templeton (1989) has defended a "cohesion" conception of species in which gene flow and barriers to it are just a couple of the factors that make a population a species, so we cannot rule out the possibility that some version of the phylogenetic species definition applies to all or almost all phenomenological species. Such a species definition would demonstrate the existence of an evolutionary kind coextensive with phenomenological species, and it would explain our capacity to recognize those phenomenological species.

However, there is a persuasive argument for doubting whether any species definition can be universal *and* at the same time treat the species category as a biological kind. I suspect some version of minimalism is inescapable for those who pursue a universal species concept—that is, an account of species in which (1) all organisms are part of some species, and (2) there is a unitary definition of what it is to be a species. To see this argument, consider a parallel case, the evolution of biological individuality. We cannot both think that all living things are organisms and have a powerful conception of what it is to be an organism. Leo Buss has shown that biological individuality, as exemplified by multicelled organisms, is not a mode of biological organization we can take for granted (Buss 1987). The evolution of the organism required the evolution both of the developmental cycle and of a complex suite of mechanisms that suppress cell-line competition. Buss has shown that those mechanisms differ in important ways between lineages. For example, in only some lineages is there a clear developmental distinction between the germline and the somatic cells. Hence "the organism" has been invented in different ways and perhaps to a different extent in, for example, plant and animal lineages. Plant physiology is more modular than that of animals: that is part of the reason why so many plants can survive being mostly eaten. Plant developmental biology is less complex than animal developmental biology because animal development requires the coordinated construction of the mechanisms for perception, movement, and behavioral integration as well as the mechanisms for digestion and physical maintenance (Buss 1985, Niklas 1997). In sum, the invention of the organism both required a long "prehistory" and brought something new and very important into the world of life (cf. Nanney, chapter 4 in this volume).

Equally, I think we have to choose between the ideas that all organisms are members of some species and that being a species is an important biological property, for just as the *organism* as a grade of biological organization had to be invented, so did the *species*. Evolutionary theory has moved close to a consensus in seeing species as historical individuals or historically defined kinds (Ghiselin 1974, Hull 1978).[4] *Numbat* names a segment of the tree of life. A particular organism is part of that species in virtue of being in that segment rather than in virtue of its intrinsic physical characteristics. A very influential subgroup within evolutionary theory sees species as of enormous significance in the evolutionary process. The reasons vary. Different theorists identify metapopulation structure; bounded gene flow; a shared mate recognition system; the variation in the gene pool; the role of the gene pool as an information store about the species' environment; and much else. Whatever one sees as the key attribute of species, however, there is a trade-off between seeing species as evolutionarily significant and espousing universalism, for if species are significant, the features that *make* them significant were *invented* over evolutionary time. Very likely (depending on the specifics of the hypothesis about species' evolutionary importance) those features will have been invented in differing ways and to differing extents in differing lineages. Vrba (1995), for example, insists that we should see species as complex systems with internal structure, not just as aggregations of organisms. If species and their properties play important roles in explaining stability and change in lineages of organisms, minimalism cannot be true. The term *species* names an explanatorily important evolutionary category (cf. Ereshefsky and Mishler, chapters 11 and 12 in this volume). But if the organization of organisms into species is a *grade*—a distinctive level of biological organization that had to be invented—universalism cannot be true, either. Living things existed before this grade came into existence, and many organisms continue to exist without being part of any such organization. In the next section, I pursue the idea that the metapopulation structure of species is salient to the role of species in evolution. The idea that species are metapopulations is not a hypothesis about all phenomenological species, but primarily one about the metazoa and perhaps not even all metazoans.

## SPECIES ARE MOSAICS

In a famous metaphor, Dobzhansky (1937) pictured closely related species as linked together through being on adjacent peaks in an adaptive landscape, with each peak representing a species niche. He wrote:

The enormous diversity of organisms may be envisaged as correlated with the immense variety of environments and of ecological niches which exist on earth. But the variety of ecological niches is not only immense, it is also discontinuous. One species of insect may feed on, for example, oak leaves, and another species on pine needles; an insect that would require food intermediate between oak and pine would probably starve to death. Hence, the living world is not a formless mass of randomly combining genes and traits, but a great array of families of related gene combinations, which are clustered on a large but finite number of adaptive peaks. Each living species may be thought of as occupying one of the available peaks in the field of gene combinations. The adaptive valleys are deserted and empty.

Furthermore, the adaptive peaks and valleys are not interspersed at random. Adjacent adaptive peaks are arranged in groups, which may be liked to mountain ranges in which the separate pinnacles are divided by relatively shallow notches. Thus, the ecological niche occupied by the species "lion" is relatively much closer to those occupied by the tiger, puma and leopard than to those occupied by wolf, coyote and jackal. The feline adaptive peaks form a group different from the group of canine "peaks". But the feline, canine, ursine, musteline and certain other groups form together the adaptive "range" of carnivores, which is separated by deep adaptive valleys from the "ranges" of rodents, bats, ungulates, primates and others. In turn, these ranges are again members of the adaptive system of mammals, which are ecologically and biologically separated, as a group, from the adaptive system of birds, reptiles, etc. The hierarchical nature of biological classification reflects the objectively ascertainable discontinuity of adaptive niches, in other words the discontinuity of ways and means by which organisms that inhabit the world derive their livelihood from the environment. (9–10)

Though striking, this metaphor is questionable in a number of ways. For example, it encapsulates a lock-and-key conception of adaptation and natural selection in which selection fashions each species to its preexisting niche. For my purposes, though, I am more interested in the oversimple relationship between phylogenetic units and ecological processes this picture projects (Vrba and Eldredge 1984, Damuth 1985, Eldredge 1989, Vrba 1993, Eldredge 1995a and 1995b). If the definition of a niche includes the biological dimensions relevant to an organism's way of life, species do not have niches.[5] Species, typically, are *ecologically fractured*: they are distributed through many different communities. The common brushtail possum is found in communities as varied as cool, temperate New Zealand rainforests, inner Sydney suburban gardens, and eucalypt woodlands. Although coyote, impala, leopards, and buffalo—to mention some of Vrba's and Eldredge's favorite examples—are exceptionally flexible and generalist animals, Vrba's and Eldredge's general point is still well made. Thus, the geographic range of plant species is a consistently good predictor of the number of insect species that attack them, for the more widely distributed a species is, the more communities in which it is found (Thompson 1994, 156–158). Temporal variation has a similar effect: the local composition of pollinating species can vary markedly from year to year (Thompson 1994, 184). Therefore, even when we consider species with restricted distributions, the biological associations of species members may change in important ways throughout their own life spans or those of their immediate descendants. Typically, species are geographic and ecological mosaics. Of course, there are exceptions; some organisms have very specific habitat requirements or very local distributions. One Australian pygmy possum is restricted to a few square miles of Australian alpine country. But Vrba and Eldredge are surely right in arguing that species are typically distributed across a number of different communities and even of community types.

Eldredge is well-known for harnessing Mayr's theory of speciation to the theory of punctuated equilibrium. He argues that the punctuated equilibrium pattern arises because adaptive change is linked to speciation (Eldredge 1989). Unless speciation takes place, adaptive change will be ephemeral as the gene combinations underlying the new adaptation will be broken up by migration from the parent population. In Mayr's view, speciation typically takes place in isolated, small, peripheral populations. Most such isolates go extinct, both because small populations are vulnerable to chance disaster and perhaps also because the outer margins of a species' range are typically suboptimal habitat.[6] In other words, peripheral isolates are often not well adapted for their specific circumstances. If they survive, they will be under strong selection pressure and are likely to differentiate from the ancestral stock.

Recent work by Vrba and Eldredge enrich this picture, though not quite in the way they may think (Gould and Eldredge 1993, Vrba 1993, Eldredge 1995a and 1995b). Gould and Eldredge stress that stasis does not mean no change; it means no net or summed change. Eldredge argues that the ecological fracturing of species explains both the fact that stasis is the predominant pattern in evolutionary history and how stasis breaks down. Moreover, if Eldredge is right, it also explains the limits on stasis: why in some species and with some traits we see a gradual shift in phenotype over time. The idea is simple. If species are typically distributed over a number of different community types, and if particular communities are ephemeral on evolutionary timescales, there is no reason to expect natural selection to be acting the same way in different communities, so selection-driven shifts in one community are apt to be undermined by Mayr's Brake. Gene flow from other communities will limit the extent of selection-driven response to purely local conditions. Such response as does take place is always vulnerable to dissolution if the local population becomes fully integrated into a larger group. Moreover, if communities disappear and are replaced by others only broadly similar, or if their composition changes in important ways over time, then even if one community type is dominant at a particular time, it is unlikely to be sufficiently persistent to produce a gradual and entrenched shift in species phenotype. The result is stasis. Eldredge clearly sees this phenomenon as a limitation on the power of natural selection: selection can produce *enduring adaptation* only when Mayr's Brake is released. But, at least to the extent in which we think of Mayr's Brake as gene flow that limits local response, this view seems questionable. It is just as natural to think of the ecological fracturing of species as a phenomenon that limits selection for specialization: parents who do not "know" where their genes are going, or whom they will be joining, should not precommit to the specifics of their environment.

According to this view, sex plays an interestingly ambiguous role in evolution. It is usually thought of as something that can accelerate evolutionary change. Vrba herself argues that the rareness of asexual phenomenological species and their lack of diversity in metazoan lineages is in itself evidence that sex is central to evolutionary change (Vrba 1995). Yet in this view, sex plays a powerful homogenizing role in limiting and destabilizing adaptation to local environments. Moreover, species—that is, *species* defined as complex systems with emergent properties—must be a relatively late evolutionary invention. Yet Eldredge and friends argue that most adaptive change is coupled with speciation. So what of the huge amount of adaptive evolution that took place before the evolution of phylogenetic species with ecological and geographic structure? If early species did not have this kind of complex metapopulation structure, and if that structure explains stasis, we would expect to see gradual phenotypic change in, say, the early metazoa. Neither Eldredge nor Gould, however, argues that the punctuated equilibrium pattern of evolutionary change is restricted to relatively recent history.

As we see in considering the "turnover pulse hypothesis," Eldredge (1995b) and Vrba (1993, 1995) emphasize the importance of the nonbiological environment in seeking to demonstrate the importance of species geographic structure. Thompson (1994) develops similar views on species structure through a consideration of coevolution; indeed, he calls his views the "geographic mosaic" theory of coevolution. He argues that species are not usually the right units of coevolutionary analysis because species ranges typically overlap rather than coincide. Moreover, even where they do coincide, the way one species has an impact upon another is typically modified by other aspects of the environment. Thompson complains that many evolutionary biologists have responded to his argument by thinking of coevolution (setting aside the relatively exceptional cases of the tight symbiotic linkage exemplified by figs and fig wasps) as "diffuse"—an amalgam of many, many species interactions. To the contrary, Thompson argues that coevolution is often highly specific, but involves a geographic mosaic: species fragments have important reciprocal interactions with one another, leading to patchwork coevolution. Populations in two metapopulations interact with one another, and through that interaction, fragments of a species diverge from one another. For example, Thompson argues that many apparently generalist parasite species—species that attack many hosts—have turned out to be metapopulations in which the individual populations specialize on one or a few hosts (pp. 128–132). Mayr's Brake does not prevent local divergence through coevolutionary interaction between species fragments: it dampens it to the extent that gene flow continues to homogenize the metapopulation, and it prevents its entrenchment. Thus, for Thompson, many species involved in coevolutionary evolution consist of largely isolated populations —incipient or potential species in the process of diverging from one another. If there is a difference between the Thompson's view on the role of fragmentation and Vrba and Eldredge's view, it is that Thompson seems to believe that a species mosaic can be stable: a species can persist as a mosaic of diverged populations.[7]

In sum, many species are not ecologically cohesive entities. To the extent then that selection pressures derive from ecological circumstances, these selection pressures will not act on a species as a whole. The fact that species are ecological mosaics explains stasis; it explains the fact that directional change in phenotype throughout the lifetime of a species is relatively infrequent amongst those species most likely to be known from the fossil record. As I explain in the next section, evolutionary change is to be expected when the relationship between evolutionary and ecological units is simplified. This simplification takes place when a mosaic metapopulation fragments into its components.

## STASIS, SPECIATION AND PHYLETIC EVOLUTION

In Eldredge's view, stasis is the result of Mayr's Brake. What releases that brake? The argument based on ecological fracturing does not predict across the board stasis. It does not apply to all aspects of natural selection; it applies to selection driven by community biological interactions, for it is these interactions that are likely to depend on the specific features of a particular community. It does not apply to species with a very small range or with very specific habitat requirements. Thus, anything that affects a few eucalyptus species in Australia will also affect koalas, given their notorious dietary specialization. Also, Eldredge's view may well not apply to many species of specialized parasites, for when we speak of an organism's environment, we must distinguish between *physical environment* and *selective environment:* the features of the environment that impinge differentially on members of the population (Brandon 1990). Some parasites are able to go through their whole life cycle on a single organism. When parasites are able to live and reproduce on a single organism, typically every member of that parasite species lives and reproduces on an organism of the same species (Thompson 1994, 124–125). This lifestyle selects for specialization. In turn, as organisms become increasingly specialized, their selective environment is increasingly dominated by a single feature. They have a very unfractured ecology, which may help explain the capacity of such parasites to evolve physiological and morphological specializations to their host. *Phyletic evolution*—gradual change in the lineage as a whole—of specialist parasites' adaptations to their host would be no surprise. Even in highly specialized and ecologically homogenous species, gene flow might break up a local coadapted gene complex that has been constructed through a period of temporary isolation, but new genes that are adaptive anywhere will be adaptive everywhere. If some speculations about species numbers in tropical rainforests are on the mark, then highly specialized, locally distributed species may yet turn out to be very numerous.[8] However, they will not be widely known, especially to paleobiology. The fracturing arguments applies to the species we are most likely to find in the fossil record—species widely distributed in space and time.

The argument from ecological fracturing, therefore, will not apply to species that are ecologically homogenous, either because they have a restricted physical distribution or because of their specialized lifestyle. We should also expect stasis to break down if selection pressures act across all or most of a species' range in ways relatively independent of the specific structure of and interactions within a community. One obvious source of such pressures is slow, widespread climate change. The browning of Australia—as it slowly became hotter, drier, and less fertile—ought to have generated selection pressures on the right temporal and physical scales, pressures fairly independent of the specific details of community structure. Perhaps the biology of invasions might produce other examples of generalized selection pressure on a broad geographic scale. In Australasia, biological invasions of rabbits, stoats, foxes, and the like seem to have the same effects in many different local habitats—though, to date, extinction or retreat to refugia seems to have been the dominant response. We might expect, therefore, to see Gondwanan paleobiology as a test of punctuated equilibrium. It seems to be an arena in which *phyletic gradualism*—slow evolutionary change within a lineage—might produce adaptation to the abiotic features of the environment.

Climate change is the focus of Vrba's work, too. She emphasizes the ecological complexity of species in explicating her "turnover pulse" hypothesis (Vrba 1993, 1995). How might stasis break down if the inference from ecological fracturing is right? Vrba thinks it typically breaks down through rapid climate changes. Organisms may not care much who their neighbours are, but they are all sensitive to key physical parameters: temperature, moisture, sunlight, and the like. She argues that each species has a distinctive habitat, defined by the range of physical parameters in which the species can survive and reproduce. This habitat should be conceived of as part of a species identity. From the very fact that species are ecologically fractured, Vrba infers that species distributions are *controlled* by these sensitivities (together with geographic barriers) rather than by their biological interactions with other species (Vrba 1993; Vrba 1995, 18–20). Though each species will have a probably unique range of environmental tolerances, these tolerances may only *constrain* its distribution. For example, many of New Zealand's tree species are still recolonizing areas from whence they were driven in the last ice age. Other plants will be absent from areas because of the phase of ecological succession in that area rather than because of their physical tolerances. From the fact that many species are ecological mosaics, it does not follow that biological factors play an unimportant role in explaining distribution.

Thus, I have reservations about the tight linkage Vrba sees between range of environmental tolerances, physical habitat requirements, and climate change. Even so, the limits of a species' tolerance do restrict its range. So what happens when the climate changes significantly? Sometimes not much: the potential space available to a species might shrink a bit, expand a bit, or shift latitudinally. If physical barriers do not intervene, the species can shift with it. Sometimes the potential space will shrink to zero, and the species will disappear with it. But stasis breaks down when environments both change (creating new selection pressures) and species' ranges fragment, at least temporarily dissolving the metapopulation, chopping it into its component populations. Thus, external shocks to the system on regional scales can both simplify the relationship between the evolutionary and the ecological units, and change the pattern of selection that operates. These shocks can create the conditions under which selection can be effective (releasing Mayr's Brake) while causing it to act. A local, isolated population is *not ecologically fragmented:* Dobzhansky's wonderful metaphor really does fit such populations. If the population is not so small that genetic variation is sharply reduced, selection can act, and it can act without counterbalance from homogenizing gene flow from neighboring populations, for there are no neighboring populations. The result is a "turnover pulse" of the kind Vrba claims to detect in the African fossil record. At around the same time, species from quite different lineages and with different ecological profiles disappear from the fossil record. In their place appear new species closely related to the departed.

What view of the nature and importance of species emerges from this conception of the interaction of speciation and selection? Let's first note that evolutionary gradualism is no threat to the objectivity of our identification of species. The skeptic supposes that if phenotypic change is typically the result of a gradual shift in a lineage over time, then species distinctions are illusions. But this view conflates features of organisms with features of populations. A limited form of evolutionary gradualism is uncontroversial. As the example of *Helianthus anomalus* shows, plant species are occasionally created in a single generation by hybridization, but in metazoan lineages, the differences between parent and viable offspring are likely to be small. Because viable offspring develop from coadapted developmental resources, any major change in those resources is likely to derail development, not generate significant change. However, there is no similar argument against fast change in *population-level properties.* A population can fragment, shrink, or change in range; disappear from some communities and become part of others; or change in genetic diversity on ecological rather than geological timescales. For example, a species hit by a climatic change can be forced through a population bottleneck that strips it of much of its previous genetic diversity. Yet populations and metapopulations are species in virtue of population-level properties. *Speciation* need not be smooth, gradual, and seamless even if *phenotypic change* is smooth, gradual, and seamless.

If Eldredge, Vrba, and those of similar views are right, two populations consist of organisms of the same species if they evolve (or more likely fail to evolve) together. The extent to which populations are evolutionarily coupled is not, of course, independent of the phenotypes of individual organisms, nor is it any simple function of those phenotypes. Populations can become permanently decoupled from one another through geological, environmental, or ecological changes that leave no obvious initial trace in individual phenotypes. An invasion can force one population to change its daily cycle; the extinction of a long distance pollinator can disjoin two plant populations. Of course, the idea that species identity and speciation is just a matter of population-level properties is controversial. Some, perhaps most, evolutionary biologists take speciation to occur only when there have been intrinsic changes in the two populations that make renewed gene flow impossible, so their test for speciation is stronger than mere de facto permanent separation of the two populations. Yet on the face of it, this criterion is puzzling, for the view that species are historically defined entities is close to the consensus view in evolutionary biology. As a consequence of such a view, the facts that make an organism a member of a particular species are now considered to be the relations of the organism to others, not its intrinsic physical characteristics.

Of course, even if we accept a thoroughly historical and relational account of what makes an organism a member of a given species, we still have good epistemic reason for an interest in intrinsic characters and changes in those characters. Often, we cannot tell of contemporary populations whether extrinsic isolating mechanisms are permanent. The extinction of a critical pollinator may be local, or a new pollinator may evolve and rejoin populations that seemed permanently separated. When the factors that have segmented a previously coupled metapopulation into disjoint parts are extrinsic, we may be unable to tell whether we are dealing with a single species that is now a geographical and ecological mosaic or a cluster of incipient species. In this sense, as O'Hara (1993) has noted, identifying species has a forward looking element to it.

However, I suspect the insistence that speciation requires intrinsic change also stems from the idea that the species evolutionary biologists identify should map closely onto phenomenological species. If Australian and New Zealand boobook owls count as separate species simply because the two populations are permanently segmented by 2,000 km of the Tasman Sea, there would be no way a taxonomist identifying a specimen in a museum could place that organism in its correct *evolutionary species* from the intrinsic physical features of the specimen alone. I think this motivation is inappropriate because it stems from a lingering adherence to the idea that there is a single and universal species category.[9] I think, therefore, we should instead see phenomenological species—identifiable clusters of organisms—as fallible clues to the existence of evolutionarily linked metapopulations. We have already been forced to accept that the concordance between phenomenological species and evolutionary species is far from perfect. Cryptic sibling species erode that correlation in one direction. Phenotypically heterogenous species and phenotypic shifts in unfragmented lineages erode it in the other direction. Thus, Niklas (1997) discusses two goldenrod "species," *Solidago rugosa* and *S. sempervirens*. These plants differ markedly, yet they "freely interbreed in nature" (p. 68). The hybrids are fertile both amongst themselves and with their parents, but are so phenotypically different from either that they are regarded as a third species. The only reason Niklas offers for thinking that there are three species here is the phenotypic differences between the individual organisms, and this reason just seems to be an undefended adherence to the idea that the species category for evolutionary biologists should match, rather than explain, the phenotypic clustering we recognize in phenomenological species. If we lump all three forms into one species, we still need an explanation of how the phenomenologically distinct populations persist, but we still have the same problem if we recognize three species, and that taxonomic decision hides the problem rather than solving it.

The distinction between the properties of organisms and the properties of the species of which they are a part is important not just for seeing the consistency of evolutionary gradualism with the view that the species category is a natural kind. Failing to note this distinction gives the cladistic species conception a false air of paradox. According to the cladistic view of species, an ancestral species becomes extinct once it fissures, creating daughter species, even if the organisms that are part of one of the daughter species are phenotypically indistinguishable from those of the parent (Ridley 1989). At first glance, this view looks like a wholly arbitrary convention whose sole purpose is to avoid the difficult problem of deciding how much the phenotypes of organisms in the daughter species must vary from those of the parent before we count the parent species as extinct.[10] Not so, however: if we think about the species properties rather than about individual phenotype properties, the fission of the lineage into two transforms the environment of the ancestor species in an extremely salient way. There is a new kid, and a potentially supplanting competitor, on the block. Moreover, the very process of fission itself will markedly change the population-level properties of the ancestral species. Its geographic range, its population structure, the range of genetic variation found within it, and the communities through which it is distributed will be different.

Eldredge (1995b) and Vrba (1993, 1995) vacillate on this issue, shackling genuinely interesting ideas about the role of species' geographic structure in speciation and about adaptive evolution to an inordinate fondness for Paterson's *recognition conception* of species (McEvey 1993). This connection seems to me to be a mistake. Paterson's account of species is a variant of the biological species concept in recognizing that species are closed systems of gene exchange, but his particular account of closure seems to me to be the wrong one for Vrba and Eldredge to endorse. Perhaps more explicitly than Dobzhansky and Mayr, Paterson treats speciation as a by-product of other evolutionary processes. To the extent that speciation depends on the previous isolation of diverging populations (i.e., to the extent that speciation is allopatric), it is clear that the mechanisms that effect separation between two gene pools must begin as by-products. In allopatry, there can be no selection for avoiding creatures that are not around anyway. A standard response is to suppose that after isolated populations diverge and hybrids cease to be as fit as individuals of either parental populations, selection can directly reinforce isolation. Paterson wants an intrinsic, nonrelational account of species, so he rejects this standard response because he supposes that the mechanisms that segment two species from one another *must be fully present* in each organism. Because selection against interpopulation mating could only act in the hybrid zone, it cannot explain the *fixation* of the isolating mechanisms in each organism in the populations. He does not seem to allow for the possibility that population-level properties, including variable features in the hybrid zone, could jointly comprise the isolating mechanisms.

Moreover, Paterson's own solution to this problem—that speciation involves the evolution of a new "specific mate recognition system" (SMRS) —leaves a major hostage to fortune. I see no reason to suppose that, in general, organisms come equipped with any such system. We know that within many species, mating is not random. Mating preferences are shaped by sexual selection, perhaps operating on inherited sensory biases, so it may well be that organisms are equipped with or develop a schemata—a stereotype—of an ideal mate. Organisms that offer themselves as mates or that are detected in an active search for mates are matched against the "ideal mate template," and their distance from the ideal is estimated. The closer to the ideal, the more willing the organism is to mate. Consider, then, those many species in which mating is not random and in which organisms do have mate assessment mechanisms. How are these mechanisms related to a specific mate recognition system? If a specific mate recognition system is an identifiable piece of behavior-guiding machinery, Paterson seems committed to a couple of idea. (1) Organisms are equipped with a threshold value of "just good enough" that defines the marginally acceptable mate; they detect whether a candidate falls above, below, or at this threshold. (2) The limits of the threshold are a developmentally normal member of the opposite sex of its species. It may be that some organisms are equipped with such a system, but it hardly seems to be a requirement for the existence of a species. New Zealand black stilts prefer to mate with a member of their own species, but if none is available, they will accept and can breed successfully with pied stilts. Famously, many male organisms have thresholds set very low, and they will attempt to mate with a wide variety of distantly related organisms. Of course, if the specific mate recognition system is not an identifiable system of behavioral machinery, but just a set of facts about how the organisms of a given population come to find their mates—taking all the normal background facts about their distribution, their environment, and the impact of their environment on their behavior into consideration—then of course black stilts do have a specific mate recognition system, for they make their first breeding effort, if possible, before pied stilts arrive at their breeding sites. But, then, despite all the rhetoric, Paterson's concept is just the biological species concept.

Though Eldredge (1995b) and Vrba (1995) use the recognition concept, it is not clear whether they think mate recognition systems are specific structures. On the one hand, they write as if they are traits subject to selection. Vrba argues that stabilizing selection is likely to ensure that specific mate recognition systems are more invariant across a species than other traits, and Eldredge (1995b) remarks that "The SMRS ... is unquestionably adaptive in the classic sense" (p. 471), which therefore supposes that the SMRS is a specific, identifiable structure. Yet he also writes: "those phenotypic attributes in the widest sense ... pertaining to reproduction constitute the SMRS" (p. 467). That huge and diverse bundle of traits is not itself a trait, a single phenotypic unit. Vrba's position is particularly difficult to assess. She endorses the recognition concept, but argues that the specific mate recognition system is a complex property of a lineage rather than a feature of individual organisms: "each unique SMRS ... is a complex emergent property ... of a particular lineage system" (Vrba 1995, 10–11). This argument seems to deny that the recognition system must be an individual trait fixed in the species, so her recognition concept is also just a version of the biological species concept. Yet she repeats Paterson's claim that the recognition concept is a nonrelational characterization of species identity.[11]

The insistence on a nonrelational account of species identity is strange, both empirically and conceptually. Empirically, the role of other species must make a difference to the way selection acts on recognition mechanisms, for these recognition mechanisms form a filtering system. They let some organisms or their genetic material through and block others. The way selection builds and stabilizes a filtering system depends both on the cost of errors and the difficulty of avoiding them. That difficulty, in turn, depends on how easy it would be to mistake an alien for a potential mate. In other words, it depends on the character of other species in the neighborhood. We would expect mate recognition systems to have a very different character in species whose range overlaps many close relatives from the systems in species that are not so afflicted. Moreover, there is a conceptual problem. Vrba herself makes the point that we cannot use the contrast between fixed and variable traits to identify species,[12] for variability is defined for a group. We have to settle the breadth of a lineage before we can tell whether a trait is fixed across it. The same seems to apply to counting recognition systems. Do black stilts have recognition systems that include pied stilts, or are black stilt–pied stilt hybrids the result of the collapse of the "complete system based on a combination of ... characters ... unique to a given cluster of related biparental organisms" (Vrba 1995, 12)? We need to identify the species before we can count the specific mate recognition systems.

Rather than buying into an exotic variant of the biological species concept, I think Vrba and Eldredge need some version of an evolutionary species concept (Simpson 1961). Species are evolutionarily linked metapopulations. Speciation is the permanent sundering of those links—whether the sundering is the result of intrinsic changes in individual phenotypes, geological fragmentation of the species habitat, or ecological changes that permanently cut populations off. In their discussions, Vrba and Eldredge assume that linkage is through gene exchange, and, of course, this is the orthodox view. It is, however, no longer the only view (see, for example, Oyama 1985, Griffiths and Gray 1994, Sterelny et al. 1996). Therefore, it is worth mentioning that the view that species are mosaics stabilized by Mayr's Brake is not committed to the idea that populations can be linked only through gene exchange. Suppose, for example, that an adaptive shift in a population is not genetic, but is instead the result of the acquisition of a more efficient symbiotic microorganism. If development is atomistic, and if the new microorganism contributes to (say) cellulose digestion in ways relatively independent of other developmental resources, nothing like Mayr's Brake will apply. The change should gradually spread, whatever the level of migration into and out of the site of the evolutionary origin. Phyletic evolution should take place in the species as better gut fauna take over. But if new developmental resources play a role in building new adaptations because of their distinctive developmental context, substantial migration from the much larger parental population is likely to disrupt the new developmental system, so Mayr's Brake would still threaten this adaptive shift in the same way that gene flow into a quasi-isolated population can disrupt new gene combinations.

In sum, I think Vrba and Eldredge's fundamental idea is right. Some phenomenological species are evolutionary units, and they have their specific evolutionary character because they are ecologically complex metapopulations. When evolutionary units have that structure, it is of central importance because it has a powerfully conservative effect on the possibilities of evolutionary change. However, Vrba and Eldredge marry this insight to an inappropriate conception of species.

## UPSHOT

It's time to draw some general morals from this discussion. Dawkins (1982) has argued that the developmental cycle is of great significance in adaptive evolution. He argues that the evolution of adaptive complexity in multicelled organisms depends on a developmental cycle from single-cell bottleneck to single-cell bottleneck. A genetic change can make important differences to the whole organism when development is funneled through this bottleneck. The change, by acting early in this process, can have global consequences. The developmental cycle matters, for only at some points in an organism's life history can a change make a global difference to the organism (Dawkins 1982, chap. 14). The idea that species have ecological structure, together with the turnover pulse hypothesis, suggests that species too have developmental cycles that are important in the same way. Adaptive phenotypic change[13] is possible only because of the metapopulation ⇒ population ⇒ metapopulation cycle, with the change tending to take place at the most simplified point in that cycle, when the incipient species is a single population. The change is entrenched only if that population both survives and is permanently isolated from the parental stock.

More generally, the importance in evolution of the geographic and ecological structure of species shows that species are real. We do not have to defend species selection to show that species have an ineliminable role in evolution. In their 1993 review of punctuated equilibrium, Gould and Eldredge identify standard modern Darwinism as having three central commitments: (1) the organism is the main unit of selection; (2) natural selection explains the existence of evolutionary novelty and hence is the "creative" force in evolution; and (3) "genealogical change at all taxonomic levels" is nothing more than the accumulation of small change in local populations (Gould 1995, 4). Eldredge, in particular, sees his opponents as "ultra-Darwinians" who accept an "extrapolationist" view of evolution,[14] and it is this final clause that Eldredge takes himself to have refuted. Macroevolutionary change is not *just* the accumulation of change in local populations, for such changes do *not* accumulate except in special circumstances. We cannot explain the evolution and entrenchment of adaptive characteristics without both explaining the role of Mayr's Brake in buffering species from change in normal times and the brake's occasional release. But Mayr's Brake is a feature of metapopulations, not populations. Hence, its existence and operation cannot be understood by extrapolating from evolution within a single population to evolution within larger temporal and geographic scales. Natural selection within a population is the only mechanism that generates adaptive phenotypic change: the first two commitments of standard Darwinism survive. That selection is only effective, however, in generating permanent phenotypic change (at least for adaptations to distinctive features of the local community) under relatively rare background conditions. The attempt, then, is to defend a modest version of the view that evolutionary species are kinds whose structure is of profound significance in evolutionary change. Not all organisms are organized into evolutionary species; not all species are complex mosaics. But where species with this structure do evolve, they are not just epiphenomenona of phenotypic change. They profoundly influence that change because they are a central feature of the environment of change.

## ACKNOWLEDGMENTS

Thanks to David Hull and Rob Wilson for their feedback on an earlier version of this paper.

## NOTES

1. Bacteria do not fall easily even into phenomenological species—at least not on the basis of the features of their morphology accessible through light microscopes. They come in a number of different shapes, but otherwise have few identifying characteristics.

2. There is gene flow between species because hybrids are fertile with both parental species: the equivalent in plants of a mule being able to breed with both donkey and horse. But there are also species formed by hybridization in which the hybrid is fertile but reproductively isolated from both parental species. The sunflower *Helianthus anomalus* seems to be an example; its parent species seem to be *Helianthus annuus* and *Helianthus petiolaris*. *Helianthus anomalus* is cross-fertile with neither of the other *Helianthus* species, but has been experimentally recreated from them by hybridization (Niklas 1997, 64–65)

3. Alternatively, the generality problem is avoided at the cost of problems equally great. Some versions of the phylogenetic species concept treat any population with a distinctive and invariant feature as a species. This treatment does avoid reviving all the problems of the biological species concept, but only at the cost of counting innumerable ephemeral populations as species. The *species category* so defined is not an evolutionary kind, in terms of either the mechanisms that produce these clumps or their evolutionary upshot.

4. As Hull (1978) points out, these ideas are interdefinable. The issue is not whether species are kinds or individuals, but whether their defining features are historical and relational or intrinsic.

5. I think this view is true according to both Elton's conception of a niche as a functional role within a community and its Hutchinsonian successor in which niches are essentially linked to their occupants (Elton 1927, Hutchinson 1965). Damuth (1985) introduced the term *avatar* to designate the fraction of a species in a particular community. Unless we define niches very thinly indeed, avatars—not species—have niches. Van Valen (1976), in his blend of an evolutionary and ecological conception of a species, may have had a very thin definition in mind in developing his notion that species have unique adaptive zones. Still, the more work one wants the niche to do in explaining the phenotypic coherence of a species, the more difficult it is to downplay the ecological differences between the different populations that make up a species. The need to define an adaptive zone thinly enough so all organisms in a species share an adaptive zone cuts across both the idea that adaptive zones explain species phenotypes and the idea that they are species specific.

6. I think we should be very cautious about accepting this idea. Notice first that this idea supposes that species' ranges are typically at equilibrium and that those ranges are set by the organism's powers of dispersal interacting with its limits of tolerance. Moreover, peripheral isolates can sometimes outrun their usual parasites and predators (Thompson 1994, 159). It may well be best, therefore, to think that the mix of selection in peripheral populations changes rather than to suppose that these populations are in suboptimal conditions compared to organisms close to the geographic center of the species' distribution.

7. Even this belief may be no more than different ideas about how to count stability. Vrba and Eldredge are palaeontologists, and Thompson is not.

8. For a very accessible introduction to this literature, see Wilson 1992, 131–141.

9. Biological taxonomy has now evolved into systematics, but the concern that species differences be marked by features identifiable in each specimen may also be a lingering vestige of traditional taxonomy and the role of the "type specimen" in that taxonomy.

10. Thus, Vrba complains, "To say that a parent species must cease to exist once it gives off a branch, and be recognised as a new species if it persists without change after branching is like saying I ceased to exist at the birth of my daughter and must be named as a new individual" (Vrba 1995, 27).

11. In contrast to the "isolation concept," Paterson's synonym for the biological species concept. The point is that identifying isolating mechanisms implicitly refers to another population, the population from which one is isolated.

12. This point is bad news for some versions of the phylogenetic species concept.

13. Less boldly, adaptive phenotypic change of a certain kind—the kind driven by community-specific ecological factors.

14. I am very skeptical of Eldredge's identification of these ultra-Darwinians. He seems to believe that accepting gene selectionism commits one to the minimalist view of species—the view that they are a merely phenomenological kind. But these are independent debates: Williams invented gene selectionism, but he defends species selection (1992).

## REFERENCES


Atran, S. (1990). *Cognitive foundations of natural history.* Cambridge: Cambridge University Press.

Berlin, B. (1992). *Ethnobiological classification: Principles of categorisation of plants and animals in traditional societies.* Princeton: Princeton University Press.

Brandon, R. (1990). *Adaptation and environment.* Cambridge, Mass.: MIT Press.

Buss, L. (1985). The uniqueness of the individual revisited. In J. Jackson, L. Buss, and R. Cook, eds., *Population biology and evolution of clonal organisms.* New Haven, Conn.: Yale University Press.

Buss, L. (1987). *The evolution of individuality.* Princeton: Princeton University Press.

Cracraft, J. (1987). Species concepts and the ontology of evolution. *Biology and Philosophy* 2, 329–346.

Damuth, J. (1985). Selection among "species": A formulation in terms of natural functional units. *Evolution* 39, 1132–1146.

Dawkins, R. (1982). *The extended phenotype.* Oxford: Oxford University Press.

Dobzhansky, T. (1937). *Genetics and the origin of the species.* New York: Columbia University Press.

Eldredge, N. (1989). *Macroevolutionary dynamics: Species, niches and adaptive peaks.* New York: McGraw-Hill.

Eldredge, N. (1995a). *Reinventing Darwin.* New York: John Wiley and Son.

Eldredge, N. (1995b). Species, selection, and Paterson's concept of the specific-mate recognition system. In D. Lambert and H. Spencer, eds., *Speciation and the recognition concept: Theory and applications.* Baltimore: Johns Hopkins Press.

Elton, C. S. (1927). *Animal ecology.* London: Sidgwick and Jackson.

Ghiselin, M. (1974). A radical solution to the species problem. *Systematic Zoology* 23, 536–544.

Gould, S. J. (1995). A task for paleobiology at the threshold of majority. *Paleobiology* 21, 1–14.

Gould, S. J., and N. Eldredge (1993). Punctuated equilibrium comes of age. *Nature* 366, 223–227.

Griffiths, P. E., and R. Gray (1994). Developmental systems and evolutionary explanation. *Journal of Philosophy* 91, 277–304.

Hull, D. (1978). A matter of individuality. *Philosophy of Science* 45, 335–360.

Hutchinson, G. E. (1965). *The ecological theater and the evolutionary play.* New Haven, Conn.: Yale University Press.

Mayr, E. (1942, second edition 1982). *Systematics and the origin of species*. New York, Columbia University Press.

Mayr, E. (1976). *Evolution and the diversity of life*. Cambridge, Mass.: Harvard University Press.

Mayr, E. (1988). *Towards a new philosophy of biology*. Cambridge, Mass.: Harvard University Press.

McEvey, S., ed. (1993). *Evolution and the recognition concept of species: The collected writing of Hugh E. H. Paterson*. Baltimore: Johns Hopkins University Press.

Niklas, K. (1997). *The Evolutionary biology of plants*. Chicago: University of Chicago Press.

O'Hara, R. J. (1993). Systematic generalization, historical fate and the species problem. *Systematic Biology* 42, 231–246.

Oyama, S. (1985). *The ontogeny of information*. Cambridge: Cambridge University Press.

Ridley, M. (1989). The cladistic solution to the species problem. *Biology and Philosophy* 4, 1–16.

Simpson, G. C. (1961). *Principles of animal taxonomy*. New York: Columbia University Press.

Sterelny, K., K. Smith, and Dickison, M. (1996). The extended replicator. *Biology and Philosophy* 11, 377–403.

Templeton, A. (1989). The meaning of species and speciation: A genetic perspective. In D. Otte and J. Endler, eds., *Speciation and its consequences*. Sunderland: Sinauer.

Thompson, J. N. (1994). *The coevolutionary process*. Chicago: University of Chicago Press.

Van Valen, L. (1976). Ecological species, multispecies and oaks. *Taxon* 25, 233–239.

Vrba, E. (1993). Turnover-pulses, the red queen and related topics. *American Journal of Science* 293A, 418–452.

Vrba, E. S. (1995). Species as habitat-specific complex systems. In D. M. Lambert and H. G. Spencer, eds., *Speciation and the recognition concept: Theory and applications*. Baltimore: Johns Hopkins University Press.

Vrba, E., and N. Eldredge (1984). Individuals, hierarchies and processes: Towards a more complete evolutionary theory. *Paleobiology* 10, 146–171.

Wiley, E. (1978). The evolutionary species concept reconsidered. *Systematic Zoology* 27, 17–26.

Williams, G. C. (1992). *Natural selection: Domains, levels and challenges*. Oxford: Oxford University Press.

Wilson, E. O. (1992). *The diversity of life*. New York: W. W. Norton.

# III Rethinking Natural Kinds

# 6 Homeostasis, Species, and Higher Taxa

Richard Boyd

## INTRODUCTION

### Overview

In this paper, I identify a class of natural kinds, properties and relations whose definitions are provided not by any set of necessary and sufficient conditions, but instead by a "homeostatically" sustained clustering of those properties or relations. It is a feature of such *homeostatic property cluster (HPC) kinds* (*properties relations,* etc.—henceforth, I'll use *kinds* as the generic term wherever it will not cause confusion) that there is always some indeterminacy or "vagueness" in their extensions.

I introduce the notion of *accommodation* between conceptual and classificatory practices and causal structures and explain why the achievement of such accommodation is necessary for successful induction and explanation. I defend the view that the naturalness (and the "reality") of natural kinds consists solely in the contribution that reference to them makes to such accommodation. In the light of this *accommodation thesis,* I explain why reference to "vague" homeostatic property cluster kinds is often essential to successful inductive and explanatory practice in the sciences.

I deploy these notions to address some aspects of the "species problem" in the philosophy of biology. I conclude that biological species are paradigmatic natural kinds, their historicality and lack of sharp boundaries notwithstanding.

Regarding the alternative conception that species are individuals, I examine the individuation of individuals in the light of considerations of accommodation and conclude that accommodation constraints operate on their individuation exactly as they do in the definition of natural kinds and categories. I conclude, in consequence, that the debate over whether species are kinds or individuals is less momentous metaphysically and methodologically than one might at first suspect, and that even those scientists who are convinced that species are individuals must conclude that they are natural kinds as well.

I draw a distinction between two equally legitimate notions of definition in science: *programmatic* definitions and *explanatory* definitions. I deploy the idea that species are homeostatic property cluster kinds together with this distinction to clarify other issues about the metaphysics of species. In the first place, I conclude that individual species have (homeostatic property cluster) essences, so that a form of "essentialism" is true for species, albeit a form of essentialism quite different from that anticipated by Mayr and others who have discussed essentialism in biology. Furthermore, I indicate how recognizing species as homeostatic property cluster phenomena and drawing the distinction between types of definitions allows us to make better sense of issues regarding "realism" and "pluralism" about species-level taxa.

I extend the application of the accommodation thesis to consideration of the question of the reality of higher taxa. I argue that some higher taxa are probably real natural kinds in the sense of the term required by the accommodation thesis—indeed, probably homeostatic property cluster natural kinds. I deploy that thesis to identify a crucial relation between judgments of arbitrariness or conventionality of representational schemes, and to show how a reference to that relation can help to clarify and to evaluate claims about the conventionality of higher taxa.

## Homeostatic Property Cluster Kinds

In the empiricist tradition since Locke, the standard conception of scientific (and everyday) kinds has been that they are defined by "nominal essences" or by other purely conventional specifications of membership conditions. Part of that conception has been a conception of linguistic precision, according to which a properly defined kind will be defined by necessary and sufficient membership conditions. Because the boundaries of kinds are, on the nominalist conception characteristic of empiricism, purely matters of convention, any failure of scientific concepts to correspond to this standard of precision could, in principle, be remedied by the adoption of more precise nominal definitions.

The realist critique of Lockean nominalism that arose with naturalistic conceptions of natural kinds and of the semantics of natural kind terms (Kripke 1971, 1972; Putnam 1972, 1975a, 1975b) was articulated around examples of a posteriori definitions of natural kinds that likewise specified necessary and sufficient membership conditions—such as natural definitions of chemical kinds by molecular formulas (e.g., "water = $H_2O$"). These critiques thus gave support to what many authors call the "traditional" essentialist conception of natural kinds, according to which, among other things, such kinds possess *real* (as opposed to nominal) essences that define them in terms of necessary and sufficient membership conditions.[1]

At the time I began thinking about these issues, philosophical conceptions of kinds and categories that did not treat definition by necessary and sufficient conditions as the relevant standard of precision were pretty much limited to Wittgensteinian and other "ordinary language" conceptions whose extrapolation to scientific cases did not seem to me very plausible.

I had the intuition, nevertheless, that the prevailing conception of linguistic precision was a holdover from logical positivism. My first foray into defending that view (Boyd 1979) focused mainly on the question of whether or not the linguistic precision appropriate in science was compatible with the use of "vague" metaphors in scientific theorizing, which has the associated risk of what Field (1973) calls "partial denotation." I concluded that partial denotation and subsequent "denotational refinement" (Field 1973) are constituents of the very phenomenon of precise reference. In the course of defending this view, I found myself advancing a conception of reference according to which certain relations between a term in use and, say, a natural kind are *constitutive* of the reference relation without any one of them being necessary for it to obtain. Thus, I became committed to the view that the relation of reference was not definable in terms of necessary and sufficient conditions.

I became convinced that this view was true of a great many scientifically and philosophically important natural kinds, categories, and relations, so in a series of papers (Boyd 1988, 1989, 1991, 1993, forthcoming b) I advanced a conception of *homeostatic property cluster kinds* to explain why there were such natural kinds.

Here's what I proposed happens in such cases. I formulate the account for monadic property terms; the account is intended to apply in the obvious way to the cases of terms for polyadic relations, magnitudes, and so on:

1. There is a family (F) of properties that are contingently clustered in nature in the sense that they co-occur in an important number of cases.

2. Their co-occurrence is, at least typically, the result of what may be metaphorically (sometimes literally) described as a sort of *homeostasis*. Either the presence of some of the properties in F tends (under appropriate conditions) to favor the presence of the others, or there are underlying mechanisms or processes that tend to maintain the presence of the properties in F, or both.

3. The homeostatic clustering of the properties in F is causally important: (theoretically or practically) important effects are produced by a conjoint occurrence of (many of) the properties in F together with (some or all of) the underlying mechanisms in question.

4. There is a kind term *t* that is applied to things in which the homeostatic clustering of most of the properties in F occurs.

5. *t* has no analytic definition; rather, all or part of the homeostatic cluster F, together with some or all of the mechanisms that underlie it, provide the natural definition of *t*. The question of just which properties and mechanisms belong in the definition of *t* is an a posteriori question—often a difficult theoretical one.

6. Imperfect homeostasis is nomologically possible or actual: some thing may display some but not all of the properties in F; some but not all of the relevant underlying homeostatic mechanisms may be present.

7. In such cases, the relative importance of the various properties in F and of the various mechanisms in determining whether the thing falls under *t*—if it can be determined at all—is an a posteriori theoretical issue rather than an a priori conceptual issue.

8. Moreover, there will be many cases of extensional indeterminacy, which are not resolvable even given all the relevant facts and all the true theories. There will be things that display some but not all of the properties in F (and/or in which some but not all of the relevant homeostatic mechanisms operate) such that no rational considerations dictate whether or not they are to be classed under *t*, assuming that a dichotomous choice is to be made.

9. The causal importance of the homeostatic property cluster F, together with the relevant underlying homeostatic mechanisms, is such that the kind or property denoted by *t* is a natural kind.

10. No refinement of usage that replaces *t* by a significantly less extensionally vague term will preserve the naturalness of the kind referred to. Any such refinement would require either that we treat as important distinctions which are irrelevant to causal explanation or to induction, or that we ignore similarities which are important in just these ways.

11. The homeostatic property cluster that serves to define *t* is not individuated extensionally. Instead, the property cluster is individuated like a (type or token) historical object or process: certain changes over time (or in space) in the property cluster or in the underlying homeostatic mechanisms preserve the identity of the defining cluster. In consequence, the properties that determine the conditions for falling under *t* may vary over time (or space), while *t* continues to have the same definition. The historicity of the individuation conditions for the definitional property cluster reflects the explanatory or inductive significance (for the relevant branches of theoretical or practical inquiry) of the historical development of the property cluster and of the causal factors that produce it, and considerations of explanatory and inductive significance determine the appropriate standards of individuation for the property cluster itself. The historicity of the individuation conditions for the property cluster is thus essential for the naturalness of the kind to which *t* refers.

## Examples

In almost any philosophical discussion about the nature of natural kinds, the author will illustrate her claims with especially persuasive illustrative examples. It will, no doubt, seem odd to readers who are biologists or philosophers of biology that in my own papers on the subject, I deployed biological species as such examples of HPC natural kinds. It is a peculiarity of the literature that in mainstream analytic philosophy, biological species are—along with chemical elements and compounds—the paradigmatic natural kinds, whereas among philosophically inclined biologists and philosophers of biology, there is almost a consensus that they are not kinds at all (see, e.g., Ghiselin 1974, Hull 1978, Ereshefsky 1991).

My aim in those papers was mainly metaphilosophical: I hoped to persuade mainstream readers that many *philosophical* categories and relations (reference, knowledge, rationality, moral goodness, and so on) might be HPC kinds. In that context, biological species served as useful illustrative examples. In the present essay, however, my aim is to establish the credibility, within the philosophy of biology, of the view that species are HPC natural kinds and to explore the implications of this conception for our understanding of the species problem in biology and of related problems about essentialism and about the reality of higher taxa.

## Strategy

I propose to address four considerations that might be thought to support the view that species are individuals and not natural kinds:

- They are not defined by necessary and sufficient conditions—specified in terms of the intrinsic properties of their members—as respectable kinds should be.
- They are necessarily restricted to particular historical periods and circumstances, whereas natural kinds are universal in the sense of not being so restricted.
- They do not fall under universal exceptionless laws as genuine natural kinds do.
- They differ from natural kinds in that what unites their members is their historical relationships to one another *rather than* their shared properties.

I maintain that the first three of these considerations draw their current plausibility from a profoundly outdated positivist conception of kinds and that the fourth participates in both this same error and in a misestimate of the explanatory role of species concepts in biology. I offer an alternative to the positivistically motivated conception of natural kinds and their essences, and explain why, in the light of this alternative, biological species properly count as natural kinds, defined by real essences, even if in some sense they are also like paradigm cases of individuals.

I then indicate how the insights of the alternative account can be extended to provide resources for the treatment of other aspects of the species problem, and even to certain issues about higher taxa.

## The Essence of Essentialism: Toward a New Understanding

One implication of the HPC conception of (some) natural kinds is that the positivist conception of natural kinds reflected in the four considerations and suggested by examples such as "water = $H_2O$" misleads us about what is *essential* to the essentialist critique of Lockean nominalism about kinds. What is essential is that the kinds of successful scientific (and everyday) practice cannot be defined by purely conventional a priori "nominal essences." Instead, they must be understood as defined by a posteriori real essences that reflect the necessity of our deferring, in our classificatory practices, to facts about causal structures in the world. What is definitely not essential to an essentialist conception of scientific (and everyday) natural kinds is that it conform to the positivist picture suggested by the four considerations. So, in defending the HPC conception and its application to the species problem, I hope to contribute to a new understanding of issues of essentialism in biology and elsewhere.

A point of clarification is in order here about the relation between my defense of a new understanding of essentialism and prominent critiques of "essentialism" in biology. Several authors (e.g., Mayr 1980, Hull 1965) point to an essentialist tradition within biology prior to the consolidation of the Darwinian revolution. According to the essentialism they have in mind, biological species, like other natural kinds, must possess definitional *essences* that define them in terms of necessary and sufficient, intrinsic, unchanging, ahistorical properties of the sort anticipated in the four given considerations. They attribute the influence of this traditional conception of species and of kinds in science, generally, to the influence of a number of philosophers, including Plato and Aristotle, and in rejecting such conceptions, they take themselves to be rejecting essentialism.

I'm offering an alternative approach to the problem of essentialism. I'll argue that species (and, probably some higher taxa) do have defining, real essences, but that those essences are quite different from the ones anticipated in the tradition that Mayr, Hull, and others criticize.

In attributing the current plausibility of the conception of natural kinds (and thus of real essences) that I criticize to the influence of recent positivism, I do not mean to dispute the claim that earlier philosophers, including ancient ones, contributed to establishing the plausibility of the sort of essentialism influential in pre-Darwinian biology. What I claim here is that what plausibility the conception of natural kinds and real essences I criticize currently enjoys among philosophers of science and philosophically sophisticated biologists derives from the legacy of recent positivist philosophy of science rather than, for example, from any lingering Platonistic or Aristotelian tendencies.[2]

## NATURAL KINDS AND ACCOMMODATION

### Accommodation and Reliable Induction

It is a truism that the philosophical theory of *natural* kinds is about how classificatory schemes come to contribute to the epistemic reliability of inductive and explanatory practices. Quine was right in "Natural Kinds"

(1969) that the theory of natural kinds is about how schemes of classification contribute to the formulation and identification of projectible hypotheses (in the sense of Goodman 1973). The naturalness of natural kinds consists in their aptness for induction and explanation; that's why (on one scientifically central notion of definition) definitions of natural kinds are reflections of the properties of their members that contribute to that aptness.

The thesis I defend here (the *accommodation thesis*) makes the further claim that what is at issue in establishing the reliability of inductive and explanatory practices, and what the representation of phenomena in terms of natural kinds makes possible, is the accommodation of inferential practices to relevant causal structures.

Here is the basic idea. Consider a simplified case in which reliable inductive practices depend on our having a suitable vocabulary of natural kind terms. Suppose that you have been conducting experiments in which you exposed various salts of sodium to flames. In each of many cases, the flame turned yellow. You conclude that always (or almost always) if a salt of sodium is heated in a flame, then a yellow flame results. You are right, and your inference is scientifically respectable.

Your inductive success in this matter is a reflection of the fact that the categories *salt of sodium*, *flame*, and *yellow* are natural categories in chemistry, and of the fact that the hypothesis you formulated with the aid of reference to these categories is a projectable one.

Now, anyone who has read Goodman (1973) can come up with indefinitely many unprojectable generalizations about such matters that fit all past data equally well, but that are profoundly false. You were able to discern the true one because your inductive practices allowed you to identify a generalization appropriately related to the causal structures of the phenomena in question. In this particular case, what distinguished the generalization you accepted from the unprojectable generalizations (which also fit the extant data) was that for any instantiation of it that makes the antecedent true, the state of affairs described by the antecedent will (in the relevant environment) cause the effect described by the consequent. Your deployment of projectable categories and generalizations allowed you to identify a *causally sustained* generalization.

What is true in this simplified example is true in general of our ability in scientific (and everyday) practice to identify true (or approximately true) generalizations: we can identify such generalizations just to the extent that we can identify generalizations that are (and will be) sustained by relevant causal structures. Things may be hairier than they are in our example; perhaps the truth makers for the antecedents of true instantiations are symptomatic effects of causes of the states of affairs described by the consequents. Perhaps the generalizations speak of causal powers and propensities rather than of determinate effects so that it is the causal sustenance of propensities rather than the causation of effects that is relevant. Perhaps the generalizations have a more complex logical form. And so forth.

Still, we are able to identify true generalizations in science and in everyday life because we are able to accommodate our inductive practices to the causal factors that sustain them. In order to do this—to frame such projectable generalizations at all—we require a vocabulary, with terms such as *sodium salt* and *flame*, which is itself accommodated to relevant causal structures. This is the essence of the accommodation thesis regarding theoretical natural kinds.

## Accommodation Demands and Two Notions of Definition

**Terminology** Some terminology will prove useful. It is widely recognized that the naturalness of a natural kind—its suitability for explanation and induction—is *discipline relative*. The states of human organisms that are natural kinds for psychology (that is, kinds reference to which facilitates accommodation of the inferential practices of psychology to relevant causal structures) may not turn out to be natural kinds in the same sense for physiology. In discussing this sort of relativity of accommodation, I prefer to speak of *disciplinary matrices* as the situations of inferential practice with respect to which accommodation is accomplished. It is characteristic of natural kind terms that, although the kinds they refer to are suited to induction and explanation in some contexts and not others, their utility for explanation and induction is rarely, if ever, circumscribed by disciplinary boundaries as these boundaries are ordinarily understood. Psychological states are natural kinds for psychology, but probably also for sociology, anthropology, intellectual history, and other disciplines. Acids form a natural kind for chemistry, but also for geology, mineralogy, metallurgy, and so on. By a *disciplinary matrix* I'll understand a family of inductive and inferential practices united by common conceptual resources, whether or not these correspond to academic or practical disciplines otherwise understood.

By the accommodation demands of a disciplinary matrix, $M$, let us understand the requirement of "fit" or accommodation between $M$'s conceptual and classificatory resources, and the relevant causal structures that would be required in order for the characteristic inductive, explanatory (or practical) aims of $M$ to be achieved. Of course, there may be basically successful disciplinary matrices, not all of whose accommodation demands can be satisfied: for some of the explanatory or inductive aims of such a disciplinary matrix, there might not be the sorts of causal structures that could sustain the sought after generalizations or regularities.

What the accommodation thesis entails is that the subject matter of the theory of natural kinds is how the use of natural kind terms and concepts (and, likewise, natural relation terms or natural magnitude terms, etc.) contributes to the satisfaction of the accommodation demands of disciplinary matrices.

**Definitions** There are two quite different but perfectly good senses of the term *definition* in play when we discuss the definitions of scientific kinds and categories. In one sense of the term, a *definition* of a natural kind is provided by specifying a certain inductive or explanatory role that the use of a natural kind term referring to it plays in satisfying the *accommodation demands* of a disciplinary matrix. Call this sort of definition of a kind a *programmatic* definition. Defining an element by the inductive/explanatory role indicated by its location in the periodic table would be an example of offering a programmatic definition for it.

There is another perfectly legitimate sense of *definition* according to which a definition of a natural kind is provided by an account of the properties shared by its members—in virtue of which, reference to the kind plays the role required by its true programmatic definitions. Call this sort of definition an *explanatory* definition. Defining a chemical element in terms of its atomic number and the associated valence structures is an example of offering an explanatory definition.

To a good first approximation (I'm ignoring here the issues of partial denotation, nonreferring expressions, subtle questions about the individuation of disciplinary matrices, translation of natural kind terms between different languages employed within the same disciplinary matrix, etc.) one can characterize true explanatory definitions in terms of the satisfaction of accommodation demands as follows:

Let $M$ be a disciplinary matrix and let $t_1, \ldots, t_n$ be the natural kind terms deployed within the discourse central to the inductive/explanatory successes of $M$. Then the families $F_1, \ldots, F_n$ of properties provide explanatory definitions of the kinds referred to by $t_1, \ldots, t_n$ just in case:

• *Epistemic access condition.* There is a systematic, causally sustained tendency—established by the causal relations between practices in $M$ and causal structures in the world—for what is predicated of $t_i$ within the practice of $M$ to be approximately true of things that satisfy $F_i$, $i = 1, \ldots, n$.

• *Accommodation condition.* This fact, together with the causal powers of things satisfying $F_1, \ldots, F_n$, causally explains how the use of $t_1, \ldots, t_n$ in $M$ contributes to accommodation of the inferential practices of $M$ to relevant causal structures: that is to the tendency for participants in $M$ to identify causally sustained generalizations and to obtain correct explanations.

To put the matter slightly differently, one can say that the explanatory definition of a natural kind is provided by an account of the family of properties shared by its members which underwrite the inductive/explanatory roles indicated by its true programmatic definitions.

**A (Sort of) Continuum of Definitions** The best-known treatments of programmatic and explanatory definitions in the philosophical literature probably lie in functionalist discussions of the definition of psychological states. The very general and abstract definitions of such states proposed by so-called *analytic* functionalists are efforts at programmatic definitions: they define psychological states in terms of very broadly characterized explanatory roles. By contrast, so-called *psychofunctionalist* accounts represent efforts at explanatory definitions of the same states. (Excellent discussions of these conceptions are to be found in Block [1980].)

There are, however, many ways in which the literature on functionalism raises issues—about the analytic-synthetic distinction and about the properties of mental states in physically impossible organisms, for example—that are irrelevant for our present purposes (for a discussion of some of them see Boyd forthcoming a). For that reason, it is probably better to take as paradigm cases of programmatic definitions the definitions of chemical elements in terms of the inductive/explanatory roles indicated by their positions in the periodic table and to take their definitions in terms of atomic number as paradigm cases of explanatory definitions.

What these examples illustrate—and what is true in general—is that both programmatic and explanatory definitions of a natural kind embody claims about the causal powers of its members. In fact, although there is an important difference between the aims of the two sorts of definitions, there is something like a continuum between the most abstractly formulated programmatic definitions of a natural kind and its explanatory definitions. Thus, for example, a chemical element might be programmatically defined in terms of the causal/explanatory role corresponding to a particular place in the periodic table, but the causal/explanatory role it occupies might equally well be spelled out in term of valence or in terms of the structure of orbitals, and so on, with ever-increasing specification of the details of its causal/explanatory role in chemistry until the characterizations in terms of causal/explanatory role converge to an account of an explanatory definition of the element in question.

Thus, the relationship between proposals for programmatic definitions, on the one hand, and proposals for explanatory definitions, on the other, is quite complex. As the literature on analytic functionalism and psychofunctionalism suggests, even when proposed programmatic and explanatory definitions for a natural kind are quite different, there *need* be no incompatibility between them. Once the "continuum" just discussed is recognized, we can see that the same can be true of two quite different programmatic definitions of the same kind, provided that they are cast at different levels of abstraction. At the same time, because programmatic definitions are a posteriori claims about the relation between the causal potentials of things and the accommodation demands of disciplinary matrices, unobvious conflicts between programmatic and explanatory definitions of the same kind, or between programmatic definitions of a kind involving different levels of abstraction, are possible.

What will prove important for our purposes in considering definitions of individual species is the simple point that programmatic formulations of species definitions in terms of explanatory roles are not, in general, rivals to explanatory definitions in terms of common factors, relations of descent, gene exchange, and so on.

## Accommodation in Inexact, Messy, and Parochial Sciences

**Kinds, Laws, and All That: The Standard Empiricist View** There is a venerable (or at least serious and admirable—depending on how inclined you are to veneration) empiricist tradition of identifying natural kinds as those kinds that *(a)* are defined by *eternal, unchanging, ahistorical, intrinsic, necessary,* and *sufficient* conditions; and that *(b)* play a role in stating laws, where laws are understood as *exceptionless, eternal,* and *ahistorical* generalizations. It is this tradition that underwrites many of the arguments that species are not natural kinds. Thus, we need to see to what extent the conclusions of this tradition can be sustained in the light of the accommodation thesis.

One thing we can point to with some certainty is the origin of the empiricist account: from three (or more) parts Hume and one part physics envy. Physics envy first. The logical empiricists' conception of precision, both of laws and of kind definitions, owes much to an idealized conception of the achievements of fundamental physics, whose laws and kinds seemed to have the properties in question.

Hume is more important here. The logical empiricist project crucially involved rationally reconstructing the notion of causation in terms of the subsumption of event sequences under laws of nature. Such a reconstruction required that the notion of a law itself have a nonmetaphysical (and, in particular, noncausal) interpretation. If by a *law* one understands just a true (or, worse yet, an approximately true) generalization, then the twentieth-century version of the Humean analysis of causation fails because there are (many) too many laws, many of them mere accidental generalizations. What empiricists needed was a syntactic (or, at any rate, a nonmetaphysical) distinction between lawlike and nonlawlike generalizations, and it was pretty clearly recognized that this distinction would have to do epistemic as well as (anti)-metaphysical work—that it would have to mark out the distinction that we would now describe as the distinction between projectable and nonprojectable generalizations.

The proposal that laws be exceptionless—that they be universally applicable (in the sense that their universal quantifiers not be restricted to any particular spatiotemporal domain)—and that they be ahistorical (in the sense that they make no reference to any particular place, time, or thing) was part of the effort to provide such a nonmetaphysical account of lawlikeness, and the characterization of natural kinds in terms of their role in such laws was a consequence of the intimate connection between lawlikeness and projectability.

Later, I address the question of whether a contemporary Humean should adopt the same conception of natural kinds and with it the implication that species cannot be kinds. (The answer will be "no.") For the present, what is important is that we recognize that the empiricist characterization of natural kinds we are considering arose not from an investigation of actual linguistic, conceptual, and inferential practices in science, but solely from an attempt to reconstruct such practices to fit an independently framed empiricist philosophical project.

**Lawlessness** According to the empiricist conception we are considering, natural kinds must figure in laws that must themselves be true and tenseless—universal generalizations that hold everywhere in space-time and that involve no references to spatiotemporal regions or to any particulars. It follows from this conception that there are no laws—and thus no natural kinds—in history, in the social sciences, in most of biology, in most of the geological sciences, in meteorology, and so on.

It should be obvious that no such conclusion about natural kinds is compatible with the account of accommodation offered here. The phenomenon that the theory of natural kinds explains—successful inductive and explanatory inferences, and the accommodation of conceptual resources to the causal structures that underwrite them—occurs no less in inductive/explanatory enterprises that seek (and achieve) more local and approximate knowledge than in fundamental physics, or whatever discipline it is whose laws are supposed to fit the empiricist conception.

The problem of projectability and the associated accommodation demands are no less real in geology, biology, and the social sciences than in (philosophers' idealization of) basic physics. What requires explanation, and what the theory of natural kinds helps to explain, is how we are able to identify *causally sustained regularities* that go beyond actually available data and how we are able to offer accurate causal explanations of particular phenomena and of such causally sustained regularities. These regularities need not be eternal, exceptionless, or spatiotemporally universal in order for our epistemic success with them to require the sort of explanation provided by the theory of natural kinds. Whatever philosophical importance (if any) there may be to the distinction between, on the one hand, causally sustained regularities and the statements that describe them, and, on the other, LAWS (Ta! Ta!), it is not reflected in the proper theory of natural kinds.

**Inexactitude** In disciplines such as geology, biology, and so on, we are largely unable to formulate exact laws. It is important to see that this fact makes the demand for accommodation of conceptual and inferential structures to relevant causal structures if anything more pressing (or, at any rate, more demanding) than it is in the case of disciplines where exact laws are available (assuming that there are any such disciplines). Here's why: the unavailability of exact laws in meteorology, for example, arises from the fact that the number of causally relevant variables with some effect on the phenomena studied is much too large to be canvassed in generalizations of the sort that practitioners (even aided by high-speed computers) can formulate. The conceptual machinery of a discipline with this feature must be adequate to the task of identifying important natural factors or parameters that correspond to causally sustained, but not exceptionless, tendencies in the phenomena being studied. That's what projectability judgments in such disciplines are about.

What this means in practice is that practitioners are faced with data that exhibit lots of discernible patterns—some, but not most, of which are in fact sustained by the sought after natural factors or parameters. Because none of these patterns comes even close to being exceptionless, researchers cannot rely on approximate exceptionlessness as a clue to projectability, as they might well in disciplines capable of discerning exact (or nearly exact) patterns. If anything, then, the task of identifying causally sustained generalizations (and explanations licensed by them) in such disciplines will be more difficult and complex than in more nearly exact disciplines. Thus, achieving accommodation between conceptual machinery and important causal structures in inexact disciplines—the task of identifying natural kinds, categories, and magnitudes—cannot possibly be less important than it is in the exact disciplines. Whatever the philosophically important differences between exact and inexact disciplines might be, they are not a matter of the unimportance of natural kinds in the latter.

**Natural Vagueness and Nonintrinsic Defining Properties** Exactly similar considerations about the task of identifying natural categories in the inexact disciplines, where taking account of all causally relevant factors is impossible, make it clear why the natural kinds in such disciplines need not (indeed cannot) be defined by necessary and sufficient membership conditions. Because, for example, a natural kind in meteorology must be defined by only a proper subset of the causally relevant factors and must participate only approximately in (only approximately) stable weather patterns, there is no prospect whatsoever that there will be absolutely determinate necessary and sufficient conditions which provide the its explanatory definition. (This is not, I should add, analytic; it's just true.) Instead, the explanatory definitions of such kinds will reflect the imperfect clustering of relevant properties that underwrites the contribution that reference to them makes to accommodation—just as the accommodation thesis requires.

It is likewise nonanalytic but true that in the inexact sciences of complex phenomena, the explanatory definitions of natural kinds often involve some relational (as opposed to intrinsic) properties. Social roles, whether in human societies or in the societies of nonhuman social animals, are clearcut examples. It is no objection to the naturalness of such kinds to say, as an ardent reductionist might, that whenever the occupier of a particular social role (alpha male, let us suppose) exhibits on a particular occasion the causal powers and dispositions characteristic of that role, there will always be intrinsic properties of other relevant organisms and of relevant features of the environment that are causally sufficient, together with intrinsic properties of that organism, to establish the causal powers and dispositions in question.

Relationally defined categories, such as social roles, are natural kinds just in case deployment of references to them contributes to the satisfaction of the accommodation demands of the disciplinary matrices in question. Their explanatory definitions include relational properties just in case the shared causal powers and dispositions among their members—upon which that contribution to accommodation depends—are causally sustained by (among other things) shared relational properties. That an imaginary and unpracticable disciplinary matrix might embody the project of, for example, predicting and explaining the behaviors of social animals by deriving them from independently formulated intrinsic physical characterizations of the animals and of their environments is irrelevant to the question of whether (partly) extrinsically defined social kinds are natural kinds in the disciplinary matrices in which we actually work.

**Historicity** It may be somewhat more difficult to see why the definitions of natural kinds need not be ahistorical and unchanging. Consider first the question of whether the explanatory definition of a natural kind can be such that members of the kind are necessarily restricted to some spatial or temporal region, or such that it involves reference to a particular space-time region or individual.

The obvious cases of natural kinds with just these properties are the historical periods recognized by an explanatorily relevant periodization of the history of some phenomena or other. Suppose for the sake of argument that important causal factors in European history are revealed if we distinguish, for any given political and economic region, between a feudal period, on the one hand, and the period of transition to recognizably modern organization of trade, production, and governance. If this is so, then the distinction in question will correspond for each region to two different natural categories of historical events and processes, such that the consequences of a historical event will tend to be significantly determined by its situation with respect to this periodization. Of course, the natural historical periods in question would have "vague" boundaries—they would possess homeostatic property cluster explanatory definitions—but as we have seen, this vagueness would not undermine their status as natural kinds in the sense appropriate to the accommodation thesis.

If an example in which the members of the kinds are historical events seems too atypical to be fully convincing, consider the (homeostatic property cluster) distinction between feudal and capitalist economic systems. It is almost certainly true that recognizing this distinction contributes fundamentally to accommodation in the disciplinary matrix that includes economic and social history.

Now, according to some economic theories (Marxist ones, for example), this distinction corresponds to quite general (inexact) "laws" of economic development such that in any suitably situated human society there would be a tendency for the means and organization of production to go through a feudal stage followed by a capitalist one. An alternative view is that the explanatory utility of the distinction rests instead on a very large number of factors peculiar to European economic history so that, although it is explanatorily important to study the transition from feudalism to capitalism in various different European countries or regions, it is important only because of factors peculiar to Europe.

What's at stake in the difference between these two conceptions is methodologically important. It is commonplace to describe China's economic organization as having been feudal until the present century. If the first conception is correct, this claim, if true, should be expected to indicate explanatorily important similarities between, say, early nineteenth-century China and fourteenth-century England. If, on the other hand, the second conception is correct, the economy of China was "feudal" only in an extended metaphorical sense of the term, and expecting to find explanatorily important similarities of the sort indicated would be a mistake.

Suppose now, for the sake of argument, that the second conception of the distinction is correct. Then deployment of the categories *feudal economy* and *capitalist economy* and of the categories employed to characterize the transition between feudal and capitalist economies will contribute to the satisfaction of the accommodation demands of economic and political history only to the extent that it is recognized that the phenomena they describe are peculiar to a particular temporal segment of European history. If this is so, then the deployment of the categories in question contributes significantly to the accommodation of the explanatory practice of economic and political historians, albeit only when they are examining economic and social developments in Europe between, for example, the tenth and twenty-first centuries.

On the assumption we are entertaining, the category *feudal economy* and the other categories in question are thus natural kinds in the sense established by the accommodation thesis. They are less widely applicable than one might have hoped—which, however, merely illustrates the claims that both programmatic and essential definitions of natural categories are a posteriori and revisable. It does not undermine the claim that these categories are natural: they do represent real achievements in the accommodation of explanatory practices in European history to relevant causal factors, and that itself is no mean feat.

My own guess is that the first of the two conceptions of the notion of a feudal economy is more likely and that this notion may well be fruitfully applicable outside the European context. Another reader might hold that the distinctions we have been discussing fail to contribute to accommodation even within the European context. What would be extraordinary, however, would be for there to be no natural kinds that exhibit historicality of the sort we are discussing.

I conclude, therefore, that we have no reason to deny that there can be genuine natural kinds that are historically delimited in the way we have been considering. Of course, if biological species are natural kinds, then almost certainly *they* are such kinds, but that is a question to which we come later.

**Noneternal Definitions** Consider now the question of whether or not the explanatory essence of a natural kind must always involve the same properties—must be in that sense *eternal* or unchanging. The obvious examples of natural kinds with noneternal definitions, if they are admitted as cogent, are those biological species whose integrity depends on gene exchange between constituent populations and reproductive isolation from closely related contraspecific populations. At any given time in the history of such a species, whatever properties operate to ensure such isolation will be constituents of its explanatory definition. With the extinction of some relevant contraspecific populations and the emergence of others, the properties that are thus parts of the species' explanatory definition can change over time.

Of course, all the elaborate machinery utilized thus far in this section is directed toward persuading the skeptical reader that biological species are natural kinds. For the reader who has not already anticipated—and been convinced by—the arguments to come, there are other examples that illustrate, albeit not so uncontroversially, the same point. Consider, for example, philosophical or scientific or religious conceptions—such as Christianity, Islam, empiricism, rationalism, behaviorism, or vitalism—considered as natural kinds in intellectual history. Such doctrines typically are motivated, molded, and sustained by a number of different factors, "internal" to the relevant discipline or practices as well as "external." Readers are now invited to consider for themselves the view (which I now advocate) that the effect of this diversity of factors is that, at any given time, such a doctrine will be characterized by a homeostatic cluster of particular doctrines, methods, explanatory and argumentative strategies, and so on.

It seems evident that the intellectual historian will treat these homeostatically defined conceptions as persisting social phenomena whose historical development forms a central part of the subject matter of her discipline. Accommodation to the complex causal factors that underwrite and change the homeostatic unity of the conceptions she studies will require that she individuate such conceptions in such a way that the doctrines, methods, and so on that constitute their definitions will change over time. This is, I suggest, exactly what historians in fact do and what they should do. So, conceptions of this sort are natural homeostatic property cluster phenomena with (in the relevant sense) noneternal definitions.

Similar considerations suggest that other categories defined in terms of causally important but evolving historical phenomena will have noneternal homeostatic property cluster definitions, at least with respect to those disciplinary matrices concerned with historical developments as well as with static situations. Social structures such as feudalism or capitalism, or monarchy and parliamentary democracy, are probable examples. I conclude that the best available conception of natural kinds implies that noneternal definitions are a perfectly ordinary phenomenon in disciplinary matrices concerned with the history of complex phenomena.

## Homeostasis, Compositional Semantics, and Disciplinary Matrices

The accommodation thesis has one more consequence of that we need to examine before we turn to issues about biological species. Disciplinary matrices are themselves HPC phenomena. What establishes the coherence of an intellectual discipline is a certain commonality of methods, explanatory strategies, relevant findings, and the like. We may see how this sort of commonality results in disciplinary coherence by recognizing that, within any disciplinary matrix, very, very many accommodation demands arise from the enormous range of quite particular phenomena for which explanations and/or predictions are sought. What we recognize as an intellectual discipline is the phenomenon manifested when a cohesive set of laws, generalizations, conceptual resources, technical and inductive methods, and explanatory strategies contributes to the satisfaction of a very wide spectrum of accommodation demands.

The conditions of satisfaction of these accommodation demands are thus themselves homeostatically related: the satisfaction of various demands tends systematically to contribute to the satisfaction of many other demands. In typical disciplines, this homeostasis is in large measure a matter of widely applicable causal knowledge: the commonalities among or systematicity in the significant causal interactions between the factors that produce the phenomena under study are such that the knowledge of such factors necessary to solve one disciplinary problem will conduce to the solution of a great many other problems.

This homeostatic tendency is reflected in the very phenomenon of natural kinds. What we recognize as a natural kind is a multipurpose category, reference to which facilitates the satisfaction of a great many accommodation demands within a disciplinary matrix. Here, then, is a particular aspect of the homeostasis just mentioned: typically, the kind distinctions central to meeting one of the accommodation demands of a disciplinary matrix will facilitate the satisfaction of many of its other accommodation demands.

What is important for our purpose is the way in which this particular aspect of disciplinary homeostasis is related to the compositional semantics of natural kind terms. We are used to the idea that natural kinds are the kinds that are the subjects of natural laws—not perhaps eternal, ahistorical, exceptionless laws, but at least explanatorily significant causal generalizations of some sort. It is important to note that even this concession to the positivist tradition overstates the connection between natural kinds and laws. The naturalness of many natural kinds is indicated not by their being the subjects of natural laws, but by the fact that reference to them is crucial for the formulation of laws with more specific subject matters. Goodman's (1973) contrast between *green* and *grue* illustrates this point. There are no interesting laws about green things generally, but references to colors like green are important in formulating explanatorily important psychological generalizations.

More scientifically important examples of the same phenomenon are provided by, for example, the categories *acid, element, ion,* and *compound* in chemistry. Few explanatorily important generalizations apply to all of the members of any of these categories, but reference to them is central to the formulation of important laws. The contribution that recognition of these categories makes to the satisfaction of accommodation demands in chemistry depends on the compositional roles of the terms *acid, element, ion,* and *compound* in specifying the subject matters of important generalizations.

Even when a natural kind exhibits its naturalness by being the subject matter of explanatorily important causal generalizations, the homeostatic contribution that its recognition makes to the satisfaction of accommodation demands in the relevant disciplinary matrix will typically depend to a great extent on the compositional role of natural kind terms referring to it. The paradigmatic natural kinds (species excepted)—chemical elements—provide a spectacular illustration of this point. There are, to be sure, laws regarding each of the elements. Nevertheless, the overwhelming majority of chemical natural kinds are compounds rather than elements, so the overwhelming majority of chemical laws do not have elements as their subject matter. Thus, the main contribution that the use of terms referring to elements makes to the satisfaction of accommodation demands in chemistry arises from the use of such terms in formulas for chemical compounds.

## The "Reality" of Natural Kinds

Two related points follow that are important for the later discussion of the metaphysics and epistemology of the species category. In the first place, the naturalness of a natural kind is not a matter of its being somehow *fundamental,* with less fundamental kinds being somehow less natural than more fundamental ones. Thus, for example, with the discovery of the phenomenon of chemical isotopes, there was no methodologically or philosophically significant problem about the true or real "*elemental* level" in chemistry, with conflicting positions regarding the question of whether the *true* or *more fundamental* elemental level consisted of categories defined just by atomic number or of categories defined by atomic number and atomic weight. The decision to adopt the practice of using the term *element* for categories of the first sort was a matter of convenience, not a matter of fundamental metaphysics or fundamental chemistry. What was important—and not just a matter of convenience or convention—was that either choice would result in the establishment of a vocabulary for chemistry in which the same class of causally and explanatorily relevant distinctions could be drawn. The naturalness of a natural kind is a matter of the contribution that reference to it makes to the satisfaction of the accommodation demands of a disciplinary matrix, in the context of a system of a compositional linguistic resources for the representation of phenomena.

This fact, in turn, constrains how we should interpret questions of "realism about" particular (allegedly) natural kinds or questions about which kinds exist or are "real." What the accommodation thesis indicates is that the metaphysical achievement that the deployment of kind terms and concepts may or may not represent is the accommodation of inferential practices to relevant causal structures, so the "reality" of a kind consists in the contribution that reference to it makes to such accommodation. What we have just seen is that—strictly speaking—questions of "realism" or "reality" are, in the first instance, questions about a family of classificatory practices incorporated into the inferential practices of a disciplinary matrix, rather than questions about particular kinds or even about families of kinds abstracted from the context of disciplinary practices.

When we ask about the "reality" of a kind or of the members of a family of kinds—or when we address the question of "realism about" them—what we are addressing is the question of what contribution, if any, reference to the kind or kinds in question makes to the ways in which the classificatory and inferential practices in which they are implicated contribute to the satisfaction of the accommodation demands of the relevant disciplinary matrix. Claims to the effect that some kind or kinds are not "real," or (equivalently) "antirealist" claims about kinds, are best understood as claims to the effect that reference to the kind or kinds in question fails to play an appropriate role in such accommodation, where the role in question is often tacitly indicated by the context in which such "antirealist" claims are made.

It is thus always preferable for such claims to be spelled out explicitly in terms of the relevant sort of contribution to accommodation, rather than by misleading reference to issues regarding the "reality of" or of "realism about" the kind(s) in question. It's important to note in this regard that what is misleading about these less precise formulations is not that they suggest that what is at issue are metaphysical questions about the kinds in question: questions about the accommodation of representational and inferential practices to real causal structures in the world are at issue, and these questions are paradigmatically metaphysical. Instead, what is misleading about formulations in terms of the "reality" or "unreality" of kinds, or of the "realism" or "antirealism" about them, is that they wrongly suggest that the issue is one regarding the metaphysical status of the families consisting of the members of the kinds in question—considered by themselves—rather than one regarding the contributions that reference to them may make to accommodation. Issues about "reality" or "realism about" are always issues about accommodation (see Boyd 1990).

## Disciplinary Relativism and Promiscuous Realism

It follows from the account developed in the preceding section that the naturalness of a natural kind will ordinarily be a matter of the role that reference to it plays in some particular family of inductive or explanatory practices. A kind may be natural "from the point of view of" some discipline or disciplinary matrix, but not "from the point of view of" another. Perhaps *jade* is a natural kind in gemology or the history of art, but not in geology (because some jade is jadite, and some is nephrite, and these two minerals are chemically quite different). This relativity to a discipline or disciplinary matrix does not compromise the naturalness or the "reality" of a natural kind. Natural kinds simply are kinds defined by the ways of satisfying the accommodations demands of particular disciplinary matrices.

Dupré (1993) makes a similar point about the relativity of the naturalness of kinds to particular projects. He argues for a "promiscuous realism" about natural kinds according to which, among other things:

> There is no God-given, unique way to classify the innumerable and diverse products of the evolutionary process. There are many plausible and defensible ways of doing so, and the best way of doing so will depend on both the purposes of the classification and the peculiarities of the organisms in question, whether those purposes belong to what is traditionally considered part of science or part of ordinary life. (p. 57)

The accommodation thesis—according to which the naturalness and the "reality" of a natural kind consist in the contribution that reference to it makes to the satisfaction of the accommodation demands of a *particular* disciplinary matrix—supports and provides a metaphysical rationale for this aspect of Dupré's conception (but probably not to his other critiques of unificatonist conceptions of science—critiques I confess to not fully understanding). Different disciplinary matrices and different accommodation demands within a disciplinary matrix will—given the complexity of the biological world—require reference to different and cross-classifying kinds in order to achieve accommodation, and this fact in no way demeans the naturalness or the "reality" of those kinds.

One of the criticisms of Dupré's conception offered by Wilson (1996) is that the classificatory categories of ordinary life and language are not natural kinds at all; he denies that common sense and common language are "in the business of individuating natural kinds at all" (p. 307). According to Wilson, ordinary language lacks the systematic purpose of uncovering order in nature, which governs scientific practice and language, and which makes it necessary for scientific terms (as opposed to ordinary language ones) to refer to natural kinds defined by real essences. Dupré (1993) himself indicates that the plurality of natural kind classifications in ordinary language is unsurprising because common sense aims to gather information about the world, rather than primarily to achieve a unified picture of it. Wilson agrees, but identifies the latter aim with the sciences and sees reference to natural kinds, defined by real essences, as appropriate only to the latter task.

The position I advocate allows one to "split the difference" between these two conceptions of everyday kinds. Although my choosing the term *disciplinary matrix* undoubtedly betrays my special concern with the issue of kinds in the theoretical sciences, everyday life provides disciplines or at any rate regimes of inferential and practical activity in which the accommodation of practices to causal structures is central. Consider the category *lily*, made famous (among a select few) by Dupré 1981. As it is employed in everyday life—in gardening, flower arranging, landscaping, decorating houses, and so on—the category *lily* does not, according to Dupré, contain such members of the biological family Liliaceae as onions and garlic and various tulips. Nor is there any biological taxon below Liliaceae whose members are just the lilies. So the term *lily* represents an ordinary life natural kind distinct from the kinds of scientific botany. Wilson agrees that onions and garlic are not lilies, but denies that the ordinary language category *lily* is a natural kind.

I suggest that the plants we ordinarily call lilies (excluding onions and garlic, etc.) do form a natural kind in the sense required by the accommodation thesis. Lilies share a family of causal properties and capacities (as it happens, a homeostatic cluster of such properties), and this fact is what explains why reference to lilies helps to satisfy the accommodation demands of the disciplinary matrix that involves gardening, landscaping, decorating, and the like. Lilies share aesthetically relevant features of structure and coloration, and they fall into a manageably small set of categories that characterize their horticulture-wise relevant growing conditions and blooming periods. Horticulturists' and gardners' particular deployment of the category *lily* contributes to their ability to achieve the botanical and aesthetic results they aim at precisely because categorization of flowering plants in terms of these shared properties achieves accommodation to relevant causal factors.

This example illustrates an important fact: even the affairs of everyday life require accommodation between conceptual/classificatory resources and causal structures, so everyday kinds are usually natural kinds in the sense defined by the accommodation thesis. Gruified gardening would be as unsuccessful as gruified mineralogy.

On the other hand, the accommodation demands of everyday practical disciplines may well often be quite different from the demands of theoretical disciplinary matrices. In particular, they may often involve far less deep or fundamental (although not necessarily less subtle) inductive and explanatory achievements. It is this fact that underwrites Wilson's insight that the kinds of everyday life are much less deeply implicated in projects of theoretical unification than scientific kinds.

Millikan (forthcoming) draws a distinction between natural kinds in general and those particular natural kinds that play a role in systematic and integrated scientific theorizing. I prefer this way of putting the distinction to Wilson's. In the first place, Millikan's approach helps to preserve the insight that everyday kinds are vehicles for satisfying accommodation demands, just as scientific natural kinds are. Secondly, I suspect that there is something like a continuum in degree of theoretical or integrative commitment between everyday accommodation-serving kinds and scientific natural kinds, and that this fact is reflected in our everyday linguistic practices.

I have in mind, of course, the cases in which reference to what are plainly scientific kinds (of diseases and medicines; of semiconductors and other electronic parts; of reagents for photographic development, etc.) plays a role in everyday practical or recreational endeavors. But I also have in mind a general feature of ordinary linguistic usage that seems to point toward a general recognition of the everyday relevance of theory-driven standards of classification.

Dupré (1981) launched the case for (what became known as) *promiscuous realism* by insisting that in the ordinary everyday sense of the term *lily*, onions (among other plants) aren't lilies. Although it is true that we don't ordinarily count onions as lilies because they aren't decorative, our judgments (even our ordinary ones) about whether onions are lilies are remarkably sensitive to the ways the question is put. Someone who says, "Onions are lilies," may seem to have spoken falsely or misleadingly, but someone who says, "Onions are a kind of lily," says something that many would intuitively accept. There are lots of similar cases ("Birds are a kind of dinosaur"; "The glass snake is a kind of lizard"; "Tomatoes are a kind of fruit"; "Mushrooms are not really a kind of plant") in which the expression "kind of" signals reference to (or, if you prefer, deference to) scientific and theoretical standards. The fact that ordinary language has such a semantic device for marking out and thus making available reference to scientific standards provides, I believe, further reason for recognizing that ordinary kinds and scientific natural kinds lie along a continuum. They do so precisely because they are all kinds of natural kinds—that is, resources for achieving accommodation.

## Natural Individuals

When we presently turn our attention to the famous (or infamous) question of whether biological species are kinds or individuals (see also de Queiroz, chapter 3 in this volume), we need to recognize that it is a consequence of the accommodation thesis that the question may not have as deep a metaphysical import as the literature would suggest. Once we begin to think of natural kinds as features of human inferential architectures—as artifacts rather than as Platonistic entities—as the accommodation thesis requires, the distinction between natural kinds and natural individuals becomes less important.

A number of philosophers have suggested something like this conclusion in discussing the species-as-individuals issue. Dupré (1993) concludes that the real question about whether species are individuals or kinds "is whether the same set of individuals can provide both the extension of a kind and the constituent parts of a larger individual. And the answer to this is clearly yes" (p. 58).

Ereshefsky (1991) understands the "traditional" notion of a natural kind approximately along the lines indicated in the earlier section entitled "strategy"; he therefore concludes that species are not kinds, but "historical entities." Still, he does maintain that some of them are individuals as well, whereas others are not, so he does not take the category *individual* to be incompatible with the much more kindlike category *historical entity*, which includes the higher taxa.

Finally, Wilson (1996) seems to hold that Dupré's conception, if developed in Dupré's promiscuous or pluralist style, would commit one to "the absurdity of saying that one and the same thing is a natural kind and an individual" (p. 310). But even he then goes on to say that the choice between the two conceptions of species is "merely pragmatic," suggesting, I believe, that neither has an advantage in satisfying the accommodation demands of biology. What I propose is that by seeing the similarities between the inductive and explanatory roles played by reference to natural kinds, on the one hand, and by reference to individuals, on the other, we can see why the distinction between natural kinds and (natural) individuals is, in an important way, merely pragmatic.

After all, successful induction and explanation depend just as much on the accommodation of our individuative practices for individuals to relevant causal structures as on the accommodation of those practices for kinds. A failure to be able to recognize the various stages in the maturation of an organism *as stages of the same organism* would undermine induction and explanation in biology just as much as a failure to deploy accommodated schemes of classification for the organisms themselves. The fact that it is, for certain familiar cases, easier to get this sort of thing right should not prevent our recognition that the classification of temporal stages as temporal stages of the same individual must meet just the same constraints of accommodation as the classification of individuals into natural kinds. Nor should this fact lead us to miss the point that sometimes accommodation of inferential practices for individuals is a real scientific achievement, as in the case of organisms whose larval and adult stages are so dissimilar as to appear contraspecific. If the truth be known, the spatial or temporal stages of a natural individual form something like a natural kind.

It may seem odd to think of the stages of some ordinary object—that rock over there, for example—as forming a natural kind; after all, particular rocks aren't typically explanatorily important enough to make the honorific title *natural kind* seem appropriate. This is less clearly so for some bigger rocks—the rock of Gibraltar, for example—or for other sorts of individuals—Oliver Cromwell, let's say. In these cases and many others, the accommodation that underwrites cogent explanations (I assume that historical explanations count as causal and require accommodation) depend on our capacities to individuate explanatorily important individual entities. Of course, if biological species are individuals, then they are individuals with the explanatory importance characteristic of natural kinds.

Even with respect to the cases of inconsequential (but still natural) individuals, our capacities to individuate are central to successful accommodation of inferential practices to causal structures. Thus, for example, experimental trials on ordinary (and individually explanatorily unimportant) mice, trees, mineral specimens, DNA samples, fossils, rivers, and so on depend for their inductive cogency on experimenters' abilities to properly individuate these things. Experimental studies on gruified mineral samples would represent failures of accommodation in just the same way and to just the same extent that such studies of (properly individuated) grue samples would.

Just think about a Quinean hydrologist studying river-kindred water stages. The distinction between natural kinds and natural individuals is almost just one of syntax. In particular, the metaphysics of accommodation is the same for natural kinds and for natural individuals.

### A Humean Note

I have just argued against a conception of natural kinds according to which they must be defined by unchanging necessary and sufficient membership conditions and must figure in eternal, ahistorical, exceptionless laws. I suggested that the current plausibility of this conception arises not from any important features of actual scientific practice, but from the demands (together with a bit of physics envy) of the logical empiricists' project of providing Humean rational reconstructions of causal notions.

Now, I have argued elsewhere (Boyd 1985b) that such Humean reconstructions must always fail. (Here's the argument in brief. Scientific realism is true, so we have [unreconstructed] knowledge of factors such as the charge of electrons. But charge just is a causal power, so knowledge of unreconstructed causal powers is actual.) What is important for our purposes is that a rejection of the Humean project of rational reconstruction is not necessary in order to accept the conclusions of the preceding sections of this essay.

Perhaps there is some metaphysically innocent notion of "law" or of "lawlikeness" in terms of which an antimetaphysical reconstruction of causal notions can be provided. Whether this notion exists or not, scientific (and historical and everyday) knowledge often depends on our being able to identify causally sustained generalizations that are neither eternal nor ahistorical nor exceptionless, and our ability to do so depends on our coordination of language and classificatory categories with causal phenomena involving and defined by imperfect property homeostasis. Any adequate Humean rational reconstruction, whether of science or of other areas of empirical knowledge, will need to be compatible with the recognition of these facts and will thus be compatible with (a suitably reconstructed version of) the homeostatic property cluster conception of natural kinds advanced here.

## SPECIES AS HOMEOSTATIC PROPERTY CLUSTER NATURAL KINDS

### Species as Homeostatic Phenomena

**Species-Level Homeostasis** It is, I take it, uncontroversial that biological species, whether or not they are natural kinds, are phenomena that exhibit something like the sort of property homeostasis that defines homeostatic property cluster natural kinds. A variety of homeostatic mechanisms—gene exchange between certain populations and reproductive isolation from others, effects of common selective factors, coadapted gene complexes and other limitations on heritable variation, developmental constraints, the effects of the organism-caused features of evolutionary niches, and so on—act to establish the patterns of evolutionary stasis that we recognize as manifestations of biological species. Indeed, the dispute between defenders of Mayr's biological species concept and theorists who hold that the species category properly includes asexually reproducing organisms is just a dispute over the relative power of these sorts of homeostatic mechanisms in sustaining the sort of homeostatic integrity characteristic of biological species.

**Quibbles and Refinements** The account of HPC natural kinds that I offered in earlier papers and rehearsed in the section "Homeostatic Property Cluster Kinds" requires some fine-tuning in order to capture species-level homeostasis, whether or not biological species are natural kinds. Here, I briefly indicate what is required. In the first place, the earlier account emphasizes the homeostatic unity of properties shared (imperfectly, of course) by all or almost all of the members of the relevant kind. The fact that there is substantial sexual dimorphism in many species and the fact that there are often profound differences between the phenotypic properties of members of the same species at different stages of their life histories (for example, in insect species), together require that we characterize the homeostatic property cluster associated with a biological species as containing lots of conditionally specified dispositional properties for which canonical descriptions might be something like, "if male and in the first molt, *P*," or "if female and in the aquatic stage, *Q*."

Once this requirement is recognized, and once the more general phenomenon of polytypic species is recognized, it becomes clear that an even more precise formulation of the homeostatic property cluster conception of species would, in the first instance, treat populations as their members and would describe species-level homeostasis as connecting causal factors that influence the statistical distribution of phenotypes among their members. No doubt, additional refinements would be in order, but like those just mentioned, they would elaborate rather than undermine the conception of biological species as homeostatic property cluster phenomena.

## Species and Accommodation

Species are homeostatic property cluster phenomena. Are they homeostatic property cluster natural kinds? The obvious questions to ask next are whether or not reference to species is crucial to the satisfaction of the accommodation demands of the relevant disciplinary matrix, and how closely the contributions that reference to them makes to accommodation resemble the contributions achieved by reference to uncontroversial examples of natural kinds.

I take it that it is uncontroversial that our ability to identify biological species and their members with some high level of reliability is central to our ability to obtain correct explanations and predictions in the biological sciences. In that regard, species are like natural kinds and like the natural individuals discussed earlier in that reference to them is central to the satisfaction of accommodation demands. Thus, the argument rehearsed earlier shows that biological species—whether kinds or individuals or whatever—are very much like natural kinds with respect to issue of the metaphysics of accommodation.

In fact, the resemblance is much greater. One way in which the family of stages that constitute some natural individual might be thought to differ from a paradigm natural kind lies in the way the commonality in properties between the various stages of the individual contributes to accommodation. In the case of paradigm natural kinds, the fact that its instances (tend to) share many explanatorily relevant properties in common is central to the contribution that reference to the kind makes to accommodation.

In the case of some natural individuals, this sort of commonality of properties is much less important to accommodation; instead, the nature and dynamics of the continuity between their temporal stages are overwhelmingly important. This is perhaps true, for example, of (individual) tropical storms and of individual forests, considered as objects of study in historical ecology. The explanatorily relevant respects of continuity between stages of such individuals enforce some similarities between nearby stages, but the continuity of historical development is probably more explanatorily central than these similarities.

Because biological species are historical entities, one might conjecture that the same sort of thing happens with them. They exhibit homeostatic unity of phenotypic properties over time, but the properties shared by individuals (better yet, populations, on the more sophisticated formulations just discussed) within a species might not be especially explanatorily significant. If this were so, then biological species would be like tropical storms rather than like paradigm natural kinds in that the historical continuities between their temporal realizations, rather than their shared properties, would be centrally important in their contributions to accommodation. The plausibility of this conjecture might be enhanced if one followed Mayr (1961) in distinguishing "functional biology" from "evolutionary biology" and offered the conjecture as relevant to the evolutionary (and thus historical) notion of species (I do not mean to imply that Mayr would approve of this application of his distinction).

If this conjecture could be maintained, then the objection that biological species differ from natural kinds in that what unites their members is their historical relationships to one another rather than their shared properties would be sustained for the case of species as objects of evolutionary theorizing.

Of course, this objection cannot be sustained. All of the standard sorts of evolutionary explanations, either for speciation or for the phenotypic properties species exhibit, tacitly (if not explicitly) presuppose that members of each of the various species in question exhibit a very wide range of shared phenotypic characters of the sort sustained by mechanisms of property homeostasis, and they ordinarily presuppose the action of many of these homeostatic mechanisms. Readers are invited to examine evolutionary explanations in terms of individual selection, kin selection, genetic drift, or founder effects, for example, to determine whether or not they fundamentally presuppose approximately static background property commonalities among the members of the relevant species, even while explaining changes in other particular properties.

**Species as Homeostatic Property Cluster Natural Kinds**

Species are at least very much like natural kinds: they reflect solutions to the accommodation demands of biology. Moreover, the ways in which reference to them contributes to satisfying these demands makes them resemble paradigmatic natural kinds as opposed to the least kindlike natural individuals (which are themselves very much like natural kinds).

I propose that biological species simply are HPC natural kinds. What is interesting is that the best arguments in favor of the alternative view—that they are individuals rather than kinds—actually support the thesis I am proposing. When the residual positivist conception of kinds is stripped away, what the best arguments that species are individuals rather than kinds come down to, at least to a good first approximation, is that organisms in the same biological species must *(a)* be members of some initial population of that species or descendants of its members (so that a species cannot become temporarily extinct and then reevolve) and *(b)* if contemporaneous, be members either of the same population or of populations that are relevantly reproductively integrated (so that the constituents of species have important internal relations with each other, as constituents of paradigm individuals do).

The more cogent reasons for insisting that species must have the two characteristics just mentioned do not depend on outdated philosophy of science, but on biology. When a family of populations of organisms satisfies *(a)* and *(b)*, the fact of their common descent and reproductive integration is a source of a tendency toward evolutionary unity. The biologically serious arguments for *(a)* and *(b)* rest on the scientific claim that without the operation of the factors they require, a family of populations will not possess the evolutionary unity characteristic of species-level taxa. (Considerations of this sort are explicit in, for example, Hull [1978] and in Ghiselin [1974].)

Let's suppose, for the sake of argument, that the considerations in favor of *(a)* and *(b)* are correct. Then common descent and reproductive integration of the sort they require are essential to establish the *homeostatic* evolutionary unity of biological species: the unity anticipated by inferences and explanations in evolutionary biology and thus required for accommodation. But, as we have seen, the unity anticipated by such inferences and explanations is the unity appropriate to HPC kinds. Both species-as-individuals theorists and their opponents are tacitly treating biological species as HPC natural kinds. That's what they are.

## Programmatic Definitions of Individual Species

It is important to reply to one possible rebuttal to the homeostatic property conception of species just defended. Someone who was persuaded that species are natural kinds and that the homeostatically unified properties their members (imperfectly) share are crucial to the satisfaction of accommodation demands in biology might still hold that, strictly speaking, a biological species is not defined by the associated homeostatic property cluster. She might reason as follows: "My favorite candidate for a programmatic definition of the species level in taxonomy is *P*. For any given species, *S*, the proper definition of *S* is provided by the formula 'the *P* that is instantiated in *T*,' rather than by the associated homeostatic property cluster (where *P* is some functional characterization of the species level in taxonomy, like Mayr's biological species concept, and *T* denotes the type specimen(s) of *S* or some other suitable representative[s])."

Such a proposal might seem attractive. After all, most extant proposed programmatic definitions of the species level are not more than a couple of paragraphs long, whereas it may be impossible to survey all the members of a species-level homeostatic property cluster, so only if something like the proposal in question were right would we ever be able to state the definition of any biological species.

What the proposal fails to take into account, however, is the distinction between programmatic and explanatory definitions. If we have an adequate programmatic definition of the species level (good luck!), then we can indeed offer programmatic definitions of individual species in the way indicated. But such programmatic definitions would not be competitors with the *explanatory* definitions provided by the relevant homeostatic property clusters (see the section "Continuum of Definitions"). This conclusion is easy to see by reflecting on the fact that the programmatic definition, "stuff that ..." (where the ellipses specifies the role of gold in the periodic table of the elements), is not a competitor for the definition of gold as the element with atomic number 79.

## Biological Species are Paradigmatic Natural Kinds (After All)

A number of philosophers have argued that the taxonomic claims put forward by species-as-individuals theorists are better and more naturally put by the claim that biological species are historically delimited natural kinds (see, e.g., Kitcher 1984). I agree, of course, but the arguments presented here do more than indicate why this is a better or more natural way of formulating taxonomic claims.

In the first place, I have offered a general theory of the nature of natural kinds (the accommodation thesis) that affords a rebuttal to the more philosophical (and positivist) arguments against the thesis that species are natural kinds. It does more than that however. The category *natural kind* is itself a natural kind in metaphysics and epistemology, and the accommodation thesis is a thesis about its essential or explanatory definition. It follows from this definition that biological species are natural kinds and not marginal examples either. Their homeostatic property cluster structure is perfectly ordinary for natural kinds; they are deeply important to the satisfaction of the accommodation demands of a very, very successful disciplinary matrix; and their departures from the positivists' conception of natural kinds are all essential to the accommodation that reference to them helps to achieve.

In fact, just as philosophers have usually thought, biological species are paradigmatic natural kinds. The natural kinds that have unchanging definitions in terms of intrinsic necessary and sufficient conditions and that are the subjects of eternal, ahistorical, exceptionless laws are an unrepresentative minority of natural kinds (perhaps even a minority of zero). Every sort of practical or theoretical endeavor that engages with the world makes accommodation demands on the conceptual and classificatory resources it deploys. Recognition of the sorts of kinds beloved by positivists can meet the demands for very few (perhaps none) of these endeavors. Instead, the sort of kinds (many of them homeostatic property cluster kinds) required for the inexact, messy, and parochial sciences are the norm. Of these kinds, biological species are entirely typical, indeed paradigmatic, examples.

## SPECIES AMONG THE TAXA

### Pluralistic Realism

**Realism** A number of authors (Dupré 1981, and chapter 1 in this volume; Mishler and Brandon 1987; Mishler and Donoghue 1982; Kitcher 1984; Ereshefsky 1992) advocate the "pluralist" view that there are different but equally legitimate strategies for sorting organisms into species. The pluralisms they advocate all seem to agree that for different groups of organisms, different standards for defining conspecificity are appropriate to the explanatory demands of evolutionary biology so that, for example, interbreeding between populations might define conspecificity in the case of one species, but not in the case of another.

For Dupré, Kitcher, and Ereshefsky (but apparently not for Mishler and Brandon or for Mishler and Donoghue) there is another dimension to the pluralism they advocate. Depending on what explanatory project is to be served, the groups of organisms assigned to the species-level taxa may be different so that, for example, a family of populations might constitute a species for the purposes of one explanatory project, but be classified into different species within the same genus for the purposes of another project (Ereshefsky proposes eliminating the "superfluous" term *species* in favor of terms such as *biospecies* and *ecospecies*, which reflect the different types of lineages reference to which is appropriate to different explanatory projects).

(There are other important differences—Mishler and Brandon as well as Mishler and Donoghue require that species be monophyletic, whereas the others do not; Kitcher differs from Ereshefsky in countenancing non-historical, nonevolutionary uses of the term *species*, but these differences are not important here.)

Each of these two dimensions to species pluralism is plausible in light of the proposal defended here that species are HPC natural kinds. The first is dictated by the reasonable assumption (defended by all the authors cited) that the homeostatic mechanisms important to the integrity of a species vary from species to species. The second is plausible in the light of the project or discipline relativity of kind definitions. What I want to indicate in the present essay is how the resources developed here can help to articulate and defend pluralistic realism. There are two obvious questions here: (1) if species taxa are properly defined by reference to different sorts of projects, in what sense are they real entities in nature? and (2) if the species category is heterogeneous in this way, what makes it the *species* category?

Kitcher's answer to the first question is that various approaches to the demarcation of species taxa correspond to features of the *objective structure* of nature, which exists independently of human thought even though different objective interests corresponding to different research programs may require demarcation by reference to different objective structures. What is important here to the *pluralist* realism Kitcher defends is that it explains realism about species in terms of the correspondence between species-level classificatory practices and objective structures, rather than in terms of some sort of unique metaphysical fundamentality of one or another of the ways of demarcating species. Different ways of demarcating species can correspond to different objective structures and thus define species categories that are equally real.

I suggest that the accommodation thesis provides us with just the machinery required to make the relevant notion of realism precise. As I suggested earlier, any talk about the "reality" of kinds or regarding "realism about" some kind or family of kinds is best understood as an imprecise way of addressing the question of the nature of the contributions (if any) that reference to those kinds makes to the satisfaction of the accommodation demands of the relevant disciplinary matrix. The objective structures existing independently of human practice are causal structures, and the "reality" of a kind consists in the contribution that reference to it makes—within the context of disciplinary practices—to the accommodation of those practices to the relevant causal structures. The sort of realist pluralism about the ways of demarcating species we are considering amounts to the insight that a plurality of species-level classificatory schemes contribute significantly to achieving (different aspects of) the accommodation of inferential practices in biology to relevant causal structures.

**The Species Level** Let us now turn to the question of why, if the species category is heterogeneous, it is appropriate to describe it as the *species category*. I have already remarked that disciplinary matrices are themselves homeostatic phenomena: the satisfaction of some of the accommodation demands of a disciplinary matrix generally tends to contribute to the satisfaction of lots of others.

What makes it possible to speak of taxa at the *species* level or of different ways of demarcating *species* is, I believe, a particular way in which homeostasis—among ways of satisfying accommodation demands—happens to work in biology. Defenders of the claim that different explanatory projects require different species definitions argue that species-level categories are deployed in biology in the service of significantly different sorts of explanatory projects and that there are different, but equally legitimate ways of demarcating species corresponding to various explanatory projects. In the terminology introduced here, they argue that these different projects place somewhat different accommodation demands on the conceptual and classificatory resources deployed by biologists—including demands on species-level classifications.

Now, there is in general a homeostatic relationship between the conditions for the satisfaction of different accommodation demands within biology. What I propose is that the category *species-level taxa* is fairly well defined, despite pluralism, because of an especially close homeostatic relation between the classificatory practices that satisfy the accommodation demands associated with the identification of the (different) primary subject matters of functional and evolutionary biology. A basic scheme of classification of (populations of) organisms that satisfies the accommodation demands of one set of projects within functional biology will come very close to satisfying the demands not only of other functional biological projects, but of the different explanatory projects in evolutionary biology, and vice versa. This second-order (or is it third-order?) homeostatic clustering of accommodation demand satisfactions is, of course, no accident. It obtains just because the sorts of stable phenomena that are the subject matter of various species-level biological explanations get their stability via a number of relatively closely (homeostatically) related evolutionary mechanisms (Wilson [1996, section 7, and chapter 7 in this volume] makes a very similar point).

Thus, the existence of a (pluralistic) species level among taxa, if there is such a level, is an artifact of an especially robust instance of the sort of homeostasis that characterizes disciplinary matrices in general (cf. Mishler, chapter 12 in this volume).

**Why There Is a "Species Problem"** The "species problem" is the problem of defining the nature of species taxa. Pluralists of the sort we are considering propose that there is no such nature—that, instead, there are many different, (approximately) equally methodologically important ways of demarcating species, each corresponding to a different legitimate way of understanding species-level taxa. If the solution is so easy, why does it represent a fairly recent proposal?

One reason, no doubt, has been the admirably motivated but (in the light of the complexity of homeostatic mechanisms) ultimately fruitless effort to establish something like a universally applicable "operational definition" of conspecificity (or at least a unitary formula that determines the relevant definition for any group of organisms) and thereby to establish consistency and uniformity of classificatory and nomenclatural practice. Arguably, the articulation of the species-as-individuals conception contributed to the plausibility of this project. If species are thought of as unique among the taxa in being evolutionary individuals in nature rather than human constructs (as many believe), then perhaps it is more plausible that a single unitary conception of conspecificity—defined in terms of the relevant notion of individuality—will be forthcoming.

What I suspect, however, is that the main source of the species problem is practical. Many disciplines are like biology in that there are schemes of classification that—by themselves—are almost adequate for the satisfaction of a wide variety of different accommodation demands—for example, the classification of the elements in chemistry and the standard classification of (what are called) mineral species in geology. In each of these disciplines, the compositional character of natural kind terms is exploited to "fine-tune" these almost adequate categories to fit more particular accommodation demands. Thus, we speak, for example, of the isotopes of chemical elements, the different physical forms of elemental sulfur, and the different varieties of quartz in order to achieve more nearly complete accommodation. There is no persisting "elements problem" in chemistry and there is no "species problem" in geology precisely because by using suitable natural adjectival terms to modify other natural terms, we can achieve accommodation, and it's merely a matter of convenience just how we do this. This is just the point I made earlier—that the compositional semantics of natural kind terms is important to the ways in which the accommodation demands of disciplinary matrices get satisfied.

Why can't we do this in biological taxonomy as well? The answer, I suggest, is that the compositional semantic structure of the standard Linnaean system of taxonomic nomenclature is inadequately flexible. Thus, for example, one might hope to take advantage of the tight homeostasis between the factors sustaining homeostasis within each particular species by settling (it might not matter exactly how) on some one reasonable way of defining the species-level taxa and then satisfying the accommodation demands of explanatory programs not perfectly served by this classification by deploying additional natural adjectival terms to differentiate further between groups of organisms or populations. (Wilson [1996 and chapter 7 in this volume] suggests that the HPC conception of natural kinds might be used to formulate a more unified conception of species. This might be one way of carrying out his project.)

The problem with such a proposal is not that it would be unworkable in the abstract—after all, that's how things are done in lots of disciplines. The problem is specific to the Linnaean hierarchy and the ways it constrains the compositional semantics of taxonomic names. Different fine-tuning would no doubt be required for different explanatory projects, but the Linnaean system of nomenclature does not have devices, for example, to distinguish between subspecies from the point of view of ecology and subspecies from the point of view of the genetics of speciation.

This is a serious practical problem, given the overwhelming need for a uniform system of biological classification and the entrenchment of the Linnaean nomenclatural scheme (cf. Dupré, Ereshefsky, chapters 1 and 11 in this volume), but there is no reason to mistake it for a metaphysical problem about fundamental entities in nature—or even about the "reality" of species in the sense defined by the accommodation thesis. Instead, it is a metaphysical problem about the lack of fit between the Linnaean hierarchy's representational resources and the causal structures important in biology (for an important account of other such metaphysical problems with the Linnaean hierarchy, see Ereshefsky 1994[3]).

## Higher Taxa and Species

**A Dubious Contrast** One of the standard themes in the metaphysics of biology is that species, being individuals, are real entities existing independently of human practice, whereas higher taxa are merely human concepts that reflect facts about the history of life and hence are largely unreal or arbitrary or merely conventional or something of the sort. The considerations we have rehearsed so far suggest that there is something seriously wrong with this approach to the metaphysics of higher taxa. In the first place, species probably aren't individuals, but they seem quite real enough nonetheless. Second, the contrast between individuals, on the one hand, and *conceptual* entities like kinds, on the other, is compromised by the fact that natural individuals are very much like kinds anyway. In particular, the correct individuation conditions (or persistence conditions) for a natural individual are a matter of how reference to it contributes to the satisfaction demands of a disciplinary matrix—a conceptual phenomenon if there ever was one. Finally, if, as pluralist realists maintain, there are different but equally legitimate ways of demarcating species, that answer to different demands for the accommodation of conceptual resources arising from different explanatory projects, then species—whether they are individuals or natural kinds—are in some sense project dependent and are thus, in yet an additional way, conceptual (or at least, concept involving) entities, so they can't contrast with higher taxa on that score.

I suggested earlier that the question of the reality of a kind should be understood as a question about the contribution that reference to it makes to accommodation, rather than as a question about its metaphysical fundamentality or anything of that sort. What I propose to do now is to explore the consequences of that approach for the issue of the metaphysics of higher taxa.

**Locke** Kitcher (1984) says that the reality of species consists in a correspondence between species classifications and the objective structure of nature. I agree, and I have proposed that the relevant objective structure is causal structure and that the relevant correspondence is a matter of the satisfaction of accommodation demands. It is tempting to articulate this claim further by saying that the realist about species believes that species are natural kinds that exist independently of scientific practice. Call this latter conception the "practice independence of natural kinds" (henceforth, *pink*) conception of realism about kinds. (There's an initially unintended pun here. I take the version of realism developed in this paper to be a natural extension of dialectical materialism in the Red tradition. I here defend that tradition against a merely pink alternative.)

If one's conception of realism about kinds is pink, then it will be tempting to treat higher taxa as (much) less real than individual organisms or species. After all, it might be thought difficult to see how Mammalia could exist independently of classificatory practice. I propose to rebut the pink conception.

Locke maintained that whereas Nature makes things similar and different, kinds are "the workmanship of men." I believe that, gender bias aside, he was right. Indeed, I think that the lesson we should draw from the accommodation thesis is that the theory of natural kinds just is (nothing but) the theory of how accommodation is (sometimes) achieved between our linguistic, classificatory, and inferential practices and the causal structure of the world. A natural kind just is the implementation—in language and in conceptual, experimental, and inferential practice—of a (component of) a way of satisfying the accommodation demands of a disciplinary matrix. Natural kinds are features not of the world outside our practice, but of the ways in which that practice engages with the rest of the world. Taxonomists sometimes speak of the "erection" of higher taxa, thus treating such taxa as, in a sense, human constructions. They are right, and the same thing is true of natural kinds in general.

Locke said that "each abstract idea, with a name to it, makes a distinct Species." His conception was that kinds are established by a sort of *unicameral* linguistic legislation: people get to establish definitions of kind(s) by whatever conventions (nominal essences) for the use of general terms they choose to adopt.

According to the accommodation thesis, we should, instead, see natural kinds as the product of *bicameral* legislation in which the (causal structure of the) world plays a heavy legislative role. A natural kind is nothing (much) over and above a natural kind term together with its use in satisfying accommodation demands. ("What else?" you ask. Well, there's whatever is necessary to accommodate translations that preserve satisfaction of accommodation demands and to accommodate phenomena such as reference failure and partial denotation.) Or, better yet, the establishment of a natural kind (remember that natural kinds are legislative achievements—that is, artifacts) consists solely in the deployment of a natural kind term (or of a family of such terms connected by practices of translation) in satisfying the accommodation demands of a disciplinary matrix. Given that the task of the philosophical theory of natural kinds is to explain how classificatory practices contribute to reliable inferences, that's all the establishment of a natural kind could consist in: natural kinds are the workmanship of women and men.

The causal structures in the world to which accommodation is required are, of course, independent of our practices (except when our practices are [part of] the subject matter; see Boyd 1989, 1990, 1991, and 1992 for better formulations). Still, natural kinds are social artifacts. That's why asking whether a kind exists independently of our practice is the wrong way to inquire about its reality. No natural kinds exist independently of practice. The kind *natural kind* is itself a natural kind in the theory of our inferential practice. That's why the reality of kinds needs to be understood in terms of the satisfaction of the accommodation demands of the relevant disciplinary matrix.

**Natural Individuals, Again** The very same points can be made about natural individuals, such as organisms. The relations of causal continuity, similarity, or whatever that unite the temporal stages of an organism exist independently of our practices, and they have the causal effects that make reference to that organism important to the satisfaction of accommodation demands independently of our practice. But the grouping of those temporal stages under a common linguistic or conceptual heading—treating them as constituting an organism—is just as much a matter of social practice in service of accommodation as the establishment of a natural kind.

It's tempting to argue that this view can't be right because even if we become extinct, dogs might continue to exist, so they must be organisms that exist independently of us. Of course, dogs might continue to exist: the persistence conditions (properly) associated with the notion of an individual dog might continue to be satisfied, but the fact that these persistence conditions are natural ones—the fact that persisting dogs are individuals "in nature," as one might say—is a fact not about nature alone, but about how biological practices are accommodated to nature. After all, some organisms would be in Mammalia even if we became extinct, and they would continue to occupy places in the relevant continuing historical lineages: in that sense Mammalia too exists independently of us.

Nature makes temporal stages similar and different, continuous and discontinuous, but things are the workmanship of women and men.

## Realism about Higher Taxa

**Higher Taxa and Accommodation** Neither for kinds nor even for individuals is the question of their reality best understood as a question about independence from our practices. That's why questions of "reality" or "realism" about them are best understood as questions about the accommodation of disciplinary matrices to causal structures. Thus, no simple contrast between species and higher taxa with respect to their independence of practice can establish the unreality (or diminished reality) of higher taxa. They may yet be unreal (or less real), but this unreality is not a matter of their being the results of human conception and practice. If they are unreal, it will be a matter of their failure to contribute effectively to accommodation.

**Assessing Accommodation: Methodological Spectra and the Equifertility Principle** I want to make a proposal about how we might fruitfully approach issues concerning the contribution that reference to higher taxa (or to any other kinds) makes to accommodation. Let's say that the choice between two alternative classificatory schemes within the context of a disciplinary matrix is arbitrary, just in case neither scheme reflects accommodation-relevant causal structures better than the other. When such a choice is arbitrary, the disciplinary matrix would (from the point of view of accommodation) be equally well served by either scheme.

Now, one measure of the extent to which a classificatory scheme contributes to accommodation—one measure of its "reality"—is given by the range of alternative schemes with respect to which a choice would be arbitrary. Philosophers or biologists who differ about the reality of higher taxa will differ about which choices between higher taxonomic schemes are arbitrary ones. How are we to assess competing claims about such arbitrariness?

It will help to answer this question if we consider the methodological import of such claims. By the *substantive conception* reflected in a disciplinary matrix at a particular time, let us understand the theories, doctrines, putative insights, and so on regarding the relevant subject matters accepted at that time. Of course, in any actual case, there will be issues and controversies of varying degrees of importance within a disciplinary matrix, so referring to the theories and so on that are accepted at a particular time involves some degree of idealization, but nothing in what I argue here depends on any subtleties about how the idealization is understood. I do intend that substantive conceptions be thought of as conceptual entities: as representations of phenomena deploying the conceptual resources of the matrix—rather than, for example, as sets of propositions understood as nonconceptual entities. The substantive conception $C_M$ of a disciplinary matrix $M$ is thus the representation within $M$ of the causal knowledge putatively achieved in $M$.

The inferential practices within a disciplinary matrix $M$ will be (except in cases where practitioners reason badly) justified by the substantive conception $C_M$. That's how the accommodation of inferential practices to causal structures is implemented (Boyd 1982, 1985a, 1990, 1991). Now, in every case—real or imaginary—there will be some arbitrary or conventional elements to the representational resources deployed within *M*. By a conventionality estimate $E_M$, for a disciplinary matrix *M*, let us understand an account of what the arbitrary or conventional elements are in *M*'s representational resources. Because (as we shall see) the methodological import of $C_M$ depends on the nature and extent of the conventionality of *M*'s representational resources, we may think of practice within a matrix at a time as being determined in part by practitioners' tacit estimates of conventionality.

Here again some harmless idealization is involved in speaking about the tacit estimates of conventionality prevailing within a matrix at a particular time. What would not be harmless, however, would be to equate the tacit estimates with the explicit estimates of conventionality articulated by practitioners within *M*. Those explicit estimates will often be more a reflection of peculiarities of the practitioners' philosophical education than of the accommodational achievements of their practices. Instead, we should think of tacit estimates of conventionality as being reflected in inferential practice. Thus, for example, the recognition that units of distance measurement are arbitrary or conventional is reflected in the fact that reference to distances in scientific laws is always in terms of distance ratios (either explicitly or via proportionality constants), whereas the nonconventionality of cardinality for sets of humans is reflected in the fact that population statistics often appear in nonratio forms in the findings of the social sciences and history.

It is important for our purposes to note a particular way in which tacit judgments of conventionality are reflected in methodological practice. Accommodation of explanatory and inferential practices to relevant causal structures is primarily achieved in mature sciences via the ways in which the substantive conception within a disciplinary matrix (formulated, of course, with the aid of reference to natural kinds, etc.) informs methodological judgments and practices—in determining projectability judgments, for example, or in determining the appropriate categories for statistical calculations. Tacit judgments of conventionality are characteristically reflected in the ways in which prevailing substantive conceptions are deployed in making such judgments.

Thus, for example, the tacit (also explicit, but that's not the point here) recognition that the assignment of negative and positive signs to the charges of electrons and protons, respectively, is conventional—rather than, say, a reflection of deficiencies or excesses—is reflected in the fact that the fact, about certain particles, that they have negative charge, whereas others have positive charge, is not taken to render projectable hypotheses to the effect that *negatively* charged particles suffer from some sort of *deficiency* in a sense in which *positively* charged particles do not.

Similarly, the recognition of the conventionality of national units of currency is reflected in the fact that no one makes use of differences in or ratios between national debts without prior conversion to some common currency or other economic measure.

These points are obvious, but important. They allow us to identify ways of specifying and assessing conventionality estimates regarding disciplinary matrices. One way of specifying an estimate of conventionality $E_M$ for a matrix $M$ with substantive content $C_M$ is to specify a range of alternatives to $C_M$ such that the choice between $C_M$ and any of these alternatives is to be understood as arbitrary or conventional in the sense that disciplinary matrices just like $M$, except that they deployed any one of these other representations, would equally well reflect facts about the relevant subject matter(s).

The examples we have just considered illustrate a quite general and fundamental methodological principle concerning conventionality and its relation to methodology—a principle that indicates another (related) way in which conventionality estimates can be specified (and sometimes assessed). According to the *equifertility principle*, when the choice between two substantive contents is arbitrary or conventional, the two substantive contents are *methodologically equifertile* in the sense that no methodological principle or practice is justified by one unless it is also justified by the other. The equifertility principle is about as obvious a methodological principle as there can be. It follows via a pretty straightforward application of the accommodation thesis—provided that one rejects the neo-Kantian view, apparently advocated by Kuhn (1970), that the adoption of a paradigm or conceptual framework can *noncausally* determine the causal structures of the relevant phenomena (see Boyd 1990, 1992).

What is especially important for the present discussion are the implications of the metaphysical innocence thesis in cases in which it is proposed that the prevailing conventionality estimate $E_M$ for a matrix $M$ is too modest and that there are alternatives to $C_M$ with respect to which the choice of $C_M$ is unexpectedly conventional. Such a proposal entails that any inference or inferential practice that would be justified (by the standards previously prevailing in the matrix) given $C_M$, but not given any one of the alternative representations, is thereby shown to be itself unjustified. No inferences that depend on conventional or arbitrary choices of representational schemes are justified.

By the *methodological spectrum* of a disciplinary matrix $M$ at a given time, let us understand the inferential strategies and methodological practices justified by $C_M$. What we have just seen is that any proposal of unexpected conventionality within a disciplinary matrix entails that the methodological spectrum of the matrix is narrower, in a systematically specifiable way, than practice within the matrix assumes. Thus, we have two ways of specifying the import of a claim of unexpected conventionality. One way characterizes the conventionality in terms of the representations with respect to which the choice of prevailing substantive content is said to be arbitrary or conventional; the other way indicates the dimensions of the narrowing of the methodological spectrum of the disciplinary matrix thereby required in the light of the equifertility thesis.

The latter characterization may be important, I suggest, in assessing the merits of proposals to revise prevailing tacit conventionality estimates. It has proven notoriously difficult for philosophers and others to achieve consensus on issues about conventionality. Sometimes, it seems to me, consensus on methodological issues is easier to achieve. When that is so, specifying the import for methodological spectra of proposals about conventionality may prove helpful.

**Extreme Cladism: A Worked Example** I propose to illustrate the way in which the equifertility principle and considerations about methodological spectra can be deployed in assessing arbitrariness claims by deploying it to criticize an extreme form of cladism about higher taxa. I do not mean to suggest that serious cladists need to hold any position close to the version I discuss or to offer a general criticism of cladistic approaches to higher taxa. Indeed, I am sympathetic to some versions of cladism. I choose the extreme version discussed here to simplify the application of the equifertility principle.

Imagine that you meet a cladist who maintains that the only scientifically legitimate constraint on the erection of taxa above the species level is that they should be strictly monophyletic. She allows that reasons of convenience might dictate the choice of one taxonomic scheme that honors strict monophyly over another, but neither choice, she claims, will more accurately reflect evolutionarily relevant features of nature.

Here's how you might reply. Consider efforts to identify and study mass extinctions. Evolutionary biologists interested in such phenomena often wish to estimate how the rate of species extinction has varied over geological time. Because the fossil record does not allow reliable distinctions to be drawn at the species level, they often compare rates of disappearance of genera or families from the fossil record by way of estimating the rate of extinction of species.

You might ask your extreme cladist colleague whether or not she finds such studies cogent. If the answer is "yes," you could point out that by choosing an alternative classificatory scheme such that the choice between it and a standard taxonomic scheme is arbitrary by her extreme cladist standards, evidence for mass extinctions could be made to disappear (just make the genus-level taxa in the new scheme correspond to, say, class-level taxa in the standard scheme). An application of the equifertility principle entails that the genus extinction data calculated with respect to the chosen scheme are no more or less indicative of evolutionary facts than the data based on more standard classificatory practices. Thus, the cladist's acceptance of the methodology of the studies in question is incompatible with her version of cladism.

A natural reply would be that given the alternative scheme in question, the relevant statistical calculations could be done with respect to appropriately chosen subgeneric categories. If your extreme cladist offered this reply, she would be acknowledging a tacit commitment to the idea that there is something natural (that is, nonarbitrary, nonconventional) about the similarity relations between species corresponding to various genus-level taxa in current classificatory practice, even if the assignment of those sorts of similarity relations to the genus level is arbitrary. She would thus be acknowledging an additional nonconventional constraint on the erection of higher taxa: they must somehow or other reflect the naturalness of those taxa assigned to the genus level in current classificatory practices (and similarly for family-level taxa if she accepts the methodological relevance of family-level statistics, and so on).

In real-life cases, resolving this issue would be more difficult, of course, but the point is this: different estimates of the degrees of arbitrariness or "reality" of classificatory schemes have quite different implications regarding the reliability of inferential methods. Often, we are in a position to evaluate these implications and thus make some headway in evaluating claims about arbitrariness.

**Homeostasis and the Reality of Higher Taxa** If some form of pluralist realism is right about taxa at the species level, then it will not do to think of nature as picking out the unique, real sort of biological taxa, with the rest being arbitrary or conventional. It does not follow, of course, that any of the levels of the Linnaean hierarchy above the species level are—given current taxonomic practice—real, in the sense provided by the accommodation thesis. Still, controversies about the species level seem to revolve around whether certain groups of similar populations should be grouped into the same subspecies, species, or genus. If pluralist realism is right, each one of the different choices from among these alternatives may, for a given family of populations, correspond to the establishment of a real taxon—which suggests, although it does not entail, that at least some subspecies and some genera (as these are ordinarily erected) are themselves real rather than arbitrary. (Ereshefsky [1991] makes the similar point that the cohesion thought by some to be distinctive of species-level taxa can be sustained by mechanisms that operate at higher taxonomic levels; cf. also Ereshefsky, chapter 11 in this volume.)

Similarly, statistical calculations like the ones mentioned in the previous section are methodologically important, which also suggests that genera are real. (Here again, there is no strict entailment. It could be, for example, that genera are real enough for such calculations to be indicative of extinction rates, but sufficiently arbitrary otherwise so that the slogan that they are "unreal" is basically right.) What I propose to do in this section is to explore the metaphysics of the proposal that some higher taxa are real.

Of course, the reality of a higher taxon would consist in the contribution that reference to it makes to accommodation. What sorts of contributions might one expect? One clue is provided by the view, characteristic of mainstream evolutionary systematics before the triumph of cladism, that higher taxa are to be thought of as defined by adaptive evolutionary innovations that constrain future courses of evolutionary development. According to this conception, species within a higher taxon—like populations within a species—share common evolutionary tendencies. In the case of higher taxa, these tendencies are derived from the constraints on evolutionary development produced by shared evolutionary innovations or novelties. Higher taxa are defined, in other words, by novel adaptations understood as sources of evolutionary tendencies toward stasis. Reference to higher taxa contributes to accommodation in evolutionary theory because the stasis-inducing factors in terms of which they are defined are important in the explanation of macroevolutionary patterns.

An important criticism of this conception of higher taxa has been that it rests on an overestimate of the extent of the role of natural selection in macroevolution. According to this criticism, many of the patterns discernible in the fossil record and reflected in the evolutionary systematists' erection of higher taxa are not products of systematic evolutionary tendencies at all, but merely the effects of historical phenomena that are random from the point of view of evolutionary theory.

It seems reasonable to extend the evolutionary systematists' conception of higher taxa as (representations of) loci of evolutionary stasis in order to claim that the reality of such a taxon consists in a distinctive configuration of stasis-enhancing factors that define it—whether these factors are matters of adaptive evolutionary innovation, developmental constraints, coevolved gene complexes, niche-organism interactions, or other sources of "phyletic inertia." According to this extended conception as well, reference to real higher taxa would contribute to accommodation because their defining properties would be crucially involved in explaining macroevolutionary patterns.

If this conception were right about some higher taxa, these taxa would, like species, be homeostatic property cluster kinds (perhaps with exceptional cases in which a single evolutionary novelty—situated, of course, within the context of other homeostatically related properties—established the relevant tendency toward stasis). The conception that some higher taxa are real in just this way would not be so deeply committed to an "adaptationist" strategy of evolutionary explanation as would more traditional evolutionary systematics, but it would be vulnerable—both in theory and in application—to the concern that many patterns in the history of life may lack altogether the sorts of explanations it anticipates.

(My understanding of traditional evolutionary systematics may have been too strongly influenced by critics of "adaptationism." Perhaps what I here present as an extension of the evolutionary systematists' conception may instead represent what they have believed all along, free from antiadaptationist caricature. If it is an extension, so much the better for the points I am making here.)

This is not the only way in which some higher taxa might turn out to be real in the sense required by the accommodation thesis, but it is a very important way. There are very good reasons to believe that at least some genera are real in this way. I have already indicated why pluralist realism about species suggests that some genus-level categories are real. If, as many authors have suggested, there are cases in which homeostasis at approximately the species level obtains in families of populations between which gene exchange is minimal or nonexistent (in the case of asexually reproducing reptilian or amphibian "species," for example), we have reasons to believe that the same sort of homeostasis might obtain in at least some recognized genera, perhaps in most.

Moreover, if some higher taxa are real kinds that are important in evolutionary theorizing, it is difficult (although, no doubt, not impossible) to see what their importance could be except as (representations of) stasis-producing factors. If that's what real higher taxa are, then it's equally difficult, given the complexity of evolutionarily relevant causal factors, to see how the contribution to stasis in any particular case could fail to involve homeostasis of several different factors. I propose, therefore, that insofar as some higher taxa are real and important categories in evolutionary theory (above and beyond their important role in representing patterns of ancestry and descent), they are probably, like species, homeostatic property cluster kinds.

If there are higher taxa that are real in this way, it is important to note that there is no particular reason to believe that their homeostatic property cluster definitions will honor strict monophyly, which is not to deny that the homeostasis linking the members of such a taxon might always crucially involve facts about the effects of their common ancestry. Thus, even if a requirement of strict monophyly is appropriate for some other higher taxa, it need not be so for taxa in question.

**Modest Cladism** Suppose, for the sake of argument, that some higher taxa—some genera for example—are real homeostatic property cluster kinds in the way indicated. What are we to make of the concern that efforts to discern evolutionary patterns in the fossil record—the causes of which define higher taxa—will identify patterns for which no explanation in terms of evolutionary tendencies exists?

The obvious answer is that this problem may arise for some higher taxa and not others. Perhaps taxa at the genus level—as taxa at that level are generally erected—are usually real in the special sense discussed here, but order-level taxa are usually not. Perhaps some such pattern obtains, but it is different across phyla, given extant practices. Perhaps taxa of shorter historical duration are more likely to reflect genuine stasis-sustaining properties. Perhaps taxa erected to account for the earlier stages in the history of life are more or less likely to be real than those taxa erected to account for later stages. Perhaps, in this regard, things are really a mess for which there is no simple characterization.

In any event, barring the extremely unlikely possibility that the standard criticisms of evolutionary systematics are somehow without force in light of the slight modification to this position we are considering, there will be some domain of higher taxa about which the cladistically inclined systematist can reasonably maintain that the only important facts about the evolution of life—which we can reflect in erecting such taxa—are historical facts about relations of ancestry and descent. About erection of taxa of this sort, the only nonconventional or nonpragmatic constraint would then be one of monophyly. This modest version of cladism is the one in which I am inclined to believe.

### Concluding Realist Postscript: Descent and Ancestry Are Real

I can't resist pointing out that the relation between a species and its daughter species is of causal significance in evolution. The erection of (at least approximately) monophyletic higher taxa does, as cladists insist, make a significant contribution to the accommodation of inferential practices in evolutionary biology to relevant causal structures. Such taxa are *real natural kinds* in the only available senses of these terms. So are species. It is a tribute (if that's the right word) to the enduring influence of empiricist conceptions of language, classification, and (anti)metaphysics that scientifically and philosophically fundamental points about the limitations of platonist conceptions of taxonomy and (overly) adaptationist conceptions of macroevolution have been formulated in philosophical terms that render obscure some of their main insights.

### ACKNOWLEDGMENTS

In formulating my approach to natural kinds, I have benefited greatly from conversations with Eric Hiddleston, Barbara Koslowski, Ruth Millikan, Satya Mohanty, Satya Shoemaker, Susanna Siegel, Jason Stanley, Zoltan Szabo, and Jessica Wilson. My thinking about biological taxonomy benefited greatly from conversations with Christopher Boyd, Kristin Guyot, and Quentin Wheeler.

### NOTES

1. Wilson (1996) goes so far as to make such conception of natural kinds part of what he calls "traditional scientific realism." It seems to me that the tradition of scientific realism was centered on the issue of refuting empiricist-verificationist arguments against knowledge of "unobservables" rather than on the issue of whether or not scientific kinds are individuated by essences that specify necessary and sufficient membership conditions. Early on, the traditional realist turn in the philosophy of science gave rise to a critique of behaviorism and to realism about mental states and properties. It is implausible to hold that scientific realists who participated in this critique believed or were committed to believing that the natural kinds of psychology always have sharp boundaries determined by necessary and sufficient membership conditions.

It is likewise implausible that traditional essentialist views always incorporated such a conception of kind definitions. Those biologists who have held that human races, as they are ordinarily recognized, have different biological essences should not be understood to have held the additional absurd position that such races always have such sharp boundaries.

2. I thank Professor David Hull for suggesting this clarification.

3. One metaphysical commitment made by Linnaeus himself that Ereshefsky criticizes is that taxa at the levels of genus and species are defined by mind-independent essences whereas taxa above these levels are subject to only pragmatic constraints. Ereshefsky denies the distinction on the grounds that there are no taxon-specific essences at any level. If the conception of essences defended here is correct, then species and probably many taxa above the species level do have essences (albeit not of the sort Ereshefsky has in mind), but all biological taxa are, in a certain sense of the term, mind dependent, or at least practice dependent.

## REFERENCES


Block N. (1980). *Readings in the philosophy of psychology* (vol. 1). Cambridge, Mass.: Harvard University Press.

Boyd, R. (1979). Metaphor and theory change. In A. Ortony, ed., *Metaphor and thought*. New York: Cambridge University Press.

Boyd, R. (1982). Scientific realism and naturalistic epistemology. In P. D. Asquith and R. N. Giere, eds., *PSA 1980*, vol. 2. East Lansing, Mich.: Philosophy of Science Association.

Boyd, R. (1983). On the current status of the issue of scientific realism. *Erkenntnis* 19, 45–90.

Boyd, R. (1985a). Lex orandi est lex credendi. In P. M. Churchland and C. Hooker, eds., *Images of science: Scientific realism versus constructive empiricism*. Chicago: University of Chicago Press.

Boyd, R. (1985b). Observations, explanatory power, and simplicity. In P. Achinstein and O. Hannaway, eds., *Observation, experiment, and hypothesis in modern physical science*. Cambridge, Mass.: MIT Press.

Boyd, R. (1988). How to be a moral realist. In G. Sayre McCord, ed., *Moral realism*. Ithaca, N.Y.: Cornell University Press.

Boyd, R. (1989). What realism implies and what it does not. *Dialectica* 43, No 1–2, 5–29.

Boyd, R. (1990). Realism, conventionality, and "realism about." In A. Boolos, ed., *Meaning and method*. Cambridge: Cambridge University Press.

Boyd, R. (1991). Realism, anti-foundationalism and the enthusiasm for natural kinds. *Philosophical Studies* 61, 127–148.

Boyd, R. (1992). Constructivism, realism, and philosophical method. In J. Earman, ed., *Inference, explanation and other philosophical frustrations*. Berkeley: University of California Press.

Boyd, R. (1993). Metaphor and theory change (second version). In A. Ortony, ed., *Metaphor and thought*, 2d ed. New York: Cambridge University Press.

Boyd, R. (forthcoming a). Kinds, complexity and multiple realization: Comments on Millikan's "historical kinds and the special sciences." *Philosophical Studies*.

Boyd, R. (forthcoming b). Kinds as the "workmanship of men": Realism, constructivism, and natural kinds. *Proceedings of the Third International Congress, Gesellschaft für Analytische Philosophie*. Berlin: de Gruyter.

Dupré, J. (1981). Natural kinds and biological taxa. *Philosophical Review* 90, 66–90.

Dupré, J. (1993). *The disorder of things*. Cambridge, Mass.: Harvard University Press.

Ereshefsky, M. (1991). Species, higher taxa, and units of evolution. *Philosophy of Science* 58, 184–101.

Ereshefsky, M. (1992). Eliminative pluralism. *Philosophy of Science* 59, 671–690.

Ereshefsky, M. (1994). Some problems with the Linnaean hierarchy. *Philosophy of Science* 61, 186–205.

Field, H. (1973). Theory change and the indeterminacy of reference. *Journal of Philosophy* 70, 462–481.

Ghiselin, M. (1974). A radical solution to the species problem. *Systematic Zoology* 23, 536–544.

Goodman, N. (1973). *Fact fiction and forecast*, 3rd ed. Indianapolis and New York: Bobbs-Merrill.

Guyot, K. (1987a). What, if anything, is a higher taxon? Ph.D thesis, Cornell University, Ithaca, New York.

Guyot, K. (1987b). Specious individuals. *Philosophica* 37, 110–126.

Hull, D. (1965). The effect of essentialism on taxonomy—two thousand years of stasis. *British Journal for the Philosophy of Science* 15, 314–362 (part one), and 16, 1–18 (part two).

Hull, D. (1978). A matter of individuality. *Philosophy of Science* 45, 335–360.

Kitcher, P. (1984). Species. *Philosophy of Science* 51, 308–333.

Kripke, S. A. (1971). Identity and necessity. In M. K. Munitz, ed., *Identity and individuation*. New York: New York University Press.

Kripke, S. A. (1972). Naming and necessity. In D. Davidson and G. Harman, eds., *The semantics of natural language*. Dordrecht, Netherlands: D. Reidel.

Kuhn, T. (1970). *The structure of scientific revolutions*, 2d ed. Chicago: University of Chicago Press.

Mayr, E. (1963). *Animal species and evolution*. Cambridge, Mass.: Harvard University Press.

Mayr, E. (1970). *Populations, species and evolution*. Cambridge, Mass.: Harvard University Press.

Mayr, E. (1976). *Evolution and the diversity of life*. Cambridge, Mass.: Harvard University Press.

Mayr, E. (1980). *Toward a new philosophy of biology: Observations of an evolutionist*. Cambridge, Mass.: Harvard University Press.

Millikan, R. (forthcoming). Historical kinds and the special sciences. *Philosophical Studies*.

Mishler, B., and R. Brandon (1987). Individuality, pluralism and the phylogenetic species concept. *Biology and Philosophy* 2, 397–414.

Mishler, B., and M. Donoghue (1982). Species concepts: A case for pluralism. *Systematic Zoology* 31, 491–503.

Putnam, H. (1972). Explanation and reference. In G. Pearce and P. Maynard, eds., *Conceptual change*. Dordrecht, Netherlands: Reidel.

Putnam, H. (1975a). The meaning of "meaning." In H. Putnam, ed., *Mind, language and reality*. Cambridge: Cambridge University Press.

Putnam, H. (1975b). Language and reality. In H. Putnam, ed., *Mind, language and reality*. Cambridge: Cambridge University Press.

Putnam, H. (1983). Why there isn't a ready-made world. In H. Putnam, ed., *Realism and reason*. Cambridge: Cambridge University Press.

Quine, W. V. O. (1969). Natural kinds. In W. V. O. Quine, ed., *Ontological relativity and other essays*. New York: Columbia University Press.

Wilson, R. (1996). Promiscuous realism. *British Journal for the Philosophy of Science* 47, 303–316.

# 7 Realism, Essence, and Kind: Resuscitating Species Essentialism?

Robert A. Wilson

## NATURAL KINDS AND SCIENTIFIC REALISM

The idea that there are natural kinds has a history in and an aptness for articulating realist views of science. Realists have traditionally held something like the following view of natural kinds: natural kinds are what the sciences strive to identify; they feature in laws of nature and so scientific explanation; they are individuated by essences, which may be constituted by unobservable (or "theoretical") properties; and they are conceiver-independent classifications of what there is in the world—they "carve nature at its joints."

The traditional realist view of natural kinds extends the following naive, commonsense view. There are objects and properties that exist independently of human observers. For example, suppose that we have before us a piece of rock. It has properties, such as a certain mass and constitution, and the rock and its properties exist independently of human observers. Scientists investigate such objects, uncover certain relationships between their properties, and develop taxonomies—natural kinds—that make these relationships more apparent. Suppose our rock has the property of being made of molten lava (composed, say, of 50% silica) and so has a certain melting point and various other chemical properties. By taxonomizing it as an *igneous* rock, scientists can both recognize its relationship to other kinds of rock and explore the relationships between the properties that igneous rocks have.

The traditional realist view of natural kinds goes beyond such a commonsense view, chiefly in the depth of its metaphysical commitments. Distinctive is the realist's view of *why* certain relationships between properties hold and why scientific taxonomies that identify natural kinds reveal further relationships between properties. Some properties are coinstantiated or correlate with one another because they feature in *laws of nature*, and these laws hold because of how nature is structured. In addition, the properties that feature in laws of nature are *intrinsic* properties of the entities that have them: they are properties that would be instantiated in those entities even if those entities were the only things that existed in the world. Natural kinds, then, categorize objects in terms of the intrinsic properties they have: same intrinsic properties, same kind of thing. This in turn explains why taxonomies that identify natural kinds lead to further revelations about how properties are related to one another, assuming that the most fundamental properties in the world are intrinsic properties. In moving from traditional realism in general to critiques of it within the philosophy of biology—within the literature on the "species problem" in particular—I want to focus on two further aspects of this overall metaphysical conception of natural kinds, *essentialism* and *unificationism*.

*Essentialism* is the view that natural kinds are individuated by essences, where the essence of a given natural kind is a set of intrinsic (perhaps unobservable) properties, each necessary and together sufficient for an entity's being a member of that kind. Realists thus say that scientific taxonomy proceeds by discovering the essences of the kinds of things that exist in the world and that this explains, in part, the theoretical and practical successes of science. The endorsement of essentialism provides a way of distinguishing natural kinds from arbitrary and conventional groupings of objects. Natural kinds are *kinds* (rather than mere arbitrary collections) because the entities so grouped share a set of intrinsic properties—an essence—and *natural* (rather than conventional or *nominal*) because that essence exists independent of human cognition and purpose.

The rejection of essentialism about species and, along with it, of the idea that species are natural kinds at all has been central to the claim that species are *individuals* (Ghiselin 1974, 1997; Hull 1976, 1978). According to this view, the traditional realist misconstrues the ontological nature of species: species are individuals rather than kinds individuated by essences. Essentialism about species has also been attacked independently in the philosophy of biology by Mayr (1970) and Sober (1980).

As a general thesis, *unificationism* is the view that scientific knowledge is *unified* in some way; for the traditional realist, it is the view that because natural kinds reflect preexisting order in the world, they are unified or integrated. But realists are not alone in holding some version of unificationism about scientific knowledge. The strongest versions of unificationism were held by the logical positivists as the "unity of science" thesis (e.g., Oppenheim and Putnam 1958) and came with a reductive view of the nature of "higher-level" scientific categories. More recent unificationist views have been nonreductive—cast in terms of the notions of constitution or *realization*, rather than in terms of identity. Traditional realism, whether in its reductionist or nonreductionist guise, implies views about the basis of membership in a given natural kind, the relationship between the various natural kinds and the complexities in nature, and the way in which natural kinds themselves are ordered. We might express these views as follows:

- the *commonality assumption:* there is a common, single set of shared properties that form the basis for membership in any natural kind
- the *priority assumption:* the various natural kinds reflect the complexities one finds in nature rather than our epistemic proclivities

• the *ordering assumption:* natural kinds are ordered so as to constitute a unity

For a traditional realist about species, the commonality assumption amounts to essentialism about natural kinds; the priority assumption points to the world rather than to ourselves as the source of the variety of natural kinds one finds; and the ordering assumption, typically expressed in the view that natural kinds are *hierarchically* organized, says that there is *one* way in which different natural kinds are related to one another.

Pluralists about species reject either the priority assumption or the ordering assumption or both. For example, Mishler and Donoghue (1982) reject the ordering assumption, but maintain the priority assumption when they say that "a variety of species concepts are necessary to adequately capture the complexity of variation patterns in nature" (p. 131). Dupré (1981, 1993), by contrast, would seem to reject both the priority and ordering assumptions in suggesting that "the best way of [classifying species] will depend on both the purpose of the classification and the peculiarities of the organisms in question" (1993, 57; cf. Dupré, chapter 1 in this volume). Kitcher seems to share this view when he says that "there is no unique relation which is privileged in that the species taxa it generates will answer to the needs of all biologists and will be applicable to all groups of organisms" (1984, 317).[1]

Traditional realism about species is indefensible, and in the next two sections I indicate just how this view has motivated the individuality thesis (the second section) and pluralism (the third section). But reflection on the similarities between the case of species and the case of neural taxonomy leaves me skeptical about the plausibility of the inferences to these two views about species (the fourth section). Moreover, I argue that the resources afforded by Richard Boyd's (1988, 1991, chapter 6 in this volume) homeostatic property cluster view of natural kinds provide a view of species that lies between traditional realism, on the one hand, and the individuality thesis and pluralism, on the other (the fifth section). I suggest that rather than rejecting the connection within traditional realism between realism, essence, and kind, we need to complicate those relationships in a way that leaves us closer to traditional realism than we might have expected.

## INDIVIDUALITY AND SPECIES TAXA

A natural way to apply traditional realism to species would be to hold that members of particular species share a set of morphological properties or a set of genetic properties, each necessary and together sufficient for membership in that species. Let me take the morphological and genetic versions of this view separately. For example, according to the former of these views, domestic dogs, members of *Canis familiaris,* share some set of observable properties—presumably determinate forms of phenotypes such as having four legs, hair, a tail, two eyes, upper and lower teeth—each necessary and together sufficient for their being members of that kind. These properties are the essential properties of being a member of *Canis familiaris.* According to the latter of these views, the species essence is not constituted by these morphological properties themselves, but by the genetic properties—such as having particular sequences of DNA in the genome—that are causally responsible for the morphological properties. In either case, the idea is that there is some set of intrinsic properties, the essence, that all and only members of *Canis familiaris* share—whether this essence be the sort of morphological properties that can be readily observed (and thus available to both common sense and science) or the sort of genetic properties whose detection requires special scientific knowledge of a more theoretical sort. The question answered by those theorists who posit phenotypes or genotypes as essences is this: what are the phenotypic or genotypic properties that an individual must have to be a member of a given species S? The answer to this question, in turn, allows these theorists to answer the question of what distinguishes S from other species.

The chief problem with either suggestion is empirical. In investigating the biological world, we don't find groups of organisms that are intraspecifically homogenous and interspecifically heterogenous with respect to some finite set of phenotypic or morphological characteristics. Rather, we find populations composed of phenotypically distinctive individual organisms; sexual dimorphism and developmental polymorphism are just two common forms of phenotypic variation within species. There simply is no set of phenotypes that all and only members of a given species share. This is true even if we extend the concept of a phenotype so as to include organismic behavior as potentially uniquely identifying properties that mark off species from one another. Precisely the same is true of genetic properties. The inherent biological variability or *heterogeneity* of species with respect to both morphology and genetic composition is, after all, a cornerstone of the idea of evolution by natural selection.

The emphasis on morphology and genotypic fragments as providing the foundations for a taxonomy of species is also shared by *pheneticists* within evolutionary biology, though their strident empiricism about taxonomy would make it anachronistic to see them as defending any version of realism or essentialism. In fact, we might see pheneticism as an attempt to move beyond traditional realism about species by shedding it of its distinctly realist cast. The idea of pheneticism is that individuals are conspecifics with those individuals to which they have a certain level of *overall phenetic similarity,* where this similarity is a weighted average of the individual phenotypes and genetic fragments individual organisms instantiate.

Both pheneticism and the traditional realist view of species focus on shared phenotype or genotype as the basis for species membership. The pheneticist sidesteps the problem—faced by the traditional realist—of intraspecific heterogeneity with respect to any putatively essential property in effect by doing away with essences altogether. However, the pheneticist still treats species as kinds rather than individuals, but they are *nominal* kinds rather than natural kinds because the measure of overall morphological similarity is a function of the conventional weightings we assign to particular morphological traits or DNA segments.

By contrast, proponents of the individuality thesis respond to the failure of essentialism with respect to species taxa by claiming that species are not natural kinds at all, but individuals or *particulars*—with individual organisms being not members of the species kind, but *parts* of species because a species itself is an individual. Species have internal coherence, discrete boundaries, spatiotemporal unity, and historical continuity—all properties that particulars have, but which neither natural nor nominal kinds have. Viewing species as individuals rather than as kinds allows us to understand how species can have a beginning (through speciation) and an end (through extinction); how organisms can change their properties individually or collectively and still belong to the same species; and why essentialism goes fundamentally wrong in its conception of the relationship between individual organism and species.

## PLURALISM AND THE SPECIES CATEGORY

The individuality thesis is a view of the nature of particular species taxa—for example, of *Canis familiaris*. Because I suggested that the individuality thesis was a competitor to both traditional realism and pheneticism, I also think of the latter two views as making claims about particular species taxa. But pheneticism is also often taken as a view about the species *category*—that is, as a view about what defines or demarcates species as a concept that applies to a unit of biological organization. So construed, pheneticism is the view that species are individuated by a measure of overall phenetic similarity, with organisms having a certain level of overall phenetic similarity counting as species, and higher-level and lower-level taxa having, respectively, lower and higher levels of similarity.

Apart from pheneticism, the various proposals that have been made about what characterizes the species category are often divided into two groups: (1) *reproductive* views, which emphasize reproductive isolation or interbreeding as criteria—including Mayr's (e.g., 1982) so-called biological species concept and relaxations of it, such as Paterson's (1985) recognition concept and Templeton's (1989) cohesion species concept; and (2) *genealogical* views, which give phylogenetic criteria the central role in individuating species and are typified by Cracraft (1983) and Wiley (1978). Unlike pheneticism, both of these families of views fit naturally with the individuality thesis as a view of species taxa.

The focus of both reproductive and genealogical views, as views of the species category, is on two questions: (*a*) what distinguishes species from other groupings of organisms, including varieties below and genera above, as well as more clearly arbitrary groupings? and (*b*) how are particular species distinguished from one another? The question that preoccupies pheneticists—namely, what properties of individual organisms determine species membership—receives only a derivative answer from proponents of reproductive and genealogical views. If one answers either (*a*) or (*b*) or both, one determines which species individual organisms belong to not by identifying a species essence, but by seeing which group, individuated in accord with the relevant answer to (*a*) or (*b*), those organisms belong to. Thus, "belonging to" can be understood in terms of part-whole relations, as it should according to the individuality thesis. Moreover, proponents of reproductive views conceive of species as *populations*, whereas proponents of genealogical views conceive of species as *lineages*, and both populations and lineages are easily understood as spatiotemporal, bounded, coherent individuals, rather than as kinds, be they natural or nominal.

It is widely accepted that there are strong objections to the claim that any of these proposals—pheneticism, reproductive views, or genealogical views—are adequate. These objections have, in turn, motivated *pluralism* about the species category, the idea being that each of the three views, or each of the more specific forms that they may take, provides a criterion for specieshood that is good for some, but not all purposes. The commonality assumption is false because, broadly speaking, phenetic, reproductive, and genealogical criteria focus on different types of properties for species membership, so there is no one type of property *that determines* kind membership. The priority assumption is also false because the different species concepts reflect the diverse biological interests of (for example) paleontologists, botanists, ornithologists, bacteriologists, and ecologists, so these concepts depend as much on our epistemic interests and proclivities as on how the biological world is structured. And the ordering assumption fails because where we locate the species category amongst other scientific categories depends on which research questions one chooses to pursue about the biological world.

Like pheneticism, reproductive and genealogical views of the species category recognize the phenotypic and genotypic variation inherent in biological populations, so they concede that there is no traditionally conceived essence in terms of which species membership can be defined. But even aside from viewing heterogeneity amongst conspecifics as intrinsic to species, these two views share a further feature that makes them incompatible with the sort of essentialism that forms a part of traditional realism. In contrast with the traditional view that essences are sets of *intrinsic* properties, reproductive and genealogical views of the species category imply that the properties determining species membership for a given organism are not intrinsic properties of that organism at all, but depend on the relations the organism bears to other organisms. Let me explain.

Although we are considering reproductive and genealogical views of the species category, I mentioned earlier that these views have a derivative view of what determines species membership for individual organisms. Reproductive views imply that a given individual organism is conspecific with organisms with which it can interbreed (Mayr), with which it shares a mate recognition system (Paterson), or with which it has genetic or demographic exchangeability (Templeton). Genealogical views imply that conspecificity is determined by a shared pattern of ancestry and descent (Cracraft) or by a shared lineage that has its own distinctive "evolutionary tendencies and historical fate" (Wiley 1978, 80). According to these views, conspecificity is not determined by shared intrinsic properties, but by organisms' standing in certain relations to one another. We can see this most clearly if we consider both views in conjunction with the individuality thesis, since conspecificity is then determined by an organism's being a part of a given reproductive population or evolutionary lineage, where neither of these is an intrinsic property of that organism. Here, we seem a long way from the traditional realist's conception of essentialism.

Any serious proposal for a more integrative conception of species must reflect the inherent heterogeneity of the biological populations that are species, and it is difficult to see how the traditional realist view of natural kinds can do so. Also, given the implicit commitment of both reproductive and genealogical views of the species category to an organism's relational rather than its intrinsic properties in determining conspecificity, the prospects for resuscitating essentialism look bleak.

## BETWEEN TRADITIONAL REALISM, INDIVIDUALITY, AND PLURALISM: THE CASE OF NEURAL TAXONOMY

Species is not the only biological category whose members are intrinsically heterogenous and relationally taxonomized. It seems telling that although traditional realism is rendered implausible for these other biological categories for much the same reasons that we have seen it to be implausible for species, there is little inclination in these other cases to opt either for an individuality thesis about the corresponding taxa or for pluralism about the corresponding categories. The categories I have in mind are *neural* categories, and I shall discuss two of these with an eye to pointing the way to a view of species somewhat closer to traditional realism than might seem defensible, given the discussion thus far.

The first example is the categorization of *neural crest cell* (Hall and Horstadius 1988; Le Douarin 1982, 1987).[2] In vertebrate embryology, the neural plate folds as the embryo develops, forming a closed structure called the neural tube. Neural crest cells are formed from the top of the neural tube and are released at different stages of the formation of the neural tube in different vertebrate species (figure 7.1). In neurodevelopment, cells migrate from the neural crest to a variety of locations in the nervous system, the neural crest being the source for the majority of neurons in the peripheral nervous system. Cell types derived from the neural crest include sensory neurons, glial cells, and Schwann cells; neural crest cells also form a part of many tissues and organs, including the eye, the heart, and the thyroid gland.

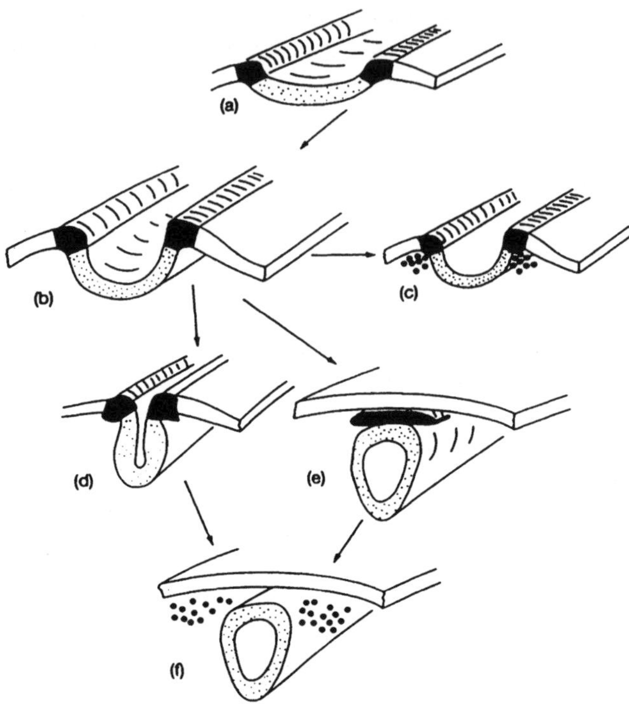


**Figure 7.1** The neural crest. A representation of the localization of the neural crest and neural crest cells (black) between neural ectoderm (stippled) and epidermal ectoderm (white) at neural plate (a), neural fold (b, c) and subsequent stages (d–f) of neural crest cell migration to illustate patterns of migration in relation to neural tube closure in various vertebrates. The time of initial migration varies between different vertebrates and can also vary along the neural axis in a single embryo. In the rat, cranial neural crest cells migrate while the neural tube is still at the open neural fold stage (c). In birds, neural crest cells remain in the neural folds until they close (d), only then beginning to migrate (f), whereas in amphibians, neural crest cells accumulate above the closed neural tube (e) before beginning their migration (f). (Reprinted with permission from Hall and Horstadius 1988.)

Neural crest cells are not taxonomized as such by any essence, as conceived by the traditional realist. The category *neural crest cells* is intrinsically heterogenous, and individual cells are individuated, in part, by one of their relational properties—their place of origin. But perhaps the category *neural crest cells* is not itself a natural kind, but rather a close to commonsense precursor to such a kind. (After all, not every useful category in science is a natural kind.) The real question, then, would be: By what criteria are refined natural kinds that derive from this category individuated?

I shall focus on the distinction that neuroscientists draw between *adrenergic* and *cholinergic* cells, both of which originate in the neural crest, because this taxonomy of neural crest cells seems initially promising as a candidate for which traditional realism is true. Adrenergic cells produce the neurotransmitter noradrenaline and function primarily in the sympathetic nervous system; cholinergic cells produce acetylcholine and function primarily in the parasympathetic nervous system. This truncated characterization of adrenergic and cholinergic cells suggests that they may fit something like the traditional realist view of natural kinds: these two types of neural crest cells are individuated by intrinsic properties or *causal powers*—their powers to produce distinctive neurotransmitters—which serve as essences that determine category membership.

Such a view of these neural categories, however, would be mistaken, a claim I substantiate in a moment. But just as mistaken would be the claim that adrenergic cells form an individual rather than a natural kind, or the claim that we should be pluralists about this category of neural cells, claims I discuss further in later sections. Standard taxonomic presentations of the two types of cells (e.g., Hall and Horstadius 1988, Le Douarin 1987) proceed by introducing a list of features that each cell type possesses, including their typical original location in the neural crest, the typical dendritic connections they make to other cells, the neural pathways they take, and their final locations and functions. Adrenergic cells are heterogenous with respect to any single one of these properties or any set of them, and it is for this reason that they do not have an essence as conceived by traditional realists. Yet in normal development, these properties tend to cluster together, and it is this feature of the *form* the heterogeneity takes that allows us, I think, to articulate a view that stops short of individuality and pluralism.

A further fact about neural crest cells dooms any attempt to individuate them in terms of their power to produce certain neurotransmitters: they are *pluripotential* in a sense that I specify in a moment. Because one goal of research into the neural crest has been to understand the paths of migration of neural crest cells, transplantation studies have played a central role in that research. In a standard paradigm, sections of the neural crest from a quail are transplanted into a chick embryo, and the phenotypic differences in development (e.g., pigmentation changes) are noted. One central and initially surprising finding from transplantation studies was that neural crest cells transplanted to a host environment tend to produce the neurotransmitter normally found in that environment, even if the cell transplanted would have produced the other neurotransmitter in its normal environment. This finding implies that factors exogenous to a given cell determine which neurotransmitter that cell produces. The best candidate we have for a traditionally conceived essence for adrenergic and cholinergic neural crest cells—the power those cells have to produce norepinephrine or acetylcholine, respectively—is not even an intrinsic property of cells. The very property we are supposing to be essential for cell type varies from cell to cell not according to facts about that cell's intrinsic properties, but according to facts about the environment in which the cell is located.

As a second example, consider the taxonomy of retinal ganglion cells. These cells receive visual information via the retina and have been extensively studied in the cat and the frog (Rowe and Stone 1980a). Chalupa

(1995) says that "we now know more about the anatomical and functional properties of retinal ganglion cells than we do about any other neurons of the mammalian brain" (p. 37), suggesting that the neural categories here are the product of relatively well-developed neuroscience. Over the last thirty years, a number of taxonomies have been proposed for retinal ganglion cells; some of these taxonomies (e.g., alpha/beta/gamma trichotomy) are based on morphological criteria, such as dendritic morphology and axon size, whereas others (e.g., the *Y*/*X*/*W* trichotomy) are based on physiological properties, such as the size of the receptive field (table 7.1). The functional distinctness of each of these kinds of retinal ganglion cell suggests that they form distinct visual channels that operate in parallel in visual processing.

As with neural crest cells and their determinate kinds, such as adrenergic and cholinergic cells, the taxonomy of retinal ganglion cells proceeds by identifying clusters of properties that each type of cell has. No one of these properties is deemed necessary or any set of them deemed sufficient for classification as a *Y*, *X*, or *W* cell; thus, there is no essence for any of these neural categories. Again, however, I want to suggest that it is implausible to see, for example, the taxa of *Y* cells as individuals rather than as a natural kind or to claim that this way of categorizing retinal ganglion cells has a pluralistic rather than a unificationist basis. The clustering of the various morphological and physiological properties in these cells again points us to a middle ground here. Large numbers of retinal ganglion cells tend to share many of a cluster of properties in their normal environments. This fact, together with the distinctness of these clusters of properties, provides the basis for individuating retinal ganglion cells into various kinds.

The biological facts in these areas of neuroscience defy philosophical views that posit traditionally conceived essences. Equally clearly they suggest an alternative to the corresponding individuality thesis and pluralism about taxonomy in the philosophy of biology more generally.

## HOMEOSTATIC PROPERTY CLUSTERS AND THE REVIVAL OF ESSENTIALISM

The middle-ground position I have in mind is based on a view introduced by Richard Boyd (1988, 1991, chapter 6 in this volume; see also Kornblith 1993), which he calls the ***homeostatic property cluster*** (hereafter, HPC) view of natural kinds. I shall adapt this view, noting explicitly where I depart from Boyd. Boyd initially introduced this view as part of his defense of a naturalistic version of realism in ethics, but from the outset he clearly intended for it to apply to natural kinds in science and to species in particular. Precursors to the HPC view include Wittgenstein's discussion of cluster concepts via the metaphor of family resemblance; Putnam's (1962) introduction of a law cluster view of scientific concepts; and Hull's (1965) argument that biologists who recognize higher taxa as cluster concepts should extend this view to species themselves. Boyd's previously published discussions have been rela-

**Table 7.1** Some properties of cat retinal ganglion cells

| | Y Cells | X Cells | W cells |
|---|---|---|---|
| Receptive field center size | large, 0.5–2.5° | small, 10′–1° | large, 0.4–2.5° |
| Linearity of center-surround summation | nonlinear | linear | not tested |
| Periphery effect | present | usually absent | absent |
| Axonal velocity | fast, 30–40 m/sec | slow, 15–23 m/sec | very slow, 2–18 m/sec |
| Soma size, peripheral retina | large, > 22 μm | medium, 14–22 μm | small, < 15 μm |
| Proportion of population | < 10% | approximately 40% | approximately 50–55% |
| Retinal distribution | concentrate near area centralis, more numerous relatively in peripheral retina | concentrate at area centralis | concentrate at area centralis and in streak |
| Central projections | to laminae A, $A_1$, and $C_{12}$ of LGN, to MIN and, via branching axon, to SC from the A-laminae of LGN to cortical areas 17 and 18, also by branching axon; and from MIN to areas 17, 18, 19 | to laminae A, $A_1$, and $C_{12}$ of LGN; thence to area 17; to midbrain (a minority), but probably not to SC | to SC, to C-laminae of LGN and thence visual cortex area 17 and/or 18, and 19 |
| Nasotemporal division | nasal cells project contralaterally; most temporal cells ipsilaterally; strip of intermingling centered slightly temporal to area centralis | nasal cells project contralaterally, temporal cells project ipsilaterally; narrow strip of intermingling centered on area centralis | nasal cells project contralaterally; most temporal cells also project contralaterally; about 40% of temporal cells project ipsilaterally |

Modified from Rowe and Stone (1977).

tively programmatic, and his current view of the implications of the HPC view for issues concerning species (see Boyd, chapter 6 in this volume) is somewhat different from the view I advocate here.

The basic claim of the HPC view is that natural kind terms are often defined by a cluster of properties, no one or particular n-tuple of which must be possessed by any individual to which the term applies, but some such n-tuple of which must be possessed by all such individuals. The properties mentioned in HPC definitions are *homeostatic* in that there are mechanisms that cause their systematic coinstantiation or clustering. Thus, an individual's possession of any one of these properties significantly increases the probability that this individual will also possess other properties that feature in the definition. This is a fact about the causal structure of the world: the instantiation of certain properties increases the chance that other particular properties will be coinstantiated because of underlying causal mechanisms and processes.

The view is a "cluster" view twice over: only a cluster of the defining properties of the kind need be present for an individual to fall under the kind, and such defining properties themselves tend to cluster together—that is, tend to be coinstantiated in the world. The first of these features of the HPC view of natural kinds allows for inherent variation among entities that belong to a given natural kind.

The second of these features distinguishes the HPC view as a *realistic* view of kinds from the Wittgensteinian view of concepts more generally to which it is indebted. On the HPC view, our natural kind concepts are regulated by information about how the world is structured, not simply by conventions we have established or language games we play. Before moving to the case of species, consider how the HPC view applies to our pair of neural kinds.

First, take the case of the individuation of neural crest cells. For a cell to be adrenergic is for it to have a certain cluster of properties that scientists have discovered; amongst other things, it is to originate in the posterior of the neural tube, to follow one of a given number of migratory paths, to function in the sympathetic nervous system, and to produce the neurotransmitter norepinephrine. Facts about the structure of the biological world—facts still being uncovered—explain why these properties tend to be (imperfectly) coinstantiated by certain kinds of cells. This clustering is the result of incompletely understood mechanisms that govern an embryo's development and is absent, either partially or wholly, just when the normal function of those mechanisms is disrupted. No single one of these properties is, however, strictly necessary for a cell to be adrenergic. The presence of *all* of them, however, is sufficient for a cell to be adrenergic, at least in the environments in which development normally occurs.[3] This feature of the HPC view marks one of the affinities between it and traditional realism, about which I say more later. On this view, adrenergic neural crest cells are a natural kind of cell, and individual cells are members of that natural kind in virtue of satisfying the homeostatic property cluster definition of that natural kind.

Second, take the case of the individuation of retinal ganglion cells. Consider in particular the physiological taxonomy of *Y*, *X*, and *W* cells. The tendency of the various physiological properties—such as the axonal velocity, soma size, and retinal distribution—to be coinstantiated by particular types of cells is no accident, but the result of underlying mechanisms governing neural development and neural functioning. Again, a determinate form of any one of these properties could be absent in a particular cell, yet the cell will still be a certain kind of cell—say, a *Y* cell—so no *one* of these properties is an essential property for being a member of that kind of retinal ganglion cell. Nevertheless, there is a general definition of what it is to be a *Y* cell, one based on the homeostatic cluster of properties that one finds instantiated in some cells and not in others. *Y* cells are a natural kind of cell with a sort of essence, albeit one different from the sort of essence characterized by traditional realism. Moreover, there is a kind of integrity to being a *Y* cell that invites a unificationist rather than a pluralistic view of it.

A more ambitious way to apply the HPC view to this example is worth noting. Although it is a substantive hypothesis that the morphological and physiological taxonomies of retinal ganglion cells are roughly coextensive, it is a hypothesis that is reasonably well confirmed (see Chalupa 1995, 40–42). The HPC view provides a natural way of integrating the two taxonomies in effect by adding together the two lists of properties in each cluster. This integration assumes, of course, that certain common mechanisms explain the presence of this new cluster of properties *qua* cluster, without which we would simply have a disjunction of two homeostatic clusters, not a new homeostatic cluster of properties.

I suggest that the HPC view applies to species taxa as follows. Particular species taxa are natural kinds defined by a homeostatic cluster or morphological, genetic, ecological, genealogical, and reproductive features. This cluster of features tend to be possessed by any organism that is a member of a given species, though no one of these properties is a traditionally defined essential property of that species, and no proper subset of them is a species essence. This clustering is caused by only partially understood mechanisms that regulate biological processes (such as inheritance, speciation, and morphological development) and the complex relations between them. More generally, the homeostatic clustering of these properties in individuals belonging to a single species is explained by facts about the structure of the biological world. For example, organisms in a given species share morphology in part because they share genetic structures, and they share these structures because of their common genealogy. This is not to suggest, however, that any one of these properties is more basic than all of the others or that there is some strict ontological hierarchy on which they can all be placed, for the dependency relations between these properties are complex and almost certainly multifarious.

Having severed the connection between the HPC view and traditional realism, let me now indicate an important affinity that the two views share. Although possession of individual properties or n-tuples of the relevant homeostatically clustered properties are not necessary for membership in the corresponding species kind, possession of all of them *is* sufficient for membership in that kind. If the homeostatic property cluster definition is sufficiently detailed, this circumstance will likely remain merely an idealization, uninstantiated in fact and approximated to a greater or lesser extent in particular cases. This in turn points to one way in which the sort of essentialism that forms a part of traditional realism is a limiting case of the sort of essentialism implicit in the HPC view of natural kinds.

The HPC view can also be applied to the species category, allowing a definition of what sorts of thing a species is that marks it off from other biological categories. First, the general nature of the cluster of properties—morphology, genetics, genealogy, and so on—will distinguish species from nonevolutionary natural kinds, such as cells (in physiology), predators (in ecology), and diseases (in epidemiology). Second, species will be distinguished from other evolutionary ranks, such as genera above and varieties below, by the particular specifications of this general cluster of properties (cf. Ereshefsky, chapter 11, and Mishler, chapter 12 in this volume). For example, for species for which reproductive criteria are applicable, reproductive isolation will distinguish a species from the mere varieties within it (because the latter are not so isolated), and interbreeding across the population will distinguish it from the genus to which it belongs. In some cases, the distinction of species as a particular rank in the biological hierarchy will be difficult to draw, but I suggest that this is a virtue, not a liability, of the HPC view because varieties sometimes *are* very like species (e.g., in cases of so-called incipient species), and species sometimes *are* very like genera (e.g., in cases of geographically isolated populations that diverge only minimally and share a recent ancestor).

That the HPC view is a *realist* view should be clear: it claims that there are natural kinds in the world individuated by properties existing independent of us and that our schemes of categorization in science track these natural kinds. Here, there seems a clear endorsement of the priority assumption from traditional realism. In addition, the properties that feature in the cluster need not be observable. For example, neither the lineages of descent nor the pathways of projection from the retina to the lateral geniculate nucleus need themselves be observable to feature in the respective HPC definitions of species and retinal ganglion cells. Lingering doubts about the realist credentials of the view should be dispelled by noting that it has traditional realism as its limiting case, one in which all of the properties in the cluster are present in all instances falling under the concept; the HPC view is a loosening of traditional realism, not an abandonment of its realist core.

Consider now the HPC view of species more explicitly vis-à-vis essentialism and unificationism. On the HPC conception, species are natural kinds, not individuals, with essentialism in the style of traditional realism a limiting case rather than a definitive feature of this type of natural kind. And just as the HPC view of species is incompatible with a traditional form of essentialism, so too is it incompatible with a traditional form of the commonality assumption, according to which all members of a natural kind must share some set of intrinsic properties. There *is*, however, a sort of common basis for membership in any given species, which can be expressed as a finite disjunction of sets of properties (and relations), and we might thus view the HPC view of species as compatible with a version of the commonality thesis that allowed such disjunctions. Likewise, because some of the criteria that define the species category may have a different level of significance in different cases—in the extreme, they may be absent altogether—simple versions of the ordering assumption are incompatible with the HPC view of natural kinds. Yet the possibility of more complicated forms of the ordering assumption would seem compatible with the HPC view because there seems to be a clear place for a unified species category amongst other (unified!) biological taxa, according to the HPC view of species.

Note how the HPC view of natural kinds preserves another idea that is a part of traditional realism: all and only members of a natural kind satisfy the corresponding definition of that kind. Anything that is a species and only things that are species will satisfy the HPC definition for species; any individual that is a member of a particular species and only such individuals will satisfy the HPC definition for a particular species—likewise for neural crest cells and retinal ganglion cells (as well as their determinate forms).

But what does it mean "to satisfy" such a definition? Thus far, I have implied that "to satisfy" is to possess "enough" of the properties specified in the HPC definition. Here, we might suspect the vagueness this implies regarding (say) the delineation of the species category and membership in particular species taxa is the Achilles' heel of the view. I want to offer two responses to this concern.

First, what counts as having "enough" of the relevant properties—as with what are the relevant properties in the first place—is an a posteriori matter determined in particular cases by the practitioners of the relevant science rather than by philosophers with a penchant for crisp universality. There need be no one answer to the question of what is "enough," but whatever answers are given in particular cases will be responsive to the clusters that one finds in the world.

Second, even once there *is* general agreement about what counts as "enough," there clearly will be cases of genuine indeterminacy with respect to both the species category and membership in particular species taxa. Yet this indeterminacy seems to me to reflect the continuities one finds in the complex biological world, whether one is investigating species, neurons, or other parts of the biological hierarchy. There will be genuine indeterminacy about the rank of given populations of organisms, and particular organisms may in some cases satisfy more than one HPC definition for particular species taxa. The former of these indeterminacies, however, is a function of the fact that under certain conditions and over time varieties *become* species, and the descendants of a given species become members of a particular genus; the latter reflects the process of speciation (and its indeterminacies) more directly.

## THE INDIVIDUALITY THESIS

Insofar as the HPC view of natural kinds embraces a form of essentialism, it presents an alternative to the individuality thesis and a revival of ideas central to traditional realism. Whether it represents a better alternative to the individuality thesis turns both on broader issues in the philosophy of science and further reflection on the nature of species in particular. Here, I simply summarize what the argument thus far has shown on this issue and what some of the options are in the issue.

Ghiselin (1974, 1997) and Hull (1976, 1978) have given multiple and diverse arguments for the individuality thesis about species—one part of their view negative (species are not natural kinds), another part positive (species are individuals). For example, it has been argued that the heterogeneity within biological populations implies that species are not natural kinds and that their status as *historical* entities within evolutionary theory supports a view of them as individuals. Insofar as the former types of argument presume a two-way conceptual connection between traditionally conceived essences and natural kinds, they carry no force against the view that species are HPC natural kinds. Thus, the view I have defended undermines negative arguments for the individuality thesis. But the HPC view of natural kinds also shows *both* types of arguments for the Ghiselin-Hull view in a new light because parity of reasoning should lead one to abandon thinking of neuronal populations as natural kinds and embrace a view of them as individuals. Of course, such parity considerations can always be undermined by the differences between how the term *species* is used within evolutionary biology and how (say) the term *retinal ganglion cell* is used within visual neuroscience. The HPC view, however, places the burden on those who think that there is something special about species talk that warrants a unique ontological view of species as individuals to show this uniqueness.

Alternatively, perhaps reflection on the neuroscientific cases should lead one to extend the individuality thesis beyond the case of species to other biological categories. Interestingly, at least some researchers in the relevant neuroscience may be amenable to this idea. For example, following Tyner (1975), Rowe and Stone (1977, 1980a, 1980b) advocate what they call a *parametric* or *polythetic* approach to the individuation of retinal ganglion cells, viewing these cells not as kinds with some type of essence, but as intrinsically heterogenous *populations* of cells that have their own internal coherence and duration. (Indeed, Rowe and Stone explicitly take their cue from the modern species concept.) The problem with such a view, it seems to me, is that central to neural taxonomy is the idea of identifying categories of cells that at least different organisms in the same species instantiate, and these instances considered together do not form an individual. For example, your adrenergic cells and my adrenergic cells considered together are not spatially bounded, occupy different temporal segments, and do not form an integrated whole. Perhaps this points the way to how the positive arguments for the individuality thesis can be sharpened in light of the parity considerations introduced with respect to the negative part of the argument for the thesis.

## PLURALISM

To remind you of what the pluralist holds about species, consider what Dupré (1993) says in articulating his version of pluralism:

There is no God-given, unique way to classify the innumerable and diverse products of the evolutionary process. There are many plausible and defensible ways of doing so, and the best way of doing so will depend on both the purposes of the classification and the peculiarities of the organisms in question.... Just as a particular tree might be an instance of a certain genus (say *Thuja*) and also a kind of timber (cedar) despite the fact that these kinds are only partially overlapping, so an organism might belong to both one kind defined by a genealogical taxonomy and another defined by an ecologically driven taxonomy. (p. 57)

In introducing pluralism as the denial of either or both of two assumptions central to traditional realism—the priority and ordering assumptions—I meant to suggest that there is some tension between pluralism and realism *punkt*. The metaphysical angst that many realists experience with pluralism concerns the extent to which one can make sense of the idea that there are incompatible but equally "natural" (i.e., real) ways in which a science can taxonomize the entities in its domain. There is at least the suspicion that, to use Dupré's terms, pluralism is driven more by the "purposes of the classification" than by the "peculiarities of the organisms in question," as Dupré's own analogy suggests. In rejecting the priority assumption, such pluralism would move one from a realist view toward a nominalist view of species (see Wilson 1996; cf. Hull, chapter 2 in this volume).

Yet the most prominent forms of pluralism about species have all labeled themselves "realist," from Dupré's "promiscuous realism" to Kitcher's "pluralistic realism." Moreover, Boyd (chapter 6 in this volume) views at least Kitcher's brand of pluralism as compatible with his own articulation of the HPC view of natural kinds—suggesting a form of realism that accepts the priority assumption, but rejects the ordering assumption. The idea that Boyd and Kitcher share is one Mishler and Donoghue express (cited earlier): the various species concepts that one can derive and thus the various orders within which one can locate species are merely a reflection of complexities within the biological world. This view has two problems—one with pluralistic realism itself, the other with viewing the HPC view of natural kinds as compatible with such pluralistic realism.

As pluralists say, one can arrive at different species concepts by emphasizing either morphological, reproductive, or genealogical criteria for the species category. Yet it is difficult to see how the choices between these sorts of alternatives could be made independently of particular research interests and epistemic proclivities, which calls into question the commitment to the priority assumption that, I claim, needs to be preserved from traditional realism in any successor version of realism. Perhaps pluralistic realists would themselves reject the priority assumption, although Boyd's own emphasis on what he calls the "accommodation demands" imposed by the causal structure of the world on inductive and explanatory projects in the sciences suggest that he himself accepts some version of the assumption.

Boyd's own view of the compatibility of the two views seems to me to fail to capitalize on the *integrationist* potential of the HPC view, one of its chief appeals. One of the striking features of the various definitions of the species category is that the properties that play central roles in each of them are not independent types of properties, but are causally related to one another in various ways. These causal relationships and the mechanisms that generate and sustain them form the core of the HPC view of natural kinds. Because the properties specified in the HPC definition of a natural kind term are homeostatically related, there is a clear sense in which the HPC view is integrationist or unificationist regarding natural kinds. By contrast, consider the view of pluralists. Kitcher (1984) says that we can think of the species concept as being a union of overlapping species concepts (pp. 336–337; cf. Hull 1965), so it is unified in some sense, but without a further emphasis on something to play the metaphysical role that underlying homeostatic mechanisms play in the HPC view, the unity to the species concept remains allusive within Kitcher's view.

Consider how the differences in views manifest themselves in a concrete case—whether asexual clonelines form species. For the pluralist, the answer to this question depends on which species concept one invokes—in particular, whether one appeals to interbreeding criteria to define the species category. By contrast, on the HPC view, asexual clonelines *are* species because they share in the homeostatic cluster of properties that defines the species category, even though they don't have at least one of those (relational) properties, interbreeding.

Likewise, consider the issue of whether there is a qualitative difference between species and other (especially higher) taxa (see Ereshefsky, chapter 11 in this volume). Again, a natural view for a pluralist to adopt is that how one construes the relationship between species and other taxa depends on which species concept one invokes. For example, on Mayr's biological species concept, species have a reality to them provided by their gene flow and its boundaries, which higher taxa lack; alternatively, pheneticists view both species and higher taxa as nominal kinds because taxa rank is determined by a conventional level of overall phenetic similarity. By contrast, on the version of the HPC view of species I have defended, although the general difference between various taxa ranks will be apparent in their different HPC definitions, there will be cases where questions of the rank of particular taxa remain unresolved by the HPC view.

## CONCLUSION

My chief aims here have been to clarify the commitments of a realist view of natural kinds and to suggest a way of modifying rather than abandoning traditional realism in light of the challenge of biological heterogeneity. Both the individuality thesis and species pluralism seem to me to be extreme reactions to the failure of traditional realism in the biological realm, but I have stopped short here of trying to make a full case for the middle-ground position I have advocated as an alternative to both of these views. That remains for another day.

Because the homeostatic property cluster view of natural kinds is in part inspired by the Wittgensteinian idea that instances falling under a common concept bear only a family resemblance to one another, it should be no surprise that it softens the contrast between natural kinds and other categories of things in the world. On the HPC view, however, two things are special about natural kinds. First, the mechanisms that maintain any given HPC are a part of the natural world, not simply our way of thinking about or intervening in the world. Second, unlike artificial or conventional kinds, natural kinds have HPC definitions that feature only properties that exist independent of us. Thus, the reason for the intermediate character of social, moral, and political categories between natural and artificial kinds is made obvious on the HPC view.

I have suggested that the concept of a natural kind needs some broadening within a realist view of science. Although I think of the resulting view as a form of essentialism, whether it is remains of less significance than the requisite broadening. But if we do think of the resulting view as a form of essentialism, then the concept of an essence need not be viewed as the concept of substance came to be viewed within modern science: as unnecessary metaphysical baggage to be jettisoned.

## ACKNOWLEDGMENTS

Precursors to this paper were presented at colloquia in the philosophy departments at the University of Western Australia, Queen's University, Canada, and the University of Colorado, Boulder in 1995–96; early thoughts about the examples discussed in the third section were first presented at the Cognitive Studies Workshop at Cornell University in May 1991. I would like to thank audiences on all four occasions for useful feedback, as well as Allison Dawe, David Hull, Ed Stein, and Kim Sterelny for helpful comments.

## NOTES

1. Later, I return to this appearance and the putative contrast between traditional realism and pluralism when I consider so-called pluralistic or promiscuous realism.

2. J. D. Trout's (1988) brief discussion of neural crest cells, including the distinction between adrenergic and cholinergic cells, stimulated my interest in this example; I also owe Barbara Finlay a note of thanks for helpful, early discussions of both this and the following example. Neither should be landed with the interpretation I give to the example.

3. Alternatively, one could make the normal developmental sequence part of the homeostatic property cluster definition itself. In either case, the sufficiency is not one that allows for a consideration of all logically or even nomologically possible cases.

## REFERENCES

Boyd, R. N. (1988). How to be a moral realist. In G. Sayre-McCord, ed., *Essays on moral realism.* Ithaca, N.Y.: Cornell University Press.

Boyd, R. N. (1991). Realism, anti-foundationalism, and the enthusiasm for natural kinds. *Philosophical Studies* 61, 127–148.

Chalupa, L. M. (1995). The nature and nurture of retinal ganglion cell development. In M. Gazzaniga, ed., *The cognitive neurosciences*. Cambridge, Mass.: MIT Press.

Cracraft, J. (1983). Species concepts and speciation analysis. In R. Johnston, ed., *Current ornithology*. New York: Plenum Press. Reprinted in Ereshefsky 1992.

Dupré, J. (1981). Natural kinds and biological taxa. *Philosophical Review* 90, 66–90.

Dupré, J. (1993). *The disorder of things: Philosophical foundations for the disunity of science*. Cambridge, Mass.: Harvard University Press.

Ereshefsky, M., ed. (1992). *The units of evolution: Essays on the nature of species*. Cambridge, Mass.: MIT Press.

Ghiselin, M. (1974). A radical solution to the species problem. *Systematic Zoology* 23, 536–544. Reprinted in Ereshefsky 1992.

Ghiselin, M. (1997). *Metaphysics and the origins of species*. Albany, N.Y.: SUNY Press.

Hall, B. K., and S. Horstadius (1988). *The neural crest*. Oxford: Oxford University Press.

Hughes, A. (1979). A rose by any other name: On "Naming of neurones" by Rowe and Stone. *Brain, Behavior, and Evolution* 16, 52–64.

Hull, D. (1965). The effect of essentialism on taxonomy: Two thousand years of stasis. *British Journal for the Philosophy of Science* 15, 314–26, and 16, 1–18. Reprinted in Ereshefsky 1992.

Hull, D. (1976). Are species really individuals? *Systematic Zoology* 25, 174–191.

Hull, D. (1978). A matter of individuality. *Philosophy of Science* 45, 335–60. Reprinted in Ereshefsky 1992.

Kitcher, P. (1984). Species. *Philosophy of Science* 51, 308–333. Reprinted in Ereshefsky 1992.

Kitcher, P. (1989). Some puzzles about species. In M. Ruse, ed., *What the philosophy of biology is*. Dordrecht, Netherlands: Kluwer

Kornblith, H. (1993). *Inductive inference and its natural ground*. Cambridge, Mass.: MIT Press.

Le Douarin, N. M. (1982). *The neural crest*. Cambridge, England: Cambridge University Press.

Le Douarin, N. M. (1987). The neural crest. In G. Adelman, ed., *Encyclopedia of neuroscience*. Boston: Birkhauser.

Mayr, E. (1970). *Populations, species, and evolution*. Cambridge, Mass.: Harvard University Press.

Mayr, E. (1982). *The growth of biological thought*. Cambridge, Mass.: Harvard University Press.

Mishler, B., and M. Donoghue (1982). Species concepts: The case for pluralism. *Systematic Zoology* 31, 491–503. Reprinted in Ereshefsky 1992.

Oppenheim, P., and H. Putnam (1958). Unity of science as a working hypothesis. In H. Feigl, M. Scriven, and G. Maxwell, eds., *Minnesota studies in the philosophy of science*. Minneapolis: University of Minnesota Press.

Paterson, H. (1985). The recognition concept of species. In E. Vrba, ed., *Species and speciation*. Pretoria: Transvaal museum monograph no. 4. Reprinted in Ereshefsky 1992.

Rowe, M. H., and J. Stone (1977). Naming of neurones: Classification and naming of cat retinal ganglion cells. *Brain, Behavior, and Evolution* 14, 185–216.

Rowe, M. H., and J. Stone (1979). The importance of knowing one's own presuppositions. *Brain, Behavior, and Evolution* 16, 65–80.

Rowe, M. H., and J. Stone (1980a). Parametric and feature extraction analyses of the receptive fields of visual neurones: Two streams of thought in the study of a sensory pathway. *Brain, Behavior, and Evolution* 17, 103–122.

Rowe, M. H., and J. Stone (1980b). The interpretation of variation in the classification of nerve cells. *Brain, Behavior, and Evolution* 17, 123–151.

Sober, E. (1980). Evolution, population thinking, and essentialism. *Philosophy of Science* 47, 350–383. Reprinted in Ereshefsky 1992.

Templeton, A. (1989). The meaning of species and speciation: A genetic perspective. In Otte and Endler, eds., *Speciation and its consequences*. Sunderland, Mass.: Sinauer. Reprinted in Ereshefsky 1992.

Trout, J. D. (1988). Attribution, content, and method: A scientific defense of commonsense psychology. Ph.D. thesis, Cornell University, Ithaca, New York.

Tyner, C. F. (1975). The naming of neurons: Applications of taxonomic theory to the study of cellular populations. *Brain, Behavior, and Evolution* 12, 75–96.

Wiley, E. O. (1978). The evolutionary species concept reconsidered. *Systematic Zoology* 27, 17–26. Reprinted in Ereshefsky 1992.

Wilson, R. A. (1995). *Cartesian psychology and physical minds*. New York: Cambridge University Press.

Wilson, R. A. (1996). Promiscuous realism. *British Journal for the Philosophy of Science* 47, 303–316.

# 8 Squaring the Circle: Natural Kinds with Historical Essences

Paul E. Griffiths

## FROM ESSENTIALISM TO INDIVIDUALISM

Chemical elements and biological species have been the two stock examples of natural kinds from Aristotle to Putnam. Samples of lead or individuals of the species *Pan troglodytes* are not only similar in various respects, but are "of the same kind" in some much deeper sense. One way to express this deeper commonality is to say that the members of a kind share an essence—a property common to all the members of a kind and responsible for each member being the kind of thing that it is. My gold watch resembles your gold navel ring in many respects, some known to us and some not, because the atoms of which both are composed share an essence: their atomic number. Essentialism in biology would suggest that my kitten Erasmus resembles Socks the cat because they too share an essence (albeit a less well understood one). Essentialism took on a new life in the 1970s, largely because of the work of the philosophers Saul Kripke (1980) and Hilary Putnam (1975). Biological species were one of the stock examples in this essentialist literature, even though by this time essentialism was regarded by many biologists as inconsistent with the basic tenets of Darwinism!

The perceived antithesis between evolution and essentialism was largely due to the work of Ernst Mayr (1959). Mayr argued that biology before Darwin was characterized by *typological thinking* in which types or kinds of organisms had ontological and explanatory priority over concrete individuals. Darwinian *population thinking* gives populations of concrete individuals ontological and explanatory priority instead. Elliott Sober has argued convincingly that the core of population thinking is the Darwinian approach to variation (Sober 1980). The typological approach explains the resemblances between the individuals in a species in terms of the underlying "natural state" of each individual, just as chemistry explains the resemblances between atoms of the same element in terms of their shared microstructure. The typological approach explains variations between the individuals in a species as perturbations of the natural state of that species. The Darwinian approach explains both resemblance and variation at the population level. Organisms resemble one another not because of something inside each of them, but because of something outside each of them: the genealogical and ecological factors that make these organisms a population or a group of related populations. The properties that differ between individuals are ontologically on a par with those properties they share. Variation is not noise obscuring the essential sameness of the members of a species, but an important, heritable property of populations consisting of the aggregate real differences between its members. Sober concludes that because these explanations of sameness and difference are central to the Darwinian tradition, Mayr is correct in concluding that Darwinism precludes identifying any phenotypic or genotypic features as a species essence. However, Sober notes that it would be quite consistent to be a Darwinian essentialist, given the right choice of essential properties (1980, 209). Population thinking excludes essential intrinsic properties, but it does not exclude essential relational properties. This paper defends just such a relational essentialism.

Mayr famously tried to characterize the relational properties that unite the members of a species. His *biological species concept* (BSC) defined species as "groups of actually or potentially interbreeding natural populations which are reproductively isolated from other such groups" (Mayr 1940, cited in Mayr 1963, 19). Underlying this formal definition of species in terms of interbreeding is the idea of a genealogical nexus. A nonreproductive worker in one beehive neither actually nor potentially interbreeds with nonreproductive workers in other hives, but that individual is united in a genealogical nexus with reproductives who actually or potentially interbreed with reproductives in other hives. These reproductives in turn are united in a genealogical nexus with nonreproductives in their hives, so the several nonreproductives are members of the same species. Attempts to extend the BSC to asexual species also rely on this underlying genealogical element in Mayr's species concept, using it as one of two criteria for specieshood. According to such proposals, an asexual species is a well-defined segment of a genealogical tree of asexual individuals that meets some other criteria, such as containing individuals roughly as morphologically similar to one another as members of a sexual species. This second criteria is designed to distinguish the species-level genealogical tree segment from the larger segments in which it is embedded and from the smaller segments that it embeds. Robert Brandon and Brent Mishler (1987) have generalized these two criteria for specieshood into *grouping criteria* and *ranking criteria*, and they have argued that any species concept must have both a grouping and a ranking criteria. In most modern species concepts, including modern versions of the BSC, the grouping criteria is genealogical. Species must be characterized by some version of monophyly—descent from a single population, a single speciation event, or any similar unique point of origin. The ranking criteria serves to distinguish species from equally monophyletic genera, families, and so forth. Although there is some disagreement over the best definition of species-level monophyly, the main disagreements between the twenty or so current species concepts are in their different ranking criteria. (For a very clear look at species-level monophyly, see Kornet 1993.)

*Individualism* about species is an idea with close links to antiessentialism, both conceptually and historically. Individualists argue that species are not kinds or types at all, but big individual objects. Organisms are not *members* of a species, but *parts* of a species. The individualist arguments of Michael Ghiselin (1974a, 1974b) and David L. Hull (1976, 1978) strongly resembled some of the earlier arguments against essentialism. They argued that species must be able to evolve and that kinds or types do not evolve. Species must be able to undergo unlimited change in any of their genetic or phenotypic characters, not only in peripheral characters. If species were kinds or types of organism, then it would not be species that evolved, but organisms or populations that changed from one species into another. The kinds or types would form a sort of biological absolute space against which evolutionary change occurred. Hull and Ghiselin also pointed to the practical failure of attempts to define species by lists of characters or statistical clusters of characters. This argument took on new force in light of the cladistic revolution in systematics, in which attempts to discern common descent replaced attempts to identify taxa by statistical clusters of characters. These and other arguments convinced the individualists that species could be defined only in terms of the pattern of ancestry and descent among organisms, but the next step in their reasoning is the most relevant to the issues of this paper. Hull and Ghiselin concluded that because species and other taxa must be defined in terms of genealogy, they must be moved from the ontological category of types or kinds to the category of individual objects. If taxa are genealogically or historically defined, then they cannot be natural kinds.

This last step in Ghiselin and Hull's argument depends on a traditional conception of natural kinds in which they are the subjects of spatiotemporally unrestricted laws of nature. If natural kinds are to figure as the subject of universal laws, they themselves must have universal applicability. Laws that make ineliminable mention of things that can exist only at a particular location in time and space are not, in the relevant sense, universal laws. If Ghiselin and Hull are correct, then biological taxa have just such a unique origin in space and time. No part of a taxon can exist outside the cone of causal influence extending from its origin event, so taxa are restricted to a particular portion of space-time and cannot be mentioned in genuine laws of nature.

The conclusion that there are no laws of nature concerning taxa has been welcomed by many theorists as part and parcel of antiessentialism. Hull (1986) has welcomed the liberatory conclusion that there is no such thing as "human nature." Attempts to distinguish normal from abnormal humans are simply misguided. John Morss (1992) has argued that there are no laws of ontogeny and particularly of child development. We should be suspicious of theories that describe a series of stages through which every child passes to reach maturity. The downside of the anomalousness of biological taxa is that it threatens the status of biology as a science. If there are no biological laws, biology is merely the study of how things happen to be around here right now (Smart 1963). The threat is not merely to laws about species and speciation. The parts of organisms and their physiological processes are standardly classified in the same way as whole organisms—using the Darwinian or evolutionary homology concept: two organs are homologous if they are copies of a single ancestral organ. Thus, the wings of pigeons are "the same" as the wings of albatrosses in a way that they are not "the same" as the wings of fruit bats. The conclusion that there are no lawlike principles of the structure and development of organs or physiological processes has not been welcomed by developmental and structural biologists.

Hull's response to these worries is interesting and has not been sufficiently discussed. He notes that there are two quite different schemes of classification in biology. Systematics and perhaps the anatomical and physiological disciplines classify by homology. If this scheme is evolutionary homology, they face the problem we have just encountered. Ecology and functional biology, however, seem to classify by analogy. Kinds such as predator, prey, digestion, thermoregulation, and so forth are not genealogically defined. Entirely unrelated organisms can share an ecological role. The wings of pigeon and fruit bat may be "the same" in the sense that they are both shaped for work amongst the branches. Likewise, genealogically unrelated DNA sequences can code for a protein with the same metabolic function. Hull suggests that it is to these categories of analogy that biology must turn in its search for laws. This suggestion is attractive when put in these abstract terms, but when we try to apply it, the results are worrying in the extreme. Developmental and structural biologists classify by homology for good reason. Functional resemblances between organs tend to be shallow. In human engineering, devices that have the same function but that were designed independently tend to be very different. In the same way, the circulatory system of an octopus is very different from that of an aquatic mammal of similar size. If developmental biology and structural biology seek only laws about functional kinds, then laws in these disciplines may be little more than performance specifications (Griffiths 1994, 1996a, 1996b). Hull's recommendation also came just as ecologists were turning from the dreams of a grand theory that had occupied them in the 1960s to a renewed interest in contingency and history (Kingsland 1985). Ecological models, it was suggested, may never achieve the status of universal laws and will remain always in need of testing and retuning for each new case. This trend in ecological thought has continued, so if biology looks to ecology for its spatiotemporally unrestricted laws of nature, it may not be pleased with what it finds.

## DOES BIOLOGY NEED NATURAL KINDS?

A number of biologists have argued that biology cannot do without natural kinds. The *process structuralist* school has suggested that biology has no real explanation of form unless it has an explanation in terms of natural kinds (Goodwin and Saunders 1989, Goodwin 1994). I have argued against this claim elsewhere (Griffiths 1996a), but it has recently been reiterated (Webster and Goodwin 1996). There is much that is correct in this new presentation, but I argue that what is correct can be accommodated by a thoroughly Darwinian and historical conception of biological kinds. Other authors have argued that understanding the nature of the characters that are candidates for evolutionary homology will require a nonevolutionary, structural concept of character identity (Müller and Wagner 1991; Wagner 1994, 1996). I consider the relations between the evolutionary homology concept and this proposed structural-developmental homology concept in my closing section.

In their 1996 book *Form and Transformation*, Brian Goodwin and Gerry Webster reiterate their claim that biology needs natural kinds of organisms, parts, and processes. If experimental biologists are to perform repeatable experiments, they must be able to say what it would be to have subjects "of the same type." If the anatomical and physiological disciplines are to extrapolate from the individuals in the laboratory to individuals elsewhere, they need to know what *sort* of things they have been examining. Goodwin and Webster argue that scientific practice only makes sense on the assumption that there are real *sorts* of things as well as real *individual* things. Individuals of the same sort share some underlying "nature," and it is the aim of science to elucidate these "natures." Goodwin and Webster agree with the anti-essentialists that the shared nature of a biological kind cannot be either a set of phenotypic characters or a set of genes. Both vary too much within the groups, such as species, about which biologists need to generalize. They infer from this variation that organisms of the same sort must have something else in common: something that must emerge as essentially the same in each individual despite differences in the developmental resources that individuals inherit. It must also be something that can be shared by many phenotypically different individuals. Differences within a species must be varied expressions of a common underlying nature. Goodwin and Webster thus reject the population-thinking model of variation in which the properties that differ between individuals are ontologically on a par with those properties that are shared. They revert to a typological model, in which variation is explained as the response of an essentially similar system to different inputs. A genetic change or a change in some other developmental input acts on the "real nature" of the organism to cause it to express a new outcome among the range of outcomes that it is, in Webster and Goodwin's phrase, "competent" to produce. Variation masks the real underlying sameness of a type of organism and it is the task of science to see through the variation to the essential sameness.

Goodwin and Webster's candidates for the real natures of organisms, parts, and processes are *morphogenetic fields*. They conceive of these fields as an emergent level of organization in the developmental process. The existence of such a level of organization explains the constancy of biological form in the face of substantial variation in all elements of the developmental matrix, including the genome. Goodwin is fond of comparing the morphogenetic field to an attractor in complex systems theory. Development from a wide range of genetic starting parameters is drawn to an attractor represented by a particular morphogenetic field (Goodwin et al. 1993). The existence of such an emergent level of organization can also explain the fact that mutations and phenocopies are often equivalent. The abnormal bithorax phenotype in *Drosophila* can be produced by a genetic change (the bithorax mutant) or by an environmental change (the bithorax phenocopy). The potential to produce the bithorax form is thus inherent in the morphogenetic field of the segment that becomes a second thorax. This potential can be triggered by several different perturbations to that field.

A revival of the morphogenetic field concept has also been advocated recently by Scott Gilbert, John Opitz, and Rudolph Raff (1996). Their conception of a morphogenetic field is much closer to the idea of a gene control circuit: a set of genes linked together by relations of feedback, excitation, and inhibition. Their field concept is directed at explaining the same phenomena as Goodwin and Webster's. By linking many genes together in this way, they hope to explain the sense in which development is an emergent phenomena: a circuit may have properties that are robust when some constituents of the circuit are changed and may be pushed into the same alternative configuration by any of several different perturbations. Despite these similarities, there is a critical difference between the two field concepts. Webster and Goodwin strongly resist the idea that a field can be reduced to the genes and other molecular machinery that underlie it. They also resist identifying the "competence" of the field with the norm of variation of those genes, mainly because their field is essentially an *invariant* across individuals of the same kind. Changing the particular genes that underly the field *makes no difference* to the field itself. When a genetic change causes a phenotypic change, according to Goodwin and Webster, we are not seeing the result of a slightly different morphogenetic field, but an identical field producing another of the outcomes within its competence. Gilbert, Opitz, and Raff make no such essentialist commitment, which demonstrates an important conceptual point. It is not necessary to postulate a theoretical entity to act as an absolute invariant in order to explain robust developmental outcomes. The field concept is entirely viable in a population-thinking form in which robust developmental outcomes are explained by the fact that many *different* (but similar) morphogenetic fields produce the same outcome.

Goodwin and Webster's case for the existence of morphological fields as developmental invariants is driven not by the need for an emergent level of developmental organization to explain canalization and mutation/phenocopy equivalences, but by the abstract methodological claim made at the beginning of this section. They do not see how extrapolation from observed to unobserved instances can be valid unless these instances share some underlying, invariant nature. They postulate invariant morphogenetic fields to meet this *epistemological* need. In the next two sections, I explain why this postulation is a misunderstanding of the natural kind concept and of how natural kinds really license such extrapolation.

## NATURAL KINDS WITHOUT TEARS

Induction and explanation presume that the world contains correlations between properties that are, to use Nelson Goodman's term, *projectable* (Goodman 1954). We can depend on these correlations holding in new cases. Theoretical categories embody current understanding of where such projectable clusters of properties are to be found. The species category, for instance, is supposed to reliably collect morphological, physiological, and behavioral properties. We can investigate these properties in the species as a whole by studying a few members of the species. That being accomplished, we can explain the fact that an individual has certain properties by citing its species: any organism that *was* of this species *would* have those properties. In Goodman's original presentation, the projectability of theoretical categories is supposed to be judged on the basis of our past experience in using the categories and others related to them—which, in practice, means that we judge projectability on the basis of our background theories of the domain to which a theoretical category applies. Our theories lead us to believe that all the chemical properties of sulphur will be reliably reproduced by future instances of that element, whereas few if any of the physical properties of Citroen cars will be reliably reproduced by future instances of that marque. Natural kinds are simply a realist interpretation of Goodman's projectable categories. The categories that figure in successful theories are projectable because the theories have some degree of versimilitude. The instances of these categories really do share an underlying nature. Therefore, from the realist perspective they adopt, Webster and Goodwin are correct in claiming that for biology to engage in induction and explanation, it must have theoretical categories that represent natural kinds.

The concept of a natural kind has a long history, stretching back at least as far as John Locke's discussion of the distinction between *real kinds* and *nominal kinds*, if not as far as Plato's famous remark about "carving nature at its joints" (Hacking 1991a). In the logical empiricist tradition, from which philosophy of science as we now know it emerged, natural kinds are envisaged as the objects of spatiotemporally unrestricted laws of nature. They are the nodes around which theories in the fundamental sciences are structured (Quine 1977). But recent decades have seen substantial changes in thought about natural kinds. Greater philosophical attention to the special sciences has led to the eclipse of the idea that these sciences are one day to be reduced to more foundational sciences such as physics and chemistry. The current received view is that the dynamics of physical systems can only be adequately captured using a hierarchy of theoretical vocabularies, each irreducible to the vocabularies below it. Irreducibility is guaranteed by the fact that descriptions in one vocabulary can be made true by indefinitely many arrangements of the structures described in lower-level vocabularies (Fodor 1974; Wimsatt 1976a, 1976b; Jackson and Pettit 1988; Lycan 1990). There are indefinitely many ways, for example, to construct instances of money, a central theoretical category of economics. An empirically successful theory with such irreducible categories cannot be eliminated without losing the knowledge embodied in its empirical generalizations. Economic generalizations about money, for example, can be made true by indefinitely many physical systems of currency and so cannot be replaced by generalizations about any category of physical systems. This idea has led to what Richard Boyd (1991) has called "the enthusiasm for natural kinds" (p. 127). Categories from any special science that enter into the generalizations of that science are now commonly regarded as natural kinds. Inflation and schizophrenia take their place alongside electrons and stars.

The generalizations of the special sciences often fail to live up to the ideal of a universal, exceptionless law of nature. Generalizations in psychology or economics are often exception-ridden or hedged with generous ceteris paribus clauses or both: decreases in the money supply *usually* lead to a contraction of the economy, *all other things being equal.* Nevertheless, the key feature of a law of nature is still present in these generalizations: they have *counterfactual force*. The idea of counterfactual force is central to the traditional idea of a law of nature because it explains how laws differ from mere widespread coincidences. It may well be true, for example, that every species with an eusocial grade of social organization has individuals that weigh less than 5,000 kg, but even if this statement turns out to be true throughout the Federation of Planets, it will not be a law of nature. Nothing in our theories licences the subjunctive conditional, "if this *were* a member of an eusocial species, it *would* weigh less than 5,000 kg." This statement lacks counterfactual force: it is not "lawlike." A key part of the conception of a natural kind is that it is a category about which there are lawlike, counterfactual-supporting generalizations. We can use induction to investigate natural kinds because we expect certain classes of properties to be connected to those kinds in a lawlike, rather than a coincidental way. For example, our background theories licence the expectation that samples of an element will possess their chemical properties in a lawlike rather than a coincidental way. Having tested the chemical properties of the samples, we can extrapolate to the chemical properties of other instances of the element.

The idea of counterfactual force is easily generalized to the exception-ridden generalizations of the special sciences. Minimally, any generalization that is a better predictor of phenomena than a suitably designed null hypothesis has some counterfactual force. This allows us to frame a minimal conception of *naturalness* for kinds. A kind is (minimally) natural if it is possible to make better than chance predictions about the properties of its instances. Suprisingly, this utterly minimal conception of a natural kind is not toothless. It does not license the conclusion that any way of classifying nature is as good as any other. Natural kinds are ways of classifying the world that correspond to some structure inherent in the subject matter being classified. They contrast to arbitrary schemes of classification about which the nominalist claim that the members of a kind share only a name is actually true. Furthermore, the minimal account of naturalness lends itself to successive restrictions that allow us to distinguish between kinds of greater or lesser naturalness and hence of greater or lesser theoretical value.

Although it is not possible in this essay to give an adequate treatment of the principles for choosing between alternative taxonomies of nature, a brief, general outline may be helpful. The value of a lawlike generalization can vary along two independent dimensions, which we might call *scope* and *force*. *Force* is a measure of the reliability of predictions made using that generalization. *Scope* is a measure of the size of the domain over which the generalization is applicable. A theoretical category about which there are generalizations of considerable scope and force is more natural than one about which generalizations tend to have more restricted scope and lesser force. For example, the claim that cladistic taxonomy is "maximally predictive" of the unobserved properties of taxa is intended to show that cladistics is superior to other systems in terms of force. There will not always be a clear winner when we compare two sets of theoretical categories on the basis of scope and force. Scope and force may trade off against one another. The scope of generalizations made with one set of categories may overlap rather than include the scope of generalizations made with the other taxonomy so that neither taxonomy can be discarded without loss of understanding.

Theoretical categories can also differ in the number of generalizations into which they enter so that one category can seem the focus of a richer scientific project than another, irrespective of comparisons of the strength of the generalizations they yield. Finally, theoretical categories are tied up in wider research programs whose relative prospects may cause us to prefer that set of categories to another despite a paucity of currently established generalizations about the preferred set of categories. None of these considerations, however, refutes the basic idea that some theoretical categories are superior to others and that some are of no foreseeable value whatever. Even if different categories are valuable for different purposes, it is still true that some are better for a particular purpose than others and that some have no foreseeable use at all. Embodying these ideas in the language of natural kinds links it to a broadly realist perspective in which the predictive and explanatory value of categories is taken to be prima facie evidence that they capture part of the structure of the world. The "enthusiasm for natural kinds" embodies the realization that there is more structure in the world than can be captured by a single taxonomy of nature.

Richard Boyd has outlined a similar conception of natural kinds using his idea of *causal homeostasis* (Boyd 1991). According to Boyd, we judge a kind to be projectable, or natural, when we have theoretical grounds for supposing (or we simply postulate) that there is a causal explanation for the property correlations we have observed. Boyd calls this underlying reason a *causal homeostatic mechanism*—something that causally explains the maintenance of the same property correlations throughout the set of instances of the kind. In my reading of Boyd, this causal homeostatic mechanism corresponds to the traditional "essence" of a natural kind. In the paradigmatic example of chemical elements, the causal homeostatic mechanism is a shared microstructure. It is because of their subatomic composition that the instances of a chemical element share their chemical properties. However, nothing in the idea of a causal homeostatic mechanism requires the mechanism to take the form of a set of intrinsic properties possessed by every member of the kind and synchronically causally producing the other properties characteristic of the kind. Money, for example, has no such microstructural essence, although it is a key node in many economic theories. The lawlike generalizations about money, such as those connecting money supply to inflation or to interest rates, hold true in an economy because of a social convention treating some class of objects as a means of exchange and because agents in that economy try to maximize their utility. Neither of these circumstances is linked to any intrinsic property of the currency units. In a similar way, if characteristic ecological successions represent natural kinds in ecology, the causal homeostatic mechanism for the kind "Fiordland rainforest succession" will include the available range of seeds and other propagules, the climate of the region, and so forth. All that is required for the existence of a natural kind is that there be some causal process in nature that links together several different properties of the objects influenced by that process. A shared microstructure is only one way of achieving this "homeostasis" of properties.

The idea of a causal homeostatic mechanism frees the idea of *essence* from many of its traditional commitments—commitments that have proved problematic in the case of biology. My interpretation of Boyd's work is that he provides a general analysis of the role that essences play in scientific reasoning about natural kinds and then redefines essence as any property that can play this role. Any state of affairs that licences induction and explanation within a theoretical category is functioning as the essence of that category. The essential property that makes particular instances members of the kind is their relation to that causal mechanism, whatever it may be. One exciting implication of this approach is that it breaks down the traditional distinction between natural kinds and kinds generated by human agency. I have exemplified this possibility by using money as an example of a natural kind. Artifactual kinds, such as kinds of tool or ceremony, can be the subject of lawlike generalizations because the sociological causes that produce them can function as essences. These sociological causes guarantee with some degree of reliability in some suitably delimited domain that instances of the kind will share a cluster of properties.

Boyd's proposal is a substantial revision of the traditional ideas of essence and natural kindhood (see also Boyd, chapter 6 in this volume). "Natural" kinds that have never been seen before can be created by social processes unique to a particular society. The fact that people think certain things form a kind can function as the essence of that kind! The justification for these conceptual revisions is that they allows insights about the formation and use of theoretical categories to be extended to the special sciences rather than restricted to a (dwindling) core of kinds with microstructural essences. The psychologist Frank C. Keil (1989) has used Boyd's ideas to argue for a continuity between category formation by developing children and category formation in science (see also Keil and Richardson, chapter 10 in this volume). I have argued that the formation of theoretical categories in psychology, including categories unique to particular cultures, is best understood as a search for causal homeostasis (Griffiths 1997).

In this section, I have tried to motivate a very general conception of natural kinds, one that discards many of the traditional associations of the natural kind concept. Natural kinds are needed for induction and explanation. They represent theoretical categories that we judge to be projectable, which requires them to enter into lawlike, counterfactual supporting generalizations. It does not require that these generalizations be universal, deterministic laws: lawlike generalizations of more limited scope and force are enough. Finally, kinds are defined by the processes that generate their instances, and for many domains of objects, these processes are extrinsic rather than intrinsic to the instances of the kind. The causal homeostatic mechanism that guarantees the projectability of a kind plays the traditional role of an essence, but it need not be a traditional, microstructural essence.

## HISTORICAL ESSENCES

Cladistic taxa and parts and processes defined by evolutionary homology have historical essences. Nothing that does not share the historical origin of the kind can be a member of the kind. Although Lilith might not have been a domestic cat,[1] as a domestic cat she is necessarily a member of the genealogical nexus between the speciation event in which that taxon originated and the speciation or extinction event at which it will cease to exist. It is not possible to be a domestic cat without being in that genealogical nexus. Furthermore, cladistic taxa and parts and processes defined by evolutionary homology have no other essential properties, which is why process structuralists such as Goodwin and Webster do not think that these categories can be adequate for developmental and structural biology. They do not see why kinds whose only essential properties are historical should be the subjects of lawlike, counterfactual-supporting generalizations about morphological and physiological properties. Yet there is a well-known Darwinian ground for expecting groups defined by common descent to share morphological and physiological characters:

> It is generally acknowledged that all organic beings have been formed on two great laws—Unity of Type and the Conditions of Existence. By unity of type is meant that fundamental agreement in structure, which we see in organic beings of the same class, and which is quite independent of their habits of life. On my theory, unity of type is explained by unity of descent. (Darwin 1859, 206)

Even in its most extreme adaptationist forms, Darwinism retains these two "great laws" as separate forces that conjointly explain biological form. The principle of heredity acts as a sort of inertial force, maintaining organisms in their existing form until some adaptive force acts to change that form. This *phylogenetic inertia* is what licenses induction and explanation of a wide range of properties—morphological, physiological, and behavioral—using kinds defined purely by common ancestry. If we observe a property in an organism, we are more likely to see it again in related organisms than in unrelated organisms. Since Darwin, this idea, much elaborated, has been the basis of comparative biology (Brooks and McLennan 1991, Harvey and Pagel 1991).

However, the mere existence of phylogenetic inertia is not the whole story. There are striking contrasts between biological traits in their tendency to persist without reference to the "conditions of life." I have argued elsewhere that it is a mistake to assume that when we have a selective explanation for the origin and fixation of a trait, there is nothing left for selection to explain (Griffiths 1992, 1996b). Many traits display a pattern of phylogenetic inertia reminiscent of the inertia of Aristotelian physics. Just as early physics expected a body with no forces acting on it to return to rest, these traits tend to atrophy when no selective forces work to maintain them. The apparently panphyletic tendency of cave-dwelling organisms to lose pigmentation and sight is a well-known example. With traits displaying this pattern, selective explanations of their maintenance are as legitimate as selective explanations of their origin. In contrast to these Aristotelian traits, other traits display an apparently Newtonian pattern of phylogenetic inertia. They are maintained over the longest geological timescales and the widest range of conditions of life, with no apparent regard for adaptive utility. Traits of this kind are the ones that make good taxonomic characters. The pattern of fused segments that marks out crustaceans among the arthropods is a well-known example, and classic morphological traits like this are not the only sort of traits that display the Newtonian pattern. Part of Konrad Lorenz's legacy was the realization that some behaviors also have a very strong phylogenetic signature.

The fact that different traits display such different patterns of phylogenetic inertia calls out for a developmental explanation. Development is the obvious place to look for something that reduces variance in certain traits and so causes them to resist atrophy or elimination as an effect of adaptive change. Proposals for developmental explanations of strong phylogenetic inertia can be divided into two types. The first type includes Rupert Riedl's (1977) concept of *burden* and William C. Wimsatt's (1986) notion of *generative entrench-*

*ment* (see also Schank and Wimsatt 1986). Both concepts draw attention to the fact that one trait may be developmentally linked to a range of other traits, making its elimination far less likely than if it were an independent developmental unit. Proposals of this type still assign a major role to selection in maintaining traits, although it acts indirectly through the structure of the development system. The second type of developmental explanation of phylogenetic inertia avoids implicating selection in any way. Examples of this type include Goodwin and Webster's concept of *generic forms* and perhaps some of Stuart Kauffman's ideas (Goodwin et al. 1993, Kauffman 1993, Webster and Goodwin 1996). These approaches take the fact that a trait is widespread in a group as a sign that this trait is an easy one for that kind of developmental system to generate. The widespread occurrence of the trait is not to be explained by its utility or its links to other useful traits, but by the structure of the developmental system. In one of Goodwin's favorite examples, the fact that there are only three patterns of phylotaxis in higher plants—patterns in which successive leaves emerge from the stem—is explained by the existence of three stable attractors that emerge when a single, continuous, quantitative developmental parameter is altered in a model of the growth of the meristem. The spiral phylotactic pattern seen in 80% of these plants emerges from the model as the outcome with the largest basin of attraction (Goodwin 1994, 116–133).

Both types of explanation of phylogenetic inertia support rather than oppose the idea that categories based on evolutionary homology will provide a natural taxonomy with which to investigate morphological and physiological characters. Past discussion of the second type of explanation has tended to give the opposite impression, however. The reasons for this tendency lie in philosophy rather than biology. Goodwin, Webster, and other process structuralists have argued that categories based on evolutionary homology do not have an underlying "nature" suitable for scientific investigation because evolutionary homologies do not have traditional, microstructural essences. They infer from this argument that if biology is to be scientific, biological kinds with such essences must exist. Furthermore, because there are some reliable taxonomic characters—the Newtonian traits just discussed—they infer that these characters must have just such underlying microstructural essences. I tried to show in the last section that the philosophical part of this process structuralist argument is mistaken. Microstructural essences are not needed to justify explanation and induction. What is left of the argument is just the postulation of developmental causes for phylogenetic inertia, with which the Darwinian can wholeheartedly agree. However, because of her general theoretical orientation, the Darwininian will have expectations very different from the process structuralist's about these developmental mechanisms. She will expect them *(a)* to have a phylogenetic pattern like other characters, and *(b)* to show variation in natural populations. The first of these expectations supports the continued use of historically defined kinds in biology, including biological investigations of the developmental basis of phylogenetic inertia. The second expectation means that even when the developmental basis of phylogenetic inertia is understood, the Darwinian will not expect to see historical kinds displaced by purely developmental definitions of taxa, parts, and processes. In the next two sections, I expand on these two points.

## WHY GHISELIN AND HULL WERE WRONG

Antiessentialists and individualists about biological taxa were wrong to suppose that there are no lawlike generalizations about these taxa. A hierachical taxonomy based on strict phylogenetic principles will collect more of the correlations between characters, from molecular to behavioral, than any other taxonomy we know how to construct. Such a taxonomy will group organisms into natural kinds because it will predict with considerable force many properties of individuals. Although such a taxonomy will predict the properties of unobserved genera or species, it will function most powerfully in predicting the properties of new members of taxa at or below the species level. A number of competing (though not necessarily exclusive) explanations of the special status of species are embodied in some of the twenty or so currently proposed species concepts. These explanations draw attention to causal processes such as gene exchange (biological species concept) or selection for the requirements of a niche (ecological species concept). These mechanisms reinforce phylogenetic inertia in keeping the members of a species clustered together in the space of biological possibility (cf. de Queiroz, chapter 3 in this volume).

Generalizations about taxa are exception-ridden. This does not, however, prevent them from being lawlike or having counterfactual force. The causal homeostatic mechanisms of taxa license the prediction that a new bird will detect its prey using visual cues or that in a new cephalopod, the blood vessels supplying the retina will lie under rather than over it. The causal homeostatic mechanisms also make it legitimate to extrapolate experimental results to other members of the same taxon, especially at the species level. The fact that such predictions and extrapolations are not absolutely reliable is simply beside the point. They are more reliable than chance, so unless there is some other way to capture the same regularities, eschewing the use of these categories would mean discarding some of our understanding of the structure of nature.

Parts and processes defined by evolutionary homology can be used for explanation and induction for the same reason that historically defined taxa can be used: phylogenetic inertia licenses the extrapolation of morphological and physiological properties in categories defined by common ancestry. Also, among these properties are the very developmental processes that are likely to explain the phenomena of phylogenetic inertia! Developmental processes, as much as other anatomical or physiological kinds, can be expected to reflect phylogeny. What lies at the bottom of all these phylogenetic patterns is, after all, the fact that related organisms inherit similar developmental resources. Plant physiology, for example, does not converge on animal physiology whenever it would be adaptively useful for it to do so because plant cells inherit a range of membrane templates, organelles, genes, and so forth that are fundamentally different from those inherited by animal cells.

If historical kinds are natural because related individuals inherit similar developmental resources, it might seem possible to define the kinds in terms of the developmental resources that underly them. This proposal would treat the shared developmental resources of a taxon as the causal homeostatic mechanism of that taxon, a mechanism that takes the form of a traditional microstructural essence possessed by all and only the members of the taxon. I suspect that this thought is at the back of several structuralist criticisms of the use of historical kinds in biology. The structural or developmental biologist sees that the processes they are investigating explain the fact that members of a taxon share a rich cluster of properties, which suggests to them that the real essence of the taxon is not its shared history, but its shared developmental processes. In the next section, I show that this very natural line of reasoning is mistaken because of the original, Darwinian considerations against essentialism outlined at the beginning of the paper.

## WHY GHISELIN AND HULL WERE RIGHT

Darwinians will resist the suggestion that taxa be defined developmentally because they expect developmental processes to be just one more product of evolution. As such, they expect developmental processes to display variation between individuals in natural populations, just as other characters do. Empirically, they do not expect to find a list of developmental properties possessed by all and only the members of a species any more than they expect to find lists of phenotypic or genotypic characters possessed by all and only members of a species. Conceptually, even if such a list of properties existed for a species, it would be an accidental not an essential matter. An individual united in a genealogical nexus with the existing members of the species, but lacking some property on the list, would still function as a member of the same evolutionary unit. The purpose of the species concept for a Darwinian is to describe the units of evolution, and essentialist species concepts fail to do this.

There are a number of reasons why Darwinians have wanted to take a more developmental perspective on evolution. A developmental perspective highlights the problems with an atomistic approach to the evolution of characters, in which each character is assumed to be optimized independently of the others (Gould and Lewontin 1979, Lewontin 1983). A developmental perspective can also draw attention to the wide range of developmental resources other than genes that can be the subject of evolutionary explanation (Griffiths and Gray 1994). Perhaps most importantly, a developmental perspective may allow a more adequate integration of phenotypic and genotypic evolution. This last motivation is at the heart of Gilbert, Opitz, and Raff's (1996) proposal to revive the morphogenetic field concept. They suggest a definition of evolution as change over time in the developmental biology of a lineage, which contrasts with the currently popular definition of evolution as change in gene frequencies in a lineage. But all these goals of a *Darwinian developmentalism* require development to be part of the process of evolution by natural selection. As such, development must be something that exhibits heritable variation. It cannot be something that is invariant across all the members of a species.

The Darwinian developmentalist is an evolutionist who focuses on development, just as a gene selectionist is an evolutionist who focuses on genes. These two views of the evolutionary process differ in important ways, but they agree on some central Darwinian themes. It is these themes that Ghiselin and Hull were right about in their insistence on a historical, anti-essentialist view of taxa and of homology. One central Darwinian themes is the ubiquity of variation. Where the Darwinian developmentalist observes a widespread phenotypic character, she will not assume that it is produced by an underlying, developmental invariant. She will be open to the idea that it is an outcome that can be produced by any of a range of different but similar developmental processes. *Canalized* developmental outcomes are precisely those that can be produced by many different configurations of developmental resources. Developmental biology illuminates how canalization occurs, but it need not do so by finding or postulating a developmental invariant other than the canalized outcome itself. Another central Darwinian theme is the value of a phylogenetic perspective in all biological investigations. The Darwinian developmentalist will expect to find a phylogenetic signature in characters of all kinds and to make extensive use of the comparative method in testing hypotheses about character associations. This phylogenetic perspective will extend to developmental biology.

## THE STRUCTURAL HOMOLOGY CONCEPT

I have defended the view that historically defined taxa are natural kinds and the corollary view that evolutionary homologues are also natural kinds. I have defended these views against some arguments associated with structuralist approaches to biology. In this closing section, I want to consider two other, recent arguments that biology needs a structural homology concept. The first argument suggests that the evolutionary homology concept is somehow unworkable without a prior conception of structural homology. This argument is mistaken, but a second, better argument points to the potential value of a structural homology concept, including its value in illuminating the basis of the evolutionary homology concept.

The mistaken argument, which we can perhaps regard as put forward by a hypothetical structuralist strawman, is that because candidates for evolutionary homology must be real characters of organisms, the identification of evolutionary homologues is parasitic on the identification of characters defined by some nonevolutionary homology concept. It is certainly true that before it can be asked whether two characters in different taxa are homologous or homoplastic, they must be identified as characters. We might, for example, measure the ratio of length to circumference of a bone, find that it was constant across a range of taxa, and use a cladistic analysis of a whole suite of characters to determine if this commonality can plausibly be identified as a homology. The first part of this procedure embodies a decision to treat the ratio as a character. It is also true that not everything that can be measured is a real character. Probably no one would bother to measure in different taxa the ratio between number of retinal receptor types and number of legs. However, it does not follow that we need to know which features of organisms are real characters before we start looking for homologies. Cladistic analysis can proceed from a list of arbitrary measurements by looking for congruences among the evolutionary trees produced by different measurements and thus "bootstrapping" itself into a reliable character set. A set of characters, different subsets of which produce similar trees, is probably a set of real units of inheritance and evolution.

The better argument for the desirability of a structural homology concept is given by Gunther P. Wagner (1994; see also Müller and Wagner 1991, Wagner 1996). The pre-Darwinian homology concept distinguished homologous resemblances among taxa from analogous ones. *Homologies* are different instances of the very same character, whereas *analogies* are different characters that happen to resemble one another. Darwin gave a specific interpretation to this idea of being *really* the same character rather than *apparently* the same character. Two characters are really the same if they are both the same as some character possessed by a common ancestor. Wagner's point is that this definition does nothing to explain the sense in which characters are "the same" by descent. Darwin has analyzed character identity horizontally, between taxa, but not vertically, between parent and offspring. It is simply assumed that some resemblances between parent and offspring amount to true character identity, just as it was previously assumed that some resemblances between taxa amount to true character identity. Wagner's point is not that the Darwinian needs to understand the vertical relation of character identity before she can begin to reconstruct phylogeny. As I have argued, the Darwinian can simply presume that there are real units of inheritance and identify good candidates for these units by trial and error. The point, rather, is that until we understand the nature and origins of the units of heritable biological form, we will not know *why* this bootstrapping procedure works. More generally, we will not understand why the historical, phylogenetic approach to biology is so useful. To explain this fundamental fact about biology we need to understand why some characters and not others display phylogenetic inertia. As Wagner (1994) puts it, "the main goal of a biological [i.e., developmental] homology concept is to explain why certain parts of the body are passed on from generation to generation for millions of years as coherent units of evolutionary change" (p. 279).

It is this sort of question that has been the focus of Wagner's more recent work on the evolution of modularization and canalization of development (Wagner 1996, Wagner et al. 1997). Wagner (1994) rejects an analysis of vertical character identity based on identical developmental origin, a modern derivative of the traditional practice of judging homology from the relative position of parts in the embryo: "Too often do we find substantial developmental variation among structurally, and presumably phylogenetically, identical body parts" (p. 276). He would presumably reject an account of vertical character identity based on identical genetic causes for the same reason: homologous characters can persist through substantial changes in the genetic inputs to their development. In place of such ideas, Wagner sets up the goal of understanding why organisms have come to have discrete, reidentifiable parts. A theory of why there are parts will tell us how those parts can be naturally taxonomized. Wagner's research program is thus (quite self-consciously) a search for natural kinds construed as the objects of lawlike generalizations.

Wagner's work is an instance of what I have described as "Darwinian developmentalism" because he looks for the origins of these units of structural homology in the evolutionary process rather than in a system of ahistorical biological types like the system postulated by the process structuralists Goodwin and Webster. Wagner and his collaborators have tried to model selective processes that favor the emergence of discrete developmental "modules" that are stabilized against various perturbations of the developmental system. Although these modules function as developmental invariants in at least some timescales, persisting with apparent disregard for the "conditions of life," they have themselves emerged as a result of the evolutionary process, and they will possess a phylogenetic signature—an association with a particular lineage.

## ACKNOWLEDGMENTS

I have benefited from the comments of Rob Wilson, Gunther Wagner, and Karola Stotz on an earlier draft of this paper.

## NOTE

1. She could exist even if domestic cats had speciated some generations ago—making her, on cladistic principles, a member of one of two new species (LaPorte 1997).

## REFERENCES

Boyd, R. (1991). Realism, anti-foundationalism and the enthusiasm for natural kinds. *Philosophical Studies* 61, 127–148.

Brooks, D. R., and D. A. McLennan (1991). *Phylogeny, ecology and behaviour.* Chicago: Chicago University Press.

Darwin, C. (1859, reprint 1964). *On the origin of species: A facsimile of the first edition.* Cambridge, Mass.: Harvard University Press.

Fodor, J. A. (1974). Special sciences. *Synthese* 28, 77–115.

Ghiselin, M. T. (1974a). *The economy of nature and the evolution of sex.* Berkeley: University of California Press.

Ghiselin, M. T. (1974b). A radical solution to the species problem. *Systematic Zoology* 23, 536–544.

Gilbert, S. F., J. M. Opitz, and R. Raff (1996). Resynthesising evolutionary and developmental biology. *Developmental Biology* 173, 357–372.

Goodman, N. (1954). *Fact, fiction and forecast.* London: Athlone Press, University of London.

Goodwin, B. C. (1994). *How the leopard changed its spots: The evolution of complexity.* New York: Charles Scribner & Sons.

Goodwin, B. C., and P. Saunders (1989). *Theoretical biology: Epigenetic and evolutionary order from complex systems.* Edinburgh: Edinburgh University Press.

Goodwin, B. C., S. A. Kauffman, and Murray, J. D. (1993). Is morphogenesis an intrinsically robust process? *Journal of Theoretical Biology* 163, 135–144.

Gould, S. J., and R. Lewontin (1979). The spandrels of San Marco and the Panglossian paradigm: A critique of the adaptationist programme. *Proceedings of the Royal Society of London* 205, 581–598.

Griffiths, P. E. (1992). Adaptive explanation and the concept of a vestige. In P. E. Griffiths, ed., *Trees of life: Essays in philosophy of biology.* Dordrecht, Netherlands: Kluwer.

Griffiths, P. E. (1994). Cladistic classification and functional explanation. *Philosophy of Science* 61(2), 206–227.

Griffiths, P. E. (1996a). Darwinism, process structuralism and natural kinds. *Philosophy of Science* 63(3), S1–S9.

Griffiths, P. E. (1996b). The historical turn in the study of adaptation. *British Journal for the Philosophy of Science* 47(4), 511–532.

Griffiths, P. E. (1997). *What emotions really are: The problem of psychological categories.* Chicago: University of Chicago Press.

Griffiths, P. E., and R. D. Gray (1994). Developmental Systems and Evolutionary Explanation. *Journal of Philosophy* 91(6), 277–304.

Hacking, I. (1991). A tradition of natural kinds. *Philosophical Studies* 61, 109–126.

Harvey, P. H., and M. D. Pagel (1991). *The comparative method in evolutionary biology.* Oxford and New York: Oxford University Press.

Hull, D. L. (1976). Are species really individuals? *Systematic Zoology* 25, 174–191.

Hull, D. L. (1978). A matter of individuality. *Philosophy of Science* 45, 335–360.

Hull, D. (1986). On human nature. *Proceedings of the Philosophy of Science Association* 1, 3–13.

Jackson, F. C., and P. Pettit (1988). Functionalism and broad content. *Mind* 97, 381–400.

Kauffman, S. A. (1993). *The origins of order: Self-organisation and selection in evolution.* New York: Oxford University Press.

Keil, F. C. (1989). *Concepts, kinds and cognitive development.* Cambridge, Mass.: Bradford Books, MIT Press.

Kingsland, S. (1985). *Modeling nature*. Chicago: University of Chicago Press.

Kornet, D. J. (1993). Permanent splits as speciation events: A formal reconstruction of the internodal species concept. *Journal of Theoretical Biology* 164, 407–435.

Kripke, S. (1980). *Naming and necessity*. Cambridge, Mass.: Harvard University Press.

LaPorte, J. (1997). Essential membership. *Philosophy of Science* 64(1), 96–112.

Lewontin, R. C. (1983). The organism as the subject and object of evolution. *Scientia* 118, 65–82.

Lycan, W. G. (1990). The continuity of levels of nature. In W. G. Lycan, ed., *Mind and cognition: A reader*. Oxford: Blackwell.

Mayr, E. (1959, reprint 1976). Typological versus populational thinking. In E. Mayr, ed., *Evolution and the diversity of life*. Cambridge, Mass.: Harvard University Press.

Mayr, E. (1963). *Animal species and evolution*. Cambridge, Mass.: Harvard University Press.

Mishler, B. D., and R. N. Brandon (1987). Individuality, pluralism and the phylogenetic species concept. *Biology and Philosophy* 2, 397–414.

Morss, J. (1992). Against ontogeny. In P. E. Griffiths, ed., *Trees of life: Essays in philosophy of biology*. Dordrecht, Netherlands: Kluwer.

Müller, G. B., and G. P. Wagner (1991). Novelty in evolution: Restructuring the concept. *Annual Review of Ecology and Systematics* 22, 229–256.

Putnam, H. (1975). The meaning of "meaning." *Mind, language and reality: Philosophical Papers*, vol. 2. Cambridge: Cambridge University Press.

Quine, W. V. O. (1977). Natural kinds. In P. Schwartz, ed., *Naming, necessity and natural kinds*. Ithaca, N.Y.: Cornell University Press.

Riedl, R. (1977). *Order in living systems*. London: Wiley.

Schank, J. C., and W. C. Wimsatt (1986). Generative entrenchment and evolution. *Philosophy of Science Association* 2, 33–60.

Smart, J. C. C. (1963). *Philosophy and scientific realism*. London: Routledge & Kegan Paul.

Sober, E. (1980, reprint 1994). Evolution, population thinking and essentialism. In *From a biological point of view*. Cambridge: Cambridge University Press.

Wagner, G. P. (1994). Homology and the mechanisms of development. In B. K. Hall, ed., *Homology: The hierarchical basis of comparative biology*. New York: Academic Press.

Wagner, G. P. (1996). Homologues, natural kinds and the evolution of modularity. *American Zoologist* 36, 36–43.

Wagner, G. P., G. Booth, and Homayoun, B. C. (1997). A population genetic theory of canalization. *Evolution* 51(2), 329–347.

Webster, G., and B. C. Goodwin (1996). *Form and transformation: Generative and relational principles in biology*. Cambridge: Cambridge University Press.

Wimsatt, W. C. (1976a). Complexity and organisation. In M. Grene and E. Mendelsohn, eds., *Boston studies in philosophy of science*, vol. 27: *Topics in philosophy of biology*. Dordrecht, Netherlands: Reidel.

Wimsatt, W. C. (1976b). Reductionism, levels of organisation and the mind/body problem. In G. Globus, G. Maxwell, and I. Savodnik, eds., *Consciousness and the brain*. New York: Plenum Press.

Wimsatt, W. C. (1986). Developmental constraints, generative entrenchment and the innate-acquired distinction. In W. Bechtel, ed., *Integrating scientific disciplines*. Dordrecht, Netherlands: Martinus Nijhoff.

# IV Species in Mind and Culture

# 9 The Universal Primacy of Generic Species in Folkbiological Taxonomy: Implications for Human Biological, Cultural, and Scientific Evolution

Scott Atran

This chapter explores the cognitive nature of folkbiology in general and of generic species in particular based in part on cross-cultural work with people living in urban areas in the United States and forest-dwelling Maya. Generic species reflect characteristics of both the scientific genus and species. A principled distinction between genus and species is not pertinent to knowledge of local environments, nor was it pertinent to the history of science until after the European Renaissance. The claim is that there is a universal appreciation of generic species as the causal foundation for the taxonomic arrangement of biodiversity and for taxonomic inference about the distribution of causally related properties that underlie biodiversity. This taxonomy is domain specific—that is, its structure does not spontaneously or invariably arise in other cognitive domains, such as the domains of substances, artifacts, or persons. It is plausibly an innately determined evolutionary adaptation to relevant and recurrent aspects of ancestral hominid environments, such as the need to recognize, locate, react to, and profit from many ambient species. Folkbiology also plays a special role in cultural evolution in general and in the development of Western biological science in particular.

Experimental results indicate that the same taxonomic rank is cognitively preferred for biological induction in two diverse populations: people raised in Michigan and Itzaj Maya of the lowland Mesoamerican rainforest. This taxonomic rank is the generic *species*—the level of *oak* and *robin*. These findings cannot be explained by domain-general models of similarity because such models cannot account for why both cultures prefer specieslike groups in making inferences about the biological world, although many people in the United States have relatively little actual knowledge or experience at this level. In fact, general relations of perceptual similarity and expectations derived from experience produce a "basic level" of recognition and recall for many of these people that corresponds to the superordinate life-form level of folkbiological taxonomy—the level of *tree* and *bird*. Still, they prefer generic species for making inductions about the distribution of biological properties among organisms and for predicting the nature of the biological world in the face of uncertainty.

A domain-specific view of folkbiology may explain the robust inductive preference for generic species across diverse cultures, regardless of actual perceptual knowledge or experience. It suggests the idea of the generic-species level as a partitioning of the ontological domains of *plant* and *animal* into mutually exclusive essences that are assumed (but not necessarily known) to have unique underlying causal natures. This partitioning may be an evolutionary design: universal taxonomic structures, centered on essence-based generic species, are arguably routine products of our "habits of mind," which may be in part naturally selected to grasp relevant and recurrent "habits of the world." This chapter explores the implications of this universal habit of mind for the evolution of human cognition, culture, and science.

## FOUR POINTS OF GENERAL CORRESPONDENCE BETWEEN FOLKBIOLOGY AND SCIENTIFIC SYSTEMATICS

In every human society, people think about plants and animals in the same special ways. These special ways of thinking, which can be termed *folk-biology*, are fundamentally different from the ways humans ordinarily think about other things in the world—such as stones, stars, tools, or even people. The science of biology also treats plants and animals as special kinds of objects, but applies this treatment to humans as well. Folkbiology, which is present in all cultures around the world, and the science of biology, whose origins are particular to Western cultural tradition, have corresponding notions of living kinds.

Consider four corresponding ways in which ordinary folk and biologists think of plants and animals as special. First, people in all cultures classify plants and animals into specieslike groups that biologists generally recognize as populations of interbreeding individuals adapted to an ecological niche. We will call such groups—for example, *redwood, rye, raccoon,* or *robin*—*generic species* for reasons that later become evident. Generic species are usually as obvious to a modern scientist as to local folk. Historically, the generic-species concept provided a pretheoretical basis for scientific explanation of the organic world in that different theories—including evolutionary theory—have sought to account for the apparent constancy of "common species" and for the organic processes that center on them (Wallace 1901, 1).

Second, there is a commonsense assumption that each generic species has an underlying causal nature or essence that is uniquely responsible for its typical appearance, behavior, and ecological preferences. People in diverse cultures consider this essence, even when hidden, responsible for the organism's identity as a complex, self-preserving entity governed by dynamic internal processes that are lawful. This hidden essence maintains the organism's integrity even as it causes the organism to grow, change form, and reproduce. For example, a tadpole and frog are in a crucial sense the same animal although they look and behave very differently and live in different places. Western philosophers, such as Aristotle and Locke, attempted to translate this commonsense notion of essence into some sort of metaphysical reality, but evolutionary biologists have rejected the notion of essence as such. Nevertheless, biologists have interpreted this conservation of identity under change as due to the fact that organisms have separate genotypes and phenotypes.

Third, in addition to the spontaneous division of local flora and fauna into essence-based species, such groups have "from the remotest period in ... history ... been classed in groups under groups. This classification [of generic species into higher- and lower-order groups] is not arbitrary like the grouping of stars in constellations" (Darwin 1872, 363).[1] The structure of these hierarchically included groups—such as white oak/oak/tree or mountain robin/robin/bird—is referred to as *folkbiological taxonomy*. Especially in the case of animals, these nonoverlapping taxonomic structures can often be scientifically interpreted in terms of speciation (that is, as related species descended from a common ancestor by splitting off from a lineage). Fourth, such taxonomies not only organize and summarize biological information, but also provide a powerful inductive framework for making systematic inferences about the likely distribution of organic and ecological properties among organisms. For example, given the presence of a disease in robins, one is "automatically" justified in thinking that the disease is more likely present among other bird species than among nonbird species. In scientific taxonomy, which belongs to the branch of biology known as *systematics*, this strategy receives its strongest expression in "the fundamental principle of systematic induction" (Warburton 1967, Bock 1973). According to this principle, given a property found among members of any two species, the best initial hypothesis is that the property is also present among all species included in the smallest higher-order taxon containing the original pair of species. For example, finding that the bacteria *Escherichia coli* share a hitherto unknown property with robins, a biologist would be justified in testing the hypothesis that all organisms share the property because *E. coli* link up with robins only at the highest level of taxonomy, which includes all organisms. This or any general-purpose system of taxonomic inference for biological kinds is grounded in a universal belief that the world naturally divides into the limited causal varieties we commonly know as (generic) species.

## FOLKBIOLOGICAL TAXONOMY

Ever since the pioneering work of Brent Berlin and his colleagues, ethnobiological evidence has been accumulating that human societies everywhere have similar folkbiological structures (Berlin, Breedlove, and Raven 1973; Hunn 1977; Hays 1983; Brown 1984; Atran 1990; Berlin 1992). These striking cross-cultural similarities suggest that a small number of organizing principles universally define systems of folkbiological classification. Folkbiological groups, or taxa, are organized into ranks that represent an embedding of distinct levels of reality. Most folkbiological systems have between three and six ranks. Taxa of the same rank are mutually exclusive and tend to display similar linguistic, biological, and psychological characteristics.

Ranks and taxa, whether in folkbiological or scientific classification, are of different logical orders, and confounding them is a category mistake. Biological ranks are second-order classes of groups (e.g., species, family, kingdom) whose elements are first-order groups (e.g., lion, feline, animal). Folkbiological ranks seem to vary little, if at all, across cultures as a function of theories or belief systems; in other words, such ranks—but not the taxa they contain—are universal. Ranks are intended to represent fundamentally different levels of reality, not convenience.[2]

The most general folkbiological rank is the folk kingdom—for example, *plant* and *animal*. Such taxa are not always explicitly named and represent the most fundamental divisions of the biological world. These divisions correspond to the notion of "ontological category" in philosophy (Donnellan 1971) and in psychology (Keil 1979). From an early age, it appears, humans cannot help but conceive of any object they see in the world as either being or not being an animal, and there is evidence for an early distinction between plants and nonliving things (Hickling and Gelman 1995, Inagaki and Hatano forthcoming). Conceiving of an object as a plant or animal seems to carry with it certain assumptions that are not applied to objects thought of as belonging to other ontological categories, such as the categories *substance* or *artifact* (Keil 1989, Mandler and McDonough 1996, Hatano and Inagaki 1998, see also Keil and Richardson, chapter 10 in this volume).

The next rank down is *life form*. Most taxa of lesser rank fall under one or another life form. Life-form taxa often have lexically unanalyzable names (simple primary lexemes), such as *tree* and *bird*, although some life-form names are analyzable, such as *quadruped*. Biologically, members of a life-form taxon are diverse. Psychologically, members of a life-form taxon share a small number of perceptual diagnostics: stem aspect, skin covering, and so forth (Brown 1984). Life-form taxa may represent adaptations to broad sets of ecological conditions, such as competition among single-stem plants for sunlight and tetrapod adaptation to life in the air (Hunn 1982, Atran 1985). Classifying by life form may occur early on: two-year-old children distinguish familiar kinds of quadruped (e.g., dog and horse) from sea animals and both of those from air animals (Mandler, Bauer, and McDonough 1991).

The core of any folk taxonomy is the generic-species level. Like life-form taxa, generic species are often named by simple lexemes, such as *oak* and *robin*. Sometimes, generic species are labeled as binomial compounds, such as *hummingbird*. On other occasions, they may be optionally labeled as binomial composites, such as *oak tree*. In both cases, the binomial makes the hierarchical relation apparent between generic species and life form.

Generic species often correspond to scientific genera (e.g., *oak*) or species (e.g., *dog*), at least for the most phenomenally salient organisms, such as larger vertebrates and flowering plants. On occasion, generic species can correspond to local fragments of biological families (e.g., *vulture*), orders (e.g., *bat*) and, especially with invertebrates, even higher-order biological taxa (Atran 1987, Berlin 1992). Generic species may also be the categories most easily recognized, most commonly named, and most easily learned by children in small-scale societies (Stross 1973). Indeed, ethnobiologists who otherwise differ in their views of folk taxonomy tend to agree that one level best captures discontinuities in nature and provides the fundamental constituents in all systems of folkbiological categorization, reasoning, and use (Bulmer 1974, Hunn 1982, Ellen 1993).

The term *generic species* is used here, rather than *folk genera/folk generic* or *folk species/folk specieme*, for four reasons:

1. Empirically, ethnobiologists and historians of systematics (as well as working biologists) mostly agree "that species come to be tolerably well defined objects ... in any one region and at any one time" (Darwin 1883, 137) and that such local species defined by ordinary people are the heart of any natural system of biological classification. Whereas zoologists and ethnozoologists generally refer to such common groups as *species* (or *speciemes*) in focusing on reproductive and geographical isolation that is more readily identified in terms of behavior (Mayr 1969, Bulmer 1970, Diamond and Bishop 1998), botanists and ethnobotanists refer to them as *genera* (or *generics*) in focusing on ease of morphological recognition without technical aids (Greene 1909/1983, Bartlett 1940, Berlin 1992). As working concepts, either term alone is likely to be more confusing for historians of systematics than the term *generic species* (see Stevens 1994), as when the zoologist George Gaylord Simpson declared that the hallmarks of priority attributed by some of his colleagues to the genus "are characteristic of the ... species, not genus" (Simpson 1961, 189).

2. Perceptually, a principled distinction between biological genus and species is not pertinent to most people around the world. For humans, the most phenomenally salient species (including most species of large vertebrates, trees, and evolutionarily isolated groups such as palms and cacti) belong to monospecific genera in any given locale. Closely related species of a polytypic genus are often difficult to distinguish locally, and no readily perceptible morphological or ecological "gap" can be discerned between them (Diver 1940).[3]

3. Historically, the distinction between genus and species did not appear until the influx of newly discovered species from around the world compelled European naturalists to sort and remember them within a worldwide system of genera built around mainly European species types (Atran 1987). The original genus concept was partially justified in terms of initially monotypic generic European species to which other species around the world might be attached (Tournefort 1694)

4. Ontologically, the term *generic species* reflects a dual character. As salient mnemonic groups, generic species are akin to genera in being those groups

most readily apparent to the naked eye (Linnaus 1751, Cain 1956). As salient causal groups, they are akin to species in being the principal centers of evolutionary processes responsible for biological diversity (Wallace 1901, Mayr 1982).

People in all cultures spontaneously partition the ontological categories *animal* and *plant* into generic species in a virtually exhaustive manner. "Virtually exhaustive" means that when people encounter an organism not readily identifiable as belonging to a named generic species, they still *expect* it to belong to one. The organism is often assimilated to one of the named taxa it resembles, but sometimes it is assigned an "empty" generic-species slot pending further scrutiny (e.g., "such and such a plant is some kind [generic-species] of tree"; cf. Berlin 1999). This partitioning of ontological categories seems to be part and parcel of the categories themselves: no plant or animal can fail in principle to belong uniquely to a generic species.

Moreover, data from developmental psychology suggests that young children presume each distinctive living kind to have an "essence," or underlying causal nature, which is responsible for the typical appearance of that kind (Gelman and Wellman 1991). At first, this presumption involves only a global understanding that the readily visible outsides of living kinds are produced by, but are perhaps different from, their initially invisible insides. Children initially lack concrete or specific pieces of knowledge about each kind (Simmons and Keil 1995). Over time, they try to flesh out the causal properties of these presumed essences as responsible for growth (Hickling and Gelman 1995), inheritance (Springer and Keil 1989), and the complementary functioning of distinct body parts in a living kind (Hatano and Inagaki 1994). Such intrinsic causal essences are universally presumed to be both teleological (unlike the mechanical causes affecting inert substances) and internally directed (unlike externally fashioned artifacts); they also appear to be unique to the cognitive domain of living kinds and primarily identified with generic species (cf. Keil and Richardson, chapter 10 in this volume).

Generic species may be further divided into folk specifics. These taxa are usually labeled binomially, with secondary lexemes. Compound names, such as *white oak* and *mountain robin*, make the hierarchical relation transparent between a generic species and its folk specifics. Folk specifics that have a tradition of high cultural salience may be labeled with primary lexemes, such as *winesap* (a kind of apple tree) and *tabby* (a kind of cat). In general, whether and how a generic species is further differentiated depends on the cultural significance of the organisms involved. Occasionally, an important folk-specific taxon will be further subdivided into contrasting folk-varietal taxa: for example, *short-haired tabby* versus *long-haired tabby*. Folk varietals are usually labeled trinomially, with tertiary lexemes that make transparent their taxonomic relationship with superordinate folk specifics and generic species —for example, *swamp white oak*.

Thus, in addition to generic species, people everywhere tend to form groups that are both subordinate and superordinate to the level of preferred groups. Cultures across the world organize readily perceptible organisms into a system of hierarchical levels designed to represent the embedded structure of life around them, with the generic-species level being the most informative. In some cultures, but not all, people may develop "theories" of life that are meant to cover all living kinds, such as Western theories of biology (Carey 1985, Atran 1990), but the very possibility of theorizing would not exist without universal construal of generic species to provide the trans-theoretical basis for scientific speculation about the biological world.

## A TAXONOMIC EXPERIMENT ON RANK AND PREFERENCE

Given these observations, cognitive studies of the "basic level" are at first sight striking and puzzling. In a justly celebrated set of experiments, Rosch and her colleagues set out to test the validity of the notion of a psychologically preferred taxonomic level (Rosch et al. 1976). Using a broad array of converging measures, they found that there is indeed a "basic level" in category hierarchies of "naturally occurring objects" such as "taxonomies" of artifacts as well as of living kinds. For artifact and living kind hierarchies, the basic level is where: (1) many common features are listed for categories, (2) consistent motor programs are used for the interaction with or manipulation of category exemplars, (3) category members have similar enough shapes so that it is possible to recognize an average shape for objects of the category, (4) the category name is the first name to come to mind in the presence of an object (e.g., *table* versus *furniture* or *kitchen table*).

There is a problem, however: the basic level that Rosch and her colleagues (1976) had hypothesized for artifacts was confirmed (e.g., *hammer, guitar*); however, the hypothesized basic level for living kinds (e.g., *maple, trout*), which Rosch initially assumed would accord with the generic-species level, was not confirmed. For example, instead of *maple* and *trout*, Rosch and the others found that *tree* and *fish* operated as basic-level categories for U.S. college students. Thus, the basic level identified for living kinds generally corresponds to the life-form level, which is superordinate to the generic-species level (see Zubin and Köpcke 1986 for findings with German).

To explore this apparent discrepancy between preferred taxonomic levels in small-scale or industrialized societies and the cognitive nature of ethnobiological ranks in general, we use inductive inference. Inference allows us to test whether or not there is a psychologically preferred rank that maximizes the strength of any potential induction about biologically relevant information and whether or not this preferred rank is the same across cultures. If a preferred level carries the most information about the world, then categories at that level should favor a wide range of inferences about what is common among members (for detailed findings under a variety of lexical and property-projection conditions, see Atran et al. 1997; Coley, Medin and Atran 1997; Coley et al. 1999).

The prediction is that inferences to a preferred category (e.g., *white oak* to *oak, tabby* to *cat*) should be much stronger than inferences to a superordinate category (*oak* to *tree, cat* to *mammal*). Moreover, inferences to a subordinate category (*swamp white oak* to *white oak, short-haired tabby* to *tabby*) should not be much stronger than or different from inferences to a preferred category. What follows is a summary of results from one representative set of experiments in two very diverse populations: midwesterners in the United States and the lowland Maya.

### Experiment Subjects and Methods

The Itzaj are Maya living in the Petén rainforest region of Guatemala. Until recently, men devoted their time to shifting agriculture, hunting, and silviculture, whereas women concentrated on the myriad tasks of household maintenance. The Itzaj were the last independent native polity to be conquered by the Spaniards (in 1697), and they have preserved virtually all ethnobiological knowledge recorded for lowland Maya since the time of the initial Spanish Conquest (Atran 1993, Atran and Ucan Ek' forthcoming). Despite the current awesome rate of deforestation and the decline of Itzaj culture, the language and ethic of traditional Maya silviculture is still very much in evidence among the generation of our informants, who range in age from fifty to eighty years old. The midwesterners in the United States were self-identified as people raised in Michigan and recruited through an advertisement in a local newspaper.

Based on extensive fieldwork with the Itzaj, we chose a set of Itzaj folkbiological categories of the kingdom (K), life-form (L), generic-species (G), folkspecific (S), and folk-varietal (V) ranks. We selected three plant life forms: *che'* = tree, *ak'* = vine, *pok~che'* = herb/bush. We also selected three animal life forms: *b'a'al~che' kuxi'mal* = "walking animal" (i.e., mammal), *ch'iich'* = birds, including bats, *käy* = fish. Three generic-species taxa were chosen from each life form so that each generic species had a subordinate folk specific, and each folk specific had a salient varietal.

Pretesting showed that participants were willing to make inferences about hypothetical diseases. The properties chosen for animals were diseases related to the heart *(pusik'al)*, blood *(k'ik'el)*, and liver *(tamen)*. For plants, there were diseases related to the roots *(motz)*, sap *(itz)*, and leaf *(le')*. Properties were chosen according to Itzaj beliefs about the essential, underlying aspects of life's functioning. Thus, the Itzaj word *pusik'al*, in addition to identifying the biological organ heart in animals, also denotes essence or heart in both animals and plants. The term *motz* denotes roots, which are considered the initial locus of the plant *pusik'al*. The term *k'ik'el* denotes blood and is conceived as the principal vehicle for conveying life from the *pusik'al* throughout the body. The term *itz* denotes sap, which functions as the plant's *k'ik'el*.

The *tamen* or liver helps to "center" and regulate the animal's *pusik'al*. The *le'* or leaf is the final locus of the plant *pusik'al*. Properties used for inferences have the form, "is susceptible to a disease of the ⟨root⟩ called ⟨X⟩." For each question, "X" was replaced with a phonologically appropriate nonsense name (e.g., *eta*) in order to minimize the task's repetitiveness.

All participants responded to a list of more than fifty questions in which they were told that all members of a category had a property (the premise) and were asked whether "all," "few," or "no" members of a higher-level category (the conclusion category) also possessed that property. The premise category was at one of four levels, either life form (e.g., $L$ = bird), generic species (e.g., $G$ = vulture), folk specific (e.g., $S$ = black vulture), or varietal (e.g., $V$ = red-headed black vulture). The conclusion category was drawn from a higher-level category, either kingdom (e.g., $K$ = animal), life form (L), generic species (G), or folk specific (S). Thus, there were ten possible combinations of premise and conclusion category levels: $L \rightarrow K$, $G \rightarrow K$, $G \rightarrow L$, $S \rightarrow K$, $S \rightarrow L$, $S \rightarrow G$, $V \rightarrow K$, $V \rightarrow L$, $V \rightarrow G$, and $V \rightarrow S$. For example, a folk specific to life form ($S \rightarrow L$) question might be, "If all black vultures are susceptible to the blood disease called eta, are all other birds susceptible?" If a participant answered "no," then the follow-up question would be, "Are some or a few other birds susceptible to disease eta, or no other birds at all?"

The corresponding life forms for the midwesterners were: mammal, bird, fish, tree, bush, and flower (on *flower* as considered a life form in the United States, see Dougherty 1979). The properties used in questions for the Michigan participants were "have protein X," "have enzyme Y," and "are susceptible to disease Z." These properties were chosen to be internal, biologically based properties intrinsic to the kind in question, but abstract enough so that rather than answering what amounted to factual questions participants would be likely to make inductive inferences based on taxonomic category membership.

## Experiment Results

Representative findings are given in figure 9.1. Responses were scored in two ways. First, we totaled the proportion of "all or virtually all" responses for each kind of question (e.g., the proportion of times respondents agreed that if red oaks had a property, all or virtually all oaks would have the same property). Second, we calculated "response scores" for each item, counting a response of "all or virtually all" as 3, "some or few" as 2, and "none or virtually none" as 1. A higher score reflected more confidence in the strength of an inference.

Figure 9.1a summarizes the results from all Itzaj informants for all life forms and diseases, and shows the proportion of "all" responses (black), "few" responses (checkered), and "none" responses (white). For example, given a premise of folkspecific (S) rank (e.g., red squirrel) and a conclusion category of generic-species (G) rank (e.g., squirrel), 49% of responses indicated

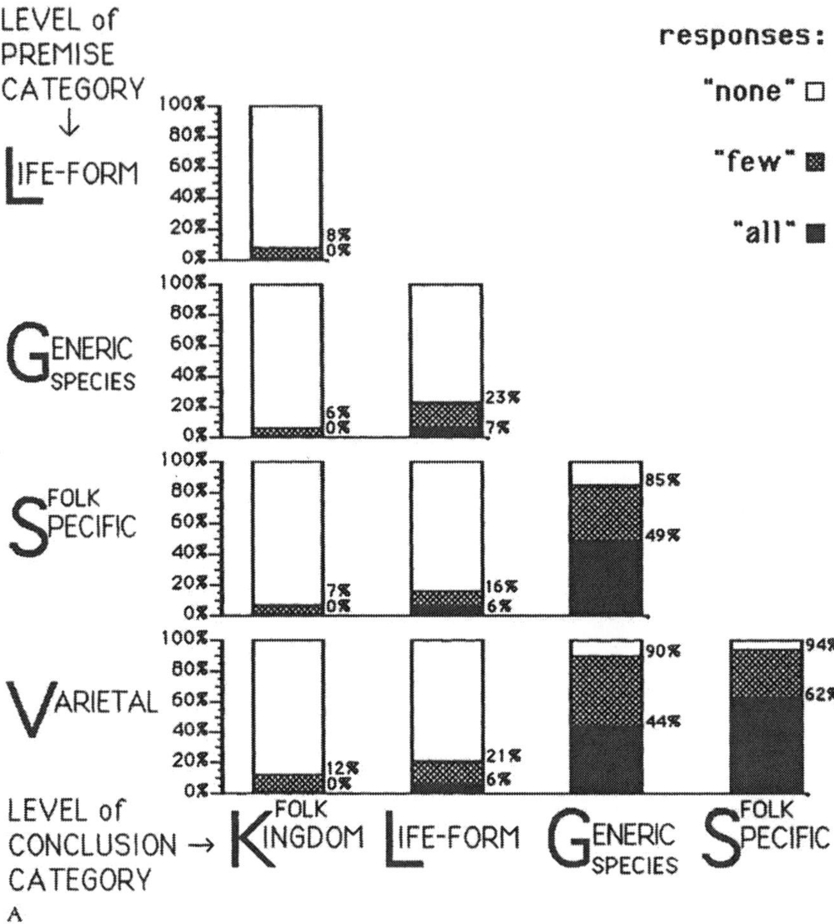


**Figure 9.1** Rank and inference: Comparing the willingness of Itzaj Maya (A) and Michigan students (B) to make inductions across folkbiological ranks Results include all life forms and biological properties, showing the proportion of "all" (black), "few" (checkered), and "none" responses (white). Main diagonals represent inferences from a given rank (premise category) to the adjacent higher-order rank (conclusion category): V(arietal) → S(pecific), S(pecific) → G(eneric species), G(eneric species) → L(ife form), L(ife form) → K(ingdom). Moving horizontally within each graph corresponds to holding the premise constant and varying the conclusion: e.g., V → S, V → G, V → L, V → K.

that "all" squirrels and not just "some" or "none" would possess a property that red squirrels have. Results were obtained by totaling the proportion of "all or virtually all" responses for each kind of question (e.g., the proportion of times respondents agreed that if red oaks had a property, all or virtually all oaks would have the same property). A higher score represented more confidence in the strength of the inductive inference. Figure 9.1b summarizes the results of Michigan participant response scores for all life forms and biological properties.

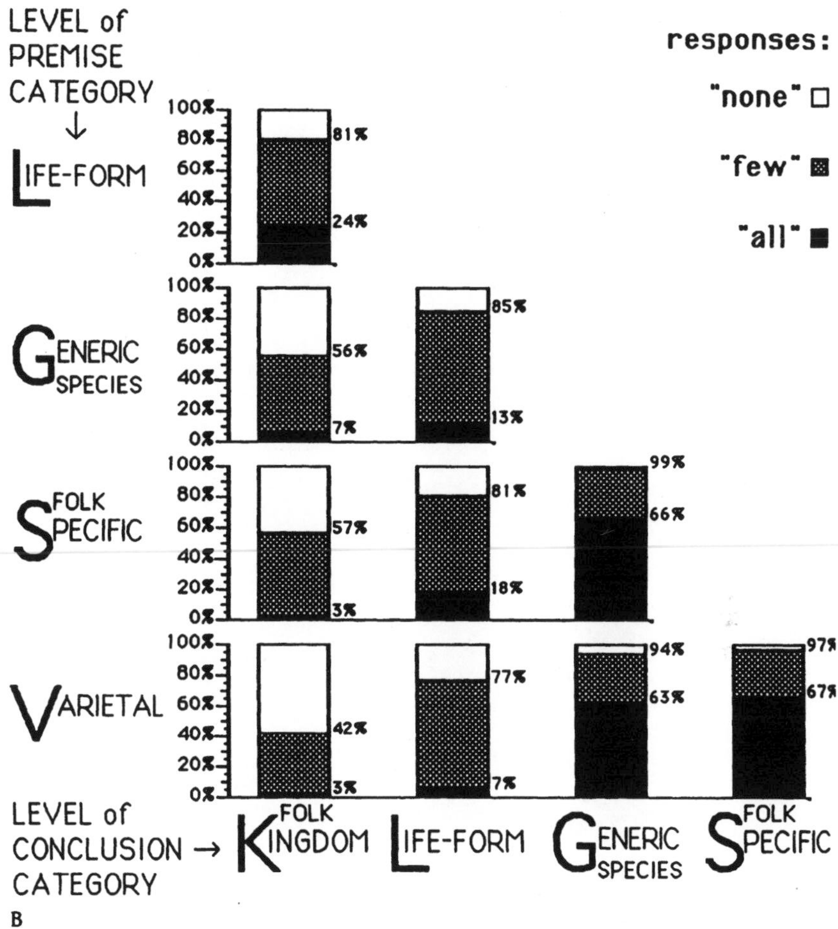


**Figure 9.1** (continued)

Response scores were analyzed using t-tests with significance levels adjusted to account for multiple comparisons. Figure 9.2 summarizes the significant comparisons ($p$-values) for "all" responses, "none" responses, and combined responses. For all comparisons, $n = 12$ Itzaj participants and $n = 21$ Michigan participants (for technical details see Atran et al. 1997).

Following The main diagonals of figures 9.1 and 9.2 corresponds to changing the levels of both the premise and conclusion categories while keeping their relative level the same (with the conclusion one level higher than the premise). Induction patterns along the main diagonal indicate a single, inductively preferred level. Examining inferences from a given rank to the adjacent higher-order rank (i.e., $V \rightarrow S$, $S \rightarrow G$, $G \rightarrow L$, $L \rightarrow K$), we find a sharp decline in strength of inferences to taxa ranked higher than generic species, whereas $V \rightarrow S$ and $S \rightarrow G$ inferences are nearly equal and similarly

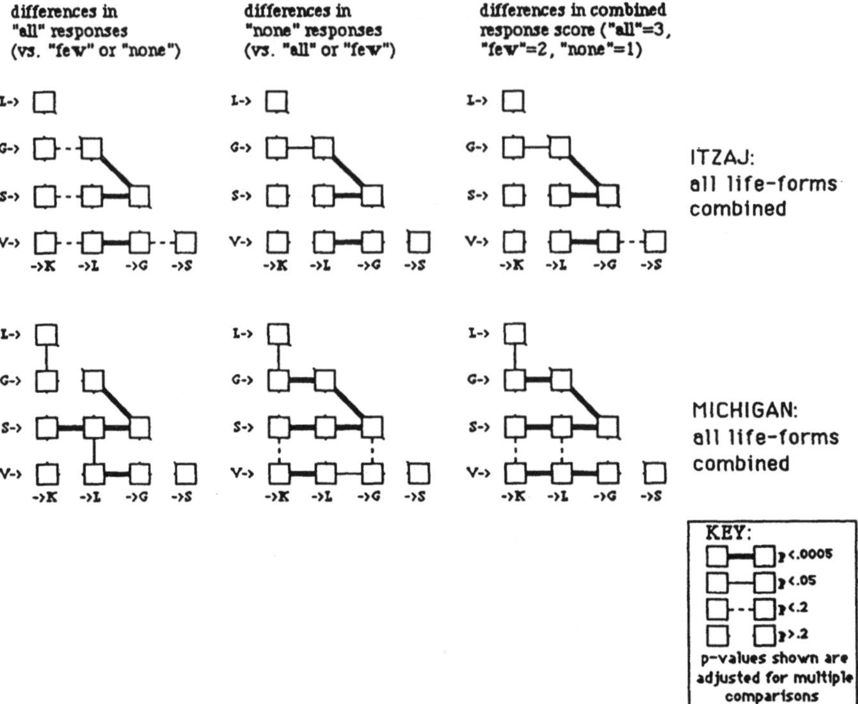


**Figure 9.2** Significant comparisons between adjacent categories in the rank and inference task for Itzaj Maya and Michigan students Results include all life forms and biological properties, showing "all" (versus "few" or "none"), "none" (versus "all" or "few"), and combined responses ("all" = 3, "few" = 2, "none" = 1). Main diagonals represent inferences from a given rank (premise category) to the adjacent higher-order rank (conclusion category): V(arietal) → S(pecific), S(pecific) → G(eneric species), G(eneric species) → L(ife form), L(ife form) → K(ingdom). Moving horizontally within each graph corresponds to holding the premise constant and varying the conclusion: e.g., V → S, V → G, V → L, V → K.

strong. Notice that for "all" responses, the overall Itzaj and Michigan patterns are nearly identical.

Moving horizontally within each graph in figures 9.1 and 9.2 corresponds to holding the premise category constant and varying the level of the conclusion.[4] Here, we find the same pattern for "all" responses for both Itzaj and Michigan participants as we did along the main diagonal. However, in the combined response scores ("all" + "few"), there is now evidence of increased inductive strength for higher-order taxa among the Michigan participants versus the Itzaj. On this analysis, both groups show the largest break between inferences to generic species versus life forms, but only the Michigan subjects also show a consistent pattern of rating inferences to life-form taxa higher than to taxa at the level of the folk kingdom: $G \rightarrow K$ versus $G \rightarrow L$, $S \rightarrow K$ versus $S \rightarrow L$, and $V \rightarrow K$ versus $V \rightarrow L$.

Finally, moving both horizontally and along the diagonal, for the Itzaj there is some hint of a difference between inductions using conclusions at the generic-species versus folk-specific levels: $V \rightarrow G$ and $S \rightarrow G$ are modestly weaker than $V \rightarrow S$. Regression analysis reveals that for the Itzaj, the folk-specific level accounts for a small but significant variance (1.4%) beyond the generic-species level. For Michigan participants, the folk-specific level is not differentiated from the generic-species level (0.2%, not significant). In fact, most of the difference between $V \rightarrow G$ and $V \rightarrow S$ inductions results from inference patterns for the Itzaj tree life form. There is evidence that the Itzaj confer some preferential status upon trees at the folkspecific level (e.g., savanna nance tree). Itzaj are forest-dwelling Maya with a long tradition of agroforestry that antedates the Spanish Conquest (Atran 1993).

### Experiment Discussion

These results indicate that both the ecologically inexperienced Michigan participants and the ecologically experienced Itzaj prefer taxa of the generic-species rank in making biological inferences; the findings go against a simple relativist account of cultural differences in folkbiological knowledge. However, the overall effects of cultural experience on folkbiological reasoning are reflected in more subtle ways that do not undermine an absolute preference for the generic species across cultures. In particular, the data point to a relative downgrading of inductive strength to higher ranks among industrialized U.S. inhabitants through knowledge attrition owing to lack of experience and a relative upgrading of inductive strength to lower ranks among silvicultural Maya through expertise.

A secondary reliance on life forms arguably owes to U.S. inhabitants' general lack of actual experience with generic species (Dougherty 1978; see Tanaka and Taylor 1991 on the effects of expertise on basic-level categorization). In one study, U.S. students used only the name *tree* to refer to more than 75% of the species they saw in a nature walk (Coley et al. 1999). Although urban people in the United States usually can't tell the difference between beeches and elms, they *expect* that biological action in the world is at the level of beeches and elms and not at the tree level. Yet without being able at least to recognize a tree, they would not even know where to begin to look for the important biological information. The Itzaj pattern reflects both overall preference for generic species and a secondary preference for lower-level distinctions, at least for kinds of trees. A strong ethic of reciprocity in silviculture still pervades the Itzaj culture; the Maya tend trees so that the forest will tend to the Maya (Atran and Medin 1997). This ethic seems to translate into an upgrading of biological interest in tree folk specifics.

These findings cannot be explained by appeals either to cross-domain notions of perceptual "similarity" or to the structure of the world "out there." On the one hand, if inferential potential were a simple function of perceptual similarity, then urban dwellers in the United States should prefer life forms for induction (in line with the conclusions drawn by Rosch and her colleagues). Yet like the Maya, they prefer generic species. On the other hand, objective reality—that is, the actual distribution of biological species within groups of evolutionarily related species—does not substantially differ in the natural environments of midwesterners and the Itzaj. Unlike the Itzaj, however, midwesterners perceptually discriminate life forms more readily than generic species. True, there are more locally recognized species of tree in the Maya area of Petén, Guatemala, than in the Midwest United States. Still, the readily perceptible evolutionary "gaps" between species are roughly the same in the two environments (most tree genera in both environments are monospecific). If anything, one might expect that having fewer trees in the U.S. environment allows each species to stand out more from the rest (Hunn 1976). For birds, the relative distribution of evolutionarily related species also seems to be broadly comparable across temperate and rainforest environments (Boster 1988).

An inadequacy in current accounts of preferred taxonomic levels may be a failure to distinguish domain-general mechanisms for best clustering stimuli from domain-specific mechanisms for best determining loci of biological information. To explain Rosch's data, it may be enough to rely on domain-general, similarity-based mechanisms. Such mechanisms may generate a basic level in any number of cognitive domains, but not the preferred level of folkbiology. Consider:

In striking contrast to the rich debate over the descriptive adequacy of accounts of folkbiological taxonomy, little attempt has been made to provide an explanatory account of the psychological mechanisms and processes that actually produce folkbiological groups. A notable exception is Hunn's (1976) "perceptual model," arguably the most influential proposal in ethnobiology (Berlin 1978). This model accords with Rosch's (1973, 1975) general account of the cognitive structure of perceptual and semantic categories in hierarchical structures. These accounts are variants of what psychologists call "similarity-based models" (Smith and Medin 1981), which organize perceptually identifiable categories on the basis of correlation or covariation of stimulus attributes. With such models, one learns to recognize a particular instance of a category by being exposed to multiple instances of the category, which implies that, as Boster (1991) puts it, "the source of biological similarity judgments is in the world, not in the brain."

To illustrate the story from a similarity-based point of view: because the attributes of having a bark, large canines, and a terrestrial habitat usually co-occur only when a dog is present, then their co-occurrence will probably figure in all and only those feature sets generally associated with the category *dog*. The mind will "automatically" tend to cluster perceptible features into "gestalts" of maximally covariant attributes, or basic-level categories, because of the "objective" discontinuities that exist in nature. Notice that for the model to work, it is not imperative that any particular feature always be necessary for defining category membership or that a given set of features always be sufficient. All that is required is that the exemplars exhibit a readily apparent "family resemblance" among a community of attributes (Rosch and Mervis 1975, Hunn 1982).

Because the processing mechanism is a general-purpose device that can pick out perceptual stimuli from whatever source, it should operate across any cognitive domain that involves separated clusters of perceptual attributes, including categories that occur naturally in everyday biological and social contexts, as well as constructed categories (e.g. artifacts). Later research has tended to confirm the findings of Rosch and her colleagues further showing that the basic level extends to artificial and natural categories as the level that people most readily recognize and that children most easily name and learn (Lassaline, Wisniewski, and Medin 1992).

The same attribute-clustering strategy can be applied recursively at higher and lower levels (Hunn 1976). Thus, the simultaneous presence of fur and live-born offspring might figure in the feature set that distinguishes the category *mammal* from other categories of superordinate-level life forms, such as *bird, fish,* and so forth. Similarly, a high body length to body height ratio, when added to the feature gestalt for *dog*, might figure in the feature set that distinguishes the subordinate-level category *dachsund* from other types of *dog*. The basic level, then, is that level without which relatively much information is lost and below which little information is gained. That is, there is a large gain in information when going from the superordinate or life-form level to the basic level, and there is only a slight gain in information going from the basic level to the subordinate or specific level.

Thus, both anthropology and psychology suggest that privilege or "basicness" could be a function of correlated features or properties producing natural clusters that are psychologically salient. These salient chunks should organize both category organization and reasoning involving categories (Anderson 1990). Compelling as this view is, however, it is inadequate to describe our findings. The challenge is to explain why the generic-species rank is preferred both by Maya, who have relatively extensive contact with the natural environment, and by Michigan students, who have relatively little. The key problem is that the linguistic and perceptual criteria for "basicness" used by Rosch and her colleagues point to the life-form level as preferred, but as we have just seen, the break point in induction appears at the more specific rank of generic species.

To explain our data may require, in addition to domain-generic perceptual heuristics, domain-specific mechanisms for the formation of biological categories that are not similarity-based. Along these lines, a "living-kind module" would involve a domain-specific sort of causal reasoning that may be called "teleo-essentialist" (Atran 1995, Keil 1995). The idea is that universal and possibly innate principles lead people to believe that visible morphotypical patterns of each readily identifiable generic species as well as nonobvious aspects of biological functioning are causally produced by an underlying essence. The nature of this essence is initially unknown, but presumed. The learner (e.g., a child) then attempts to discover how essences govern the heritable teleological relations between visible parts, how they link initially ill-perceived internal parts to morphotypical parts through canonical patterns of irreversible growth and how they determine the stable and complex functioning of visible and nonobvious parts. Virtually all people in all cultures cannot help but follow through this spontaneously triggered "research program," which compels them to deepen and extend the domain of information relevant to living kinds within a taxonomic framework that focuses attention on generic species.

Notice that although a generic species may fail to be "basic" in Rosch's sense of a maximally rich cluster of readily available perceptual information, it may still be preferred as a maximally rich bundle of anticipated biological information. In other words, domain-specific constraints on categorization and category-based reasoning may diverge from domain-general constraints. When and where they do, the expectation is that domain-specific constraints are paramount.

In small-scale societies, adults as well as children learn about generic species just by being told about them or by seeing a single instance. In our society, one need only describe a single instance in a picture book or point to an isolated example in a zoo or museum to have an adult or child instantly extend that poor and fragmentary instance of experience to an indefinitely extendible category. The taxonomic position of the category is immediately fixed as a generic species. This fixture "automatically" carries with it a complex internal structure that is partially presumed and partially inferred, but by no means directly known.

How can people conceive of a given category as a generic species without primarily relying on perception? Ancillary encyclopedic knowledge may be often crucial. Thus, one may have detailed perceptual knowledge of dogs but not of oaks. Yet a story that indicates where an oak lives or how it looks or grows or that its life is menaced may be sufficient to trigger the presumption that oaks comprise a generic species just like dogs do. But such cultural learning produces the same results under widely divergent conditions of experience in different social and ecological environments, which indicates that the learning itself is strongly motivated by cross-culturally shared cognitive mechanisms that do not depend primarily on experience.

In conjunction with encyclopedic knowledge of what is already known for the natural world, language is important in targeting preferred kinds by triggering biological *expectations* in the absence of actual experience or knowledge of those kinds (Gelman, Coley, and Gottfried 1994). Language alone, however, would not suffice to induce the expectation that, for people who live in urban areas in the United States, little or poorly known generic species are more biologically informative than better known life forms. Some other process must invest the generic-species level with inductive potential. Language alone can only signal that such an expectation is appropriate for a given lexical item; it cannot determine the nature of that expectation. Why presume that an appropriately tagged item is the locus of a "deep" causal nexus of biological properties and relationships? Why suppose at all that there is such a nexus that spontaneously justifies and motivates the expectations, inferences, and explorations relating little known or nonobvious aspects of a presumably fundamental biological reality?

It is logically impossible that such presumptions come from (repeated exposure to) the stimuli themselves. Logically, the world of stimuli (the only world we are in direct contact with) is a flux of indefinitely many associations that no structurally unbiased processing device could ever hope to order in finite time (Goodman 1972). In other words, input to the mind cannot alone cause an instance of experience (e.g., a sighting in nature or in a picture book) or any finite number of fragmentary instances to be generalized into a category that subsumes a rich and complex set of indefinitely many instances and stimuli. This projective capacity for category formation can only come from the mind, never from the world alone. The empirical question, then, is whether or not this projective capacity is simply domain general or also domain specific—that is, whether we have one general perceptually based similarity metric (built on some innate quality space of a priori phenomenal associations [see Quine 1960]) or something more. For any given category domain—say, living kinds as opposed to artifacts or substances—the process would be domain general if and only if one could generate both the categories of any number of domains from the stimuli alone and the very same cognitive mechanisms for associating and generalizing those stimuli. As we have seen, current domain-general similarity models of category formation and category-based reasoning fail to account for the taxonomic privilege of the generic-species level across cultures.

## Experiment Summary

Our findings suggest that fundamental categorization and reasoning processes in folkbiology are rooted in domain-specific conceptual presumptions and not exclusively in domain-general, similarity-based (e.g., perceptual) heuristics. People in either subsistence or industrialized cultures may differ on the level at which they most easily identify organisms, but still prefer the same absolute level of reality for biological reasoning—namely, the generic-species rank. Michigan college students have greater secondary reliance on life forms because life forms are what the students most easily recognize and know from experience in urbanized environments, whereas Itzaj Maya have greater secondary reliance on folk specifics because their silvicultural life depends on experience at that level. Despite the compelling needs established by lived experience, both the U.S. students and the Maya overwhelmingly and in nearly equal measure subordinate such influences to a preference for generic species. I have argued that they show this preference because they presume the biological world to be partitioned at that rank into nonoverlapping kinds, each with its own unique causal essence, or inherent underlying nature, whose visible products they may or may not readily perceive.[5] People anticipate that the biological information value of these preferred kinds is maximal whether or not there is also visible indication of maximal covariation of perceptual attributes. This does not mean that more general perceptual cues have no inferential value when applied to the folk-biological domain. On the contrary, the findings here point to a significant role for such cues in targeting basic-level life forms as secondary foci for inferential understanding in a cultural environment where biological awareness is poor, as among many people who live in urban areas in the United States.

## GENERIC SPECIES, NATURAL SELECTION, AND THE EVOLUTION OF HUMAN COGNITION

There may possibly be an evolutionary design for a cognitive division of labor between domain-general perceptual heuristics and domain-specific learning mechanisms: the former enabling flexible adaptation to variable conditions of experience, and the latter invariably steering us to those abiding aspects of biological reality that are both causally recurrent and especially relevant to the emergence of human life and cognition. One hallmark of adaptation is a phylogenetic history that extends beyond the species in which the adaptation is perfected: for example, ducklings crouching in the presence of hawks but not of other kinds of birds suggests dedicated mechanisms for something like species recognition. To be sure, the world itself is neither chaos or flux: species are often locally self-structuring entities that are reproductively and ecologically isolated from other species through natural selection. But there is no a priori reason for the mind to always focus on categorizing and relating species qua species unless doing so has served some adaptive function, and the adaptive functions of organisms rarely, if ever, evolve or operate in nature as all-purpose mechanisms.

All organisms must function to procure energy to survive, and they also must procure (genetic) information for recombination and reproduction (Eldredge 1986). The first requirement is primarily satisfied by other species and by an indiscriminate use of any individual of the other species (e.g., energy-wise, it does not generally matter which chicken or which spinach plant you eat). The second requirement is usually only satisfied by genetic information unique to individual conspecifics (e.g., genetically, it matters who is chosen as a mate and who is considered kin). On the one hand, humans recognize other humans by individuating them with the aid of species-specific triggering algorithms that "automatically" coordinate perceptual cues (e.g., facial recognition schemata) with conceptual assumptions (e.g., intentions). Thus, children spontaneously read the mind of a mother by her gaze in order to predict her behavior, but they also employ a similar strategy to understand and predict the behavior of other individuals who could be potential allies or enemies in life. On the other hand, people do not spontaneously individuate the members of other species in this way, but only as examplars of the (generic) species that identifies them as causally belonging to one and only one essential kind.

Natural selection basically accounts only for the appearance of complexly well-structured biological traits designed to perform important functional tasks of adaptive benefit to organisms. In general, naturally selected adaptations are structures functionally "perfected for any given habit" (Darwin 1883, 140) and that have "very much the appearance of design by an intelligent designer ... on which the wellbeing and very existence of the organism depends" (Wallace 1901, 138). Plausibly, the universal appreciation of generic species as the causal foundation for the taxonomic arrangement of biodiversity and for taxonomic inference about the distribution of causally related properties that underlie biodiversity is one such functional evolutionary adaptation.

## GENERIC SPECIES AND CULTURAL EVOLUTION

Folkbiology in general and generic-species concepts in particular represent a stable knowledge structure that is supported by high interinformant agreement and that regularly and recurrently (within and across cultures) serves as a principled basis for the transmission and acquisition of more variable and extended forms of cultural knowledge. Consider, for example, the spontaneous emergence of totemism—the correspondence of social groups with generic species—at different times and in different parts of the world. Why, as Lévi-Strauss (1963) aptly noted, are totems so "good to think"? In part, totemism uses representations of generic species to represent groups of people; however, this pervasive metarepresentational inclination arguably owes its recurrence to its ability to ride piggyback on folkbiological taxonomy.

Generic species and groups of generic species are inherently well structured, attention arresting, memorable, and readily transmissible across minds. As a result, they readily provide effective pegs on which to attach knowledge and behavior of less intrinsically well-determined social groups. In this way, totemic groups can also become memorable, attention arresting, and transmissible across minds, which are the conditions for any meme to become culturally viable (see Sperber 1996 for a general view of culture along the lines of an "epidemiology of representations"). A significant feature of totemism that enhances both memorability and its capacity to grab attention is that it violates the general behavior of biological species: members of a totem, unlike members of a generic species, generally do not interbreed, but only mate with members of other totems in order to create a system of social exchange. Notice that this violation of core knowledge is far from arbitrary. In fact, it is a pointed violation of human beings' intuitive ontology, and as such it readily mobilizes most of the assumptions people ordinarily make about biology in order to help build societies around the world (Atran and Sperber 1991).

In sum, folkbiological concepts are special players in cultural evolution, whose innate stability derivatively attaches to more complex representational forms, thus enhancing the latter's prospects for regularity and recurrence in transmission within and across cultures. This transmission includes knowledge that cumulatively enriches (e.g., to produce folk expertise), pointedly violates (e.g., to produce religious belief), or otherwise displaces (e.g., to produce science) the intuitive ontology prescribed by folkbiology.

## GENERIC SPECIES, COMMON SENSE AND SCIENCE

Much of the history of systematics has involved attempts to adapt to a more global setting the locally relevant principles of folkbiology—such as the taxonomic embedding of biodiversity, the primacy of species, and the teleo-essentialist causality that makes sense of taxonomic diversity and the life functions of species. This process has been far from uniform (e.g., initial rejection of plant but not animal life forms; recurrent but invariably failed attempts to define essential characters for species and other taxa; intermittent attempts to reduce teleological processes to mechanics; and so forth).

Historical continuity should not be confounded with the epistemic continuity or use of folk knowledge as a learning heuristic for scientific knowledge. Scientists have made fundamental ontological shifts away from folk understanding in the construal of species, taxonomy, and underlying causality. For example, biological science today rejects fixed taxonomic ranks, the primary and essential nature of species, teleological causes of species existence, and phenomenal evidence for the existence of taxa (e.g., *tree* cannot be a scientifically valid superordinate plant group, but *bacteria* almost assuredly should be).

Nevertheless, from the vantage of our own evolutionary history, it may be more important to the everyday life of our species (or at least to the aspects of everyday life that we became sensitive to as we evolved) that our ordinary concepts be adaptive than true. Relative to ordinary human perceptions and awareness, evolutionary and molecular biology's concerns with vastly extended and minute dimensions of time and space may be of only marginal value. The ontological shift required by science may be so counterintuitive and irrelevant to everyday life as to render inappropriate and maladaptive the use of scientific knowledge in grasping and responding to everyday circumstances. This situation makes untenable any uniform application of the doctrine of externalism for living kind concepts (i.e., the belief that understanding the ordinary meaning and reference of living-kind terms necessarily involves commitment or deference to scientific knowledge when available—in other words, to a likelier nomological account of the world). Scientific knowledge cannot wholly subsume or subvert folkbiological knowledge.

Reliance on folk concepts rather than on scientific concepts may depend on context. Belief in essences, for example, may greatly help people explore the world by prodding them to look for regularities and to seek explanations of variation in terms of underlying patterns. This strategy may bring order to ordinary circumstances, including those circumstances relevant to human survival. In other circumstances, such as wanting to know what is correct or true for the cosmos at large, folk concepts and beliefs may hinder more than help. For example, the essentialist bias to understand variation in terms of deviance is undoubtedly a hindrance to evolutionary thinking. Even in some everyday matters, the tendency to essentialize or to explain variation in terms of deviation from some essential ideal or norm (e.g., people as mental or biological "deviants") can be an effortlessly "natural" but wrong way to think.

Consider racism—that is, the projection of biological essences onto social groups. Although people may be endowed with distinct intuitive ontologies, they need to integrate and adapt them to the actual conditions of individual experience and cultural life. Humans are cognitively resourceful and eclectic and will tend to use whatever is readily available to make better sense of the world (Inagaki and Hatano 1991). Humans and animals are cognitively "adjacent" ontological domains that share many higher-order features of animacy and life (Sommers 1959, Keil 1979), which is a situation that favors transference of knowledge between these domains.[6] By reducing the natural variation among individuals to biologically justified social essences, people can artifactually create reliable conditions for deciding whom to trust. No matter how biologically spurious, once people are essentialized in this way, a causally efficacious "looping effect" would set in between people's expectations of in-group versus out-group behavior and of actual behaviors induced by such expectations (Hacking 1995). The end result, however, may be culturally more costly or evil than beneficial or good. As Darwin remarks in *Notebooks*, "Animals—whom we have made our slaves we do not consider equals—Do not slave holders wish to make the black man other kind? ... to consider him as other animal."

Science teaches us that we can do better than merely get by with what we are easily able to handle or to do from birth. The task of the anthropology of science is to explore the scope and limits of common sense in order to help us better understand the development and objectives of science (Atran 1990, 1998). For example, it helps us to better understand why it is so difficult to teach biology students evolutionary theory and why it is so difficult to get psychologists and philosophers to stop talking as if biological species were natural kinds with lawful natures or metaphysical essences. It may also help people to understand that although not all members of a species—including our own—are created equal, neither are any member groups essentially unequal from a biological standpoint.

## GENERIC SPECIES AND THE SPECIES CONCEPT

Darwin's own work strongly suggests that humankind's appreciation of phenomenal species does not simply produce psychologically convenient clusterings of phenotypic variation. There is the assumption and presumably to a degree the reality of an underlying causal story. Yet, in a sense, Darwin's triumph is a paradox. From a strictly cosmic standpoint, the title of his great work, *On the Origins of Species*, is ironic and misleading—much as if Copernicus had entitled his attack on the geocentric universe, *On the Origins of Sunrise*. Of course, in order to attain that cosmic understanding, Darwin could no more dispense with thinking about "common species" than Copernicus could avoid thinking about the sunrise (Wallace 1901, 1–2). Where does this paradox leave the species concept?

In philosophy of biology, the current debate over the species concept seems to center on whether or not there is a single theoretically significant level of organization that covers all organisms (Kitcher 1993). For the most part, theoretical significance is equated with significance to evolutionary theory, which at this stage in the historical development of science remains the most powerful and creative paradigm in biology (Sober 1993). Accepting the primacy of evolutionary theory seems to rule out species concepts that may be preferable on mainly pragmatic or operational grounds—such as historical primacy (Linnaean species), maximal covariance of many present and absent characters (pheneticists' basic taxonomic units), or minimally nested character-state distributions (speciation in pattern cladism).

Unfortunately for evolutionary monists, presently no one species concept is simultaneously able to deal adequately with issues of interbreeding (delimiting the boundaries of reproduction and gene flow), phylogenesis (fixing geneaological ascendance and descendance), and ecology (determining the geographical distribution of biodiversity)—all of which are thought to be fundamental to the causal patterning and development of life on Earth. This situation lets nontheoretical or at least nonevolutionary considerations of a pragmatic sort back in through the side door to weigh on the choice of which evolutionary species concept should be primary (cf. Ereshefsky 1992), which in turn risks biasing evolutionary theory itself away from a correct or true understanding of life's emergence.

An alternative to this dilemma has been a call for pluralism, yielding as many species concepts as may accord with various equal, different, or combined considerations from psychology, history, logic, metaphysics, or the several branches of biology (cf. Dupré 1993). For a bystander eyeing the multiplication of uses and abuses that pluralism seems able to generate, such an alternative could well leave not only truth but also clarity, in the abyss. There are also other alternatives. One is to leave the species concept as a general, open-textured umbrella for a converging range of significant thought and research that need only be carefully specified intermittently and in context—like the framework notion of representation in cognitive psychology. Still another alternative is to abandon the species concept altogether as outworn, hopelessly vague, or evidently false—like the notion of the ether in physics (cf. Ereshefsky, Mishler, chapters 11 and 12 in this volume). Either of these alternatives, however, severs any connection with the specific commonsense intuitions that universally allow humans beings to believe in the organic world as a causally related totality in the first place and that were a necessary (if not sufficient) condition for the development of biological and evolutionary thinking (by contrast, even without notions of representation or of the ether, human belief in the mind or the world of substance is conceivable, as are the sciences of cognitive psychology and physics).

Perhaps the species concept, like teleology, should be allowed to survive in science more as a regulative principle that enables the mind to establish a regular communication with the ambient environment than as an epistemic principle that guides the search for nomological truth. Once communication is established with the world, science may discover deeper channels or more significant overlapping networks of causality. The persistence of a species concept would function to ensure only that these diverse scientific explorations are never wholly disconnected or lost from one another or from that aspect of phenomenal reality that will always remain as evident to a Maya as to a modern scientist.

## A SYNOPSIS OF FOLKBIOLOGY

Folkbiology (FB) may be canonically described as follows:

1. FB has a cognitive structure that is culturally universal and that places a priori constraints on the ways human beings ordinarily categorize and reason inductively about the properties and relationships of organic objects. This structure consists of

   1.1. Categorical distinctions and principles in the intuitive ontology of human beings, such that

      1.1.1. Every natural object is either a living kind or not.

      1.1.2. Every living kind is either an animal or a plant; that is, it belongs to a *folk kingdom*.

      1.1.3. Each animal or plant belongs to one and only one essential grouping, or *generic species*.

      1.1.4. In addition:

         1.1.4.1. Most cultures also partition generic species among *life forms* and divide them into *folk specifics*.

         1.1.4.2. Most cultures that have life forms and folk specifics also have *intermediates* (groupings subordinate to life forms), *folk varietals* (groupings subordinate to folk specifics), and occasionally *subvarietals* (groupings subordinate to folk specifics).

   1.2. This ontological structure may be described as a *ranked taxonomy* (RT). Formally, RT may be partially characterized as follows:

1.2.1. RT is headed by the (named or unnamed) categories *animal* and *plant*, which are *folk kingdoms* (FK). A folk kingdom is a maximal taxonomic category with respect to a «kind of» relation, **K**, such that

1.2.1.1. **K** is a two-place, acyclic relation with a finite domain, $\mathbf{T}^*$ (i.e., for no sequence $x1, \ldots, xn$ of members of its domain do we have $x1\,\mathbf{K}\,x2, \ldots, xn-1\,\mathbf{K}\,xn, xn\,\mathbf{K}\,x1$);

1.2.1.2. **K** is transitive over every *taxonomic category*, **T** (i.e., any subset of its domain).

1.2.1.3. No item is of two distinct kinds unless one is a kind of the other (i.e., for any members $x$, $y$, $z$ of **T** such that $x\,\mathbf{K}\,y$ and $x\,\mathbf{K}\,z$, either $y = z$ or $y\,\mathbf{K}\,z$ or $z\,\mathbf{K}\,y$).

1.2.1.4. Every **T** consists of a head item $h$ and everything in $\mathbf{T}^*$ that is a «kind of» $h$ (i.e., for some $h \in \mathbf{T}^*$, $\mathbf{T} = \{h\} \cup \{x/x\,\mathbf{K}\,h\}$. Taxon $h$ is then called the *head* of taxonomic category **T**.

1.2.1.5. It follows that the set $\mathbf{T}^*$ of taxa with respect to **K** is partitioned into disjoint taxonomic kingdoms with respect to **K**. The head of a **K**-kingdom (i.e., plant, animal) stands in relation **K** to no member of $\mathbf{T}^*$.

1.2.2. For each FK there is a ranking **R**, such that each rank with respect to **R** has a special conceptual status within the system of folk concepts. A *ranking* of **T** with respect to **K** is a function **R** from set **T** onto a set of consecutive integers $\{m, \ldots, n\}$, with $m < 0$ and $n > 0$, which satisfies the following condition: $(\forall x, y \in \mathbf{T})$ [if $x$ **K** $y$ then $\mathbf{R}(y) > \mathbf{R}(x)$]. The integers $m, \ldots, n$ in the range R are called *ranks* with respect to **R**, and $\mathbf{R}(x)$ is the *rank* of $x$ with respect to **R**.

1.2.2.1. Rank $n$ is the rank of *folk kingdom* (FK).

1.2.2.2. Rank 0 is the rank of *generic species* (GS).

1.2.2.3. Rank $n - 1$ is the rank of *life form* (LF).

1.2.2.4. Rank $-1$ is the rank of *folk specific* (FS).

1.2.2.5. Rank $-2$ is the rank of *folk varietal* (FV).

1.2.2.6. Rank $-3$ is the rank of *folk subvarietal*.

1.2.2.7. Taxa (named or unnamed) between ranks $n - 1$ and 0 are *intermediate*.

1.2.3. In any system of folk concepts, FK and GS (i.e., ranks 0 and $n$) are mandatory in the sense that every terminal kind is a subkind of some taxon of that rank, such that

1.2.3.1. A *terminal* kind has no subkinds (i.e., $x$ is terminal for **K** if and only if $x$ is in the domain of **K** and there is no $y$ such that $y$ **K** $z$).

1.2.3.2. $\forall x$ [$x$ is terminal $\rightarrow$ ($\mathbf{R}(x) = i \vee \exists y(x \mathbf{K} y$ and $\mathbf{R}(y) = i$))].

1.2.3.2.1. It follows that if **T** is a taxonomic category, the maximal rank $n$ of the head of **T** is mandatory.

1.2.3.2.2. It also follows that if a level is mandatory, it partitions the taxa at that level or lower (into mutually exclusive groups of organisms).

1.2.4. It remains an open question, whether or not

1.2.4.1. LFs are mandatory. If so, then apparently unaffiliated generic species are in fact monotypic life forms; that is, the LF and its single GS are extensionally (perceptually) equivalent, but conceptually distinct.

1.2.4.2. Some intermediate taxa are ranked. If so, then any such intermediate taxon is a subkind of some life form, such that: $(\forall x \in \mathbf{T})[\mathbf{R}(x) = n - 2 > \exists y(x \mathbf{K}\, y$ and $\mathbf{R}(y) = n - 1)]$.

1.2.5. In the historical development of Western systematics:

1.2.5.1. Rank $n$ became the biological kingdom (Cesalpino 1583).

1.2.5.2. Rank 0 fissioned into ranks 0, the biological species, and 1, the genus (including monospecific genera) (Tournefort 1694).

1.2.5.3. Rank $n - 1$ became the biological class (Linnaeus 1738).

1.2.5.4. Rank $n - 2$ was initially formalized as the biological family (Jussieu 1789).

1.2.5.5. Taxa below rank 0 became unranked infraspecific groups (Darwin 1859).

2. FB is *domain specific* (rather than domain general) in the sense that it involves cognitive operations partially dedicated to the perceptual identification and conceptual processing of nonhuman living kinds as generic species of animals and plants:

2.1. Experiments with people in highly diverse cultures (U.S. Midwest, Mayan rainforest), suggest that

2.1.1. People everywhere categorize and reason about nonhuman organisms in terms of essence-based generic species and ranked taxonomies.

2.1.2. Domain-general, similarity-based models fail to account for these findings.

3. This domain-specific capacity for the universal and spontaneous understanding of the organic world is

   3.1. Likely grounded in a priori abilities that are independent of actual experience and thus *innate* to the human mind/brain.

   3.2. Plausibly an *evolutionary adaptation* (rather than a by-product or accident) to relevant and recurrent features of hominid ancestral environments (e.g., wide-ranging subsistence involving the understanding of potentially indefinitely many species and habitats).

4. Systematic biology culturally evolved from folkbiology. Within an ever-expanding global (and ultimately cosmic) framework, natural historians from Aristotle to Darwin sought ways to extend or overcome the limits of understanding inherent in people's spontaneous abilities to parse the local environment's biodiversity.

## ACKNOWLEDGMENTS

The comparative studies reported here were funded by the National Science Foundation (SBR 93-19798, 97-07761) and the French Ministry of Research and Education (Contract CNRS 92-C-0758). They were codirected with Douglas Medin. Participants in this project on biological understanding across cultures include Alejandro López (psychology, Max Planck), John Coley (psychology, Northeastern University), Elizabeth Lynch (psychology, Northwestern University), Ximena Lois (linguistics, Crea-Ecole Polytechnique), Valentina Vapnarsky (anthropology, Université de Paris X), Edward Smith and Paul Estin (psychology, University of Michigan), and Brian Smith (biology, University of Texas, Arlington).

## NOTES

1. Thus, comparing constellations in the cosmologies of ancient China, Greece, and the Aztec Empire shows little commonality. By contrast, herbals like the ancient Chinese *ERH YA*, Theophrastus's *Peri Puton Istorias*, and the Aztec *Badianus Codex*, share important features, such as the classification of generic species into tree and herb life forms (Atran 1990, 276).

2. Generalizations across taxa of the same rank thus differ in logical type from generalizations that apply to this or that taxon. *Termite*, *pig*, and *lemon tree* are not related to one another by a simple class inclusion under a common hierarchical node, but by dint of their common rank—in this case, the level of generic species. A system of rank is not simply a hierarchy, as some suggest (Rosch 1975, Premack 1995, Carey 1996). Hierarchy—that is, a structure of inclusive classes—is common to many cognitive domains, including the domain of artifacts. For example, *chair* often falls under *furniture* but not *vehicle*, and *car* falls under *vehicle* but not *furniture*. But there is no ranked system of artifacts: no inferential link or inductive framework spans both *chair* and *car* or *furniture* and *vehicle* by dint of a common rank, such as the artifact *species* or the artifact *family*.

3. For example, in a comparative study of Itzaj Maya and rural Michigan college students, we found that the great majority of mammal taxa in both cultures correspond to scientific species and that most also correspond to monospecific genera: 30 of 40 (75%) basic Michigan mammal

terms denote biological species, of which 21 (70%, or 53% of the total) are monospecific genera; 36 of 42 (86%) basic Itzaj mammal terms denote biological species, of which 25 (69%, or 60% of the total) are monospecific genera (López et al. 1997, Atran 1999). Similarly, a Guatemalan government inventory of the Itzaj area of the Petén rainforest indicates that 69% (158 of 229) are monospecific (AHG/APESA 1992; cf. Atran and Ucan Ek' forthcoming), the same percentage of monospecific tree genera (40 of 58) as in our study of the Chicago area (Medin et al. 1997).

4. Moving vertically within each graph corresponds to changing the premise while holding the conclusion category constant. This maneuver allows us to test another domain-general model of category-based reasoning: the similarity-coverage model (Osherson et al. 1990). According to this model, the closer the premise category is to the conclusion category, the stronger the induction should be. Our results show only weak evidence for this general reasoning heuristic, which fails to account for the various "jumps" in inductive strength that indicate absolute or relative preference.

5. By contrast, a partitioning of artifacts (including those of organic origin, such as foods) is neither mutually exclusive nor inherent: some mugs may or may not be cups; an avocado may be a fruit or vegetable depending upon how it is served; a given object may be a bar stool or a waste bin depending on the social context or perceptual orientation of its user; and so on.

6. Although there may a cognitive susceptibility to racism, the resultant social groups are by no means as evident across cultures as are notions of generic species. Neither is the intermittent ranking of social formations (e.g., armies) consistent across cultures or necessarily bound to essentialized groups. In short, the apparent features of folkbiological taxonomy found in the domain of persons and social groups appears to involve a variable transference of principles from folkbiology rather than a common ontological foundation or mode of construal. The flip side of this ontological transference is anthropomorphism—that is, the projection of human intentionality onto animals. Given the initial absence of causal knowledge about (usually furtive) animals, the use of person analogies (especially by children) may initially lead to useful and accurate predictions about entities phylogenetically similar to humans (Inagaki and Hatano 1991).

## REFERENCES

AHG/APESA (1992). *Plan de desarollo integrado de Pet(n: Inventario forestal del Departamento del Pet(n (Convenio Gobiernos Alemania y Guatemala)*. Santa Elena, Petén: SEGEPLAN.

Anderson, J. (1990). *The adaptive character of thought*. Hillsdale, N.J.: Erlbaum.

Atran, S. (1985). The nature of folk-botanical life forms. *American Anthropologist* 87, 298–315.

Atran, S. (1987). Origins of the species and genus concepts. *Journal of the History of Biology* 20, 195–279.

Atran, S. (1990). *Cognitive foundations of natural history: Towards an anthropology of science*. Cambridge, England: Cambridge University Press.

Atran, S. (1993). Itza Maya tropical agro-forestry. *Current Anthropology* 34, 633–700.

Atran, S. (1995). Causal constraints on categories and categorical constraints on biological reasoning across culture. In D. Sperber, D. Premack, and A. Premack, eds., *Causal cognition*. Oxford: Oxford University Press.

Atran, S. (1998b). Folkbiology and the anthropology of science. *Behavioral and Brain Sciences* 21, 547–611.

Atran, S. (1999). Itzaj Maya folkbiological taxonomy. In D. Medin and S. Atran, eds., *Folkbiology*. Cambridge, Mass.: MIT Press.

Atran, S., and D. Medin (1997). Knowledge and action: Cultural models of nature and resource management in Mesoamerica. In M. Bazerman, D. Messick, A. Tinbrunsel, and K. Wayde-Benzoni, eds., *Environment, ethics, and behavior*. San Francisco: New Lexington Press.

Atran, S., and D. Sperber (1991). Learning without teaching: Its place in culture. In L. Tolchinsky-Landsmann, ed., *Culture, schooling and psychological development*. Norwood, N.J.: Ablex.

Atran, S., and E. Ucan Ek' (forthcoming). Classification of useful plants among the Northern Peten Maya (Itzaj). In C. White, ed., *Ancient Maya diet*. Salt Lake City: University of Utah Press.

Atran, S., P. Estin, J. Coley, and D. Medin (1997). Generic species and basic levels: Essence and appearance in folkbiology. *Journal of Ethnobiology* 17, 22–45.

Bartlett, H. (1940). History of the generic concept in botany. *Bulletin of the Torrey Botanical Club* 47, 319–362.

Berlin, B. (1978). Ethnobiological classification. In E. Rosch and B. Lloyd, eds., *Cognition and categorization*. Hillsdale, N.J.: Erlbaum.

Berlin, B. (1992). *Ethnobiological classification*. Princeton: Princeton University Press.

Berlin, B. (1999). One Maya Indian's view of the plant world, In D. Medin and S. Atran, eds., *Folkbiology*. Cambridge, Mass.: MIT Press.

Berlin, B., D. Breedlove, and P. Raven (1973). General principles of classification and nomenclature in folkbiology. *American Anthropologist* 74, 214–242.

Berlin, B., D. Breedlove, and P. Raven (1974). *Principles of Tzeltal plant classification*. New York: Academic Press.

Bock, W. (1973). Philosophical foundations of classical evolutionary taxonomy. *Systematic Zoology* 22, 275–392.

Boster, J. (1988). Natural sources of internal category structure. *Memory and Cognition* 16, 258–270.

Boster, J. (1991). The information economy model applied to biological similarity judgment. In L. Resnick, J. Levine, and S. Teasley, eds., *Perspectives on socially shared cognition*. Washington, D.C.: American Psychological Association.

Brown, C. (1984). *Language and living things: Uniformities in folk classification and naming*. New Brunswick, N.J.: Rutgers University Press.

Bulmer, R. (1970). Which came first, the chicken or the egg-head? In J. Pouillon and P. Maranda, eds., *Echanges et communications: Mélanges offerts à Claude Lévi-Strauss*. The Hague: Mouton.

Bulmer, R. (1974). Folkbiology in the New Guinea Highlands. *Social Science Information* 13, 9–28.

Cain, A. (1956). The genus in evolutionary taxonomy. *Systematic Zoology* 5, 97–109.

Carey, S. (1985). *Conceptual change in childhood*. Cambridge, Mass.: MIT Press.

Carey, S. (1996). Cognitive domains as modes of thought. In D. Olson and N. Torrance, eds., *Modes of thought*. New York: Cambridge University Press.

Cesalpino, A. (1583). *De plantis libri XVI*. Florence: Marescot.

Coley, J., D. Medin, and S. Atran (1997). Does rank have its privilege? Inductive inferences in folkbiological taxonomies. *Cognition* 63, 73–112.

Coley, J., E. Lynch, J. Proffitt, D. Medin, and S. Atran (1999). Inductive reasoning in folkbiological thought. In D. Medin and S. Atran, eds., *Folkbiology*. Cambridge, Mass.: MIT Press.

Darwin, C. (1859). *On the origins of species by means of natural selection*. London: Murray.

Darwin, C. (1872, reprint 1883). *On the origins of species by means ofnatural selection*, 6th ed. New York: Appleton.

Darwin, C. (1960–1967). Darwin's notebooks on transmutation of species. G. de Beer, M. Rowlands, and B. Skarmovsky, eds., *Bulletin of the British Museum (Natural History)* 2, 27–200, and 3, 129–176 (notebooks B, C, D, E).

Diamond, J., and D. Bishop (1998). Ethno-ornithology of the Ketengban people, Indonesian New Guinea. In D. Medin and S. Atran, eds., *Folkbiology*. Cambridge, Mass.: MIT Press.

Diver, C. (1940). The problem of closely related species living in the same area. In J. Huxley, ed., *The new systematics*. Oxford: Clarendon Press.

Donnellan, K. (1971). Necessity and criteria. In J. Rosenberg and C. Travis, eds., *Readings in the philosophy of language*. Englewood-Cliffs, N.J.: Prentice-Hall.

Dougherty, J. (1978). Salience and relativity in classification. *American Ethnologist* 5, 66–80.

Dougherty, J. (1979). Learning names for plants and plants for names. *Anthropological Linguistics* 21, 298–315.

Dupré, J. (1993). *The disorder of things*. Cambridge, Mass.: Harvard University Press.

Eldredge, N. (1986). Information, economics and evolution. *Annual Review of Ecology and Systematics* 17, 351–369.

Ellen, R. (1993). *The cultural relations of classification*. Cambridge: Cambridge University Press.

Ereshefsky, M. (1992). Eliminative pluralism. *Philosophy of Science* 59, 671–690.

Gelman, S., and H. Wellman (1991). Insides and essences. *Cognition* 38, 214–244.

Gelman, S., J. Coley, and G. Gottfried (1994). Essentialist beliefs in children. In L. Hirschfeld and S. Gelman, eds., *Mapping the mind*. New York: Cambridge University Press.

Goodman, N. (1972). *Problems and projects*. Indianapolis: Bobbs-Merrill.

Greene, E. (1983). *Landmarks in botany*, 2 vols. Stanford: Stanford University Press (originally compiled for the Smithsonian in 1909).

Hacking, I. (1995). The lopping effects of human kinds. In D. Sperber, D. Premack, and A. Premack, eds., *Causal cognition*. Oxford: Oxford University Press.

Hatano, G., and K. Inagaki (1994). Young children's naive theory of biology. *Cognition* 50, 171–188.

Hatano, G., and K. Inagaki (1998). A developmental perspective on informal biology. In D. Medin and S. Atran, eds., *Folkbiology*. Cambridge, Mass.: MIT Press.

Hays, T. (1983). Ndumba folkbiology and general principles of ethnobotanical classification and nomenclature. *American Anthropologist* 85, 592–611.

Hickling, A., and S. Gelman (1995). How does your garden grow? Evidence of an early conception of plants as biological kinds. *Child Development* 66, 856–876.

Hull, D. (1997). The ideal species definition and why we can't get it. In M. Claridge, H. Dawah, and M. Wilson, eds., *Species: The units of biodiversity*. London: Chapman and Hill.

Hunn, E. (1976). Toward a perceptual model of folkbiological classification. *American Ethnologist* 3, 508–524.

Hunn, E. (1977). *Tzeltal folk zoology*. New York: Academic Press.

Hunn, E. (1982). The utilitarian factor in folkbiological classification. *American Anthropologist* 84, 830–847.

Inagaki, K., and G. Hatano (1991). Constrained person analogy in young children's biological inference. *Cognitive development* 6, 219–231.

Inagaki, K., and G. Hatano (forthcoming). Young children's recognition of commonalities between plants and animals. *Child Development*.

Jussieu, A-L. (1789). *Genera plantarum*. Paris: Herissant.

Keil, F. (1979). *Semantic and conceptual development: An ontological perspective*. Cambridge, Mass.: Harvard University Press.

Keil, F. (1989). *Concepts, kinds, and cognitive development*. Cambridge, Mass.: MIT Press.

Keil, F. (1995). The growth of causal understandings of natural kinds. In D. Sperber, D. Premack, and A. Premack, eds., *Causal cognition*. Oxford: Oxford University Press.

Kitcher, P. (1993). *The advancement of science*. New York: Oxford University Press.

Lassaline, M., E. Wisniewski, and D. Medin (1992). Basic levels in artificial and natural categories. In B. Burns, ed., *Percepts, concepts and categories*. New York: Elsevier.

Lévi-Strauss, C. (1963). The bear and the barber. *Journal of the Royal Anthropological Institute* 93, 1–11.

Linnaeus, C. (1738). *Classes plantarum*. Leiden: Wishoff.

Linnaeus, C. (1751). *Philosophia botanica*. Stockholm: G. Kiesewetter.

López, A., S. Atran, J. Coley, D. Medin, and E. Smith (1997). The tree of life: Universals of folk-biological taxonomies and inductions. *Cognitive Psychology* 32, 251–295.

Mandler, J., P. Bauer, and L. McDonough (1991). Separating the sheep from the goats: Differentiating global categories. *Cognitive Psychology* 23, 263–298.

Mandler, J., and L. McDonough (1996). Drinking and driving don't mix: Inductive generalization in infancy. *Cognition* 59, 307–335.

Mayr, E. (1969). *Principles of systematic zoology*. New York: McGraw-Hill.

Mayr, E. (1982). *The growth of biological thought*. Cambridge, Mass.: Harvard University Press.

Medin, D., E. Lynch, J. Coley, and S. Atran (1997). Categorization and reasoning among tree experts: Do all roads lead to Rome? *Cognitive Psychology* 32, 49–96.

Osherson, D., E. Smith, O. Wilkie, A. López, and E. Shafir (1990). Category-based induction. *Psychological Review* 97, 85–200.

Premack, D. (1995). Forward to part IV: Causal understanding in naïve biology. In D. Sperber, D. Premack, and A. Premack, eds., *Causal cognition*. Oxford: Oxford University Press.

Quine, W. (1960). *Word and object*. Cambridge, Mass.: Harvard University Press.

Rosch, E. (1973). On the internal structure of perceptual and semantic categories. In T. Moore, ed., *Cognitive development and the acquisition of language*. New York: Academic.

Rosch, E. (1975). Universals and cultural specifics in categorization. In R. Brislin, S. Bochner, and W. Lonner, eds., *Cross-cultural perspectives on learning*. New York: Halstead.

Rosch, E., and C. Mervis (1975). Family resemblances: Studies in the internal structure of natural categories. *Cognitive Psychology* 8, 382–439.

Rosch, E., C. Mervis, W. Grey, D. Johnson, and P. Boyes-Braem (1976). Basic objects in natural categories. *Cognitive Psychology* 8, 382–439.

Simmons, D., and F. Keil (1995). An abstract to concrete shift in the development of biological thought: The insides story. *Cognition* 56, 129–163.

Simpson, G. (1961). *Principles of animal taxonomy*. New York: Columbia University Press.

Smith, E., and D. Medin (1981). *Categories and concepts*. Cambridge, Mass.: Harvard University Press.

Sober, E. (1993). *Philosophy of biology*. Boulder, Colo.: Westview Press.

Sommers, F. (1959). The ordinary language tree. *Mind* 68, 160–185.

Sperber, D. (1996). *Explaining culture: A naturalistic approach*. Oxford: Blackwell.

Springer, K., and F. Keil (1989). On the development of biologically specific beliefs: The case of inheritance. *Child Development* 60, 637–648.

Stevens, P. (1994). Berlin's "Ethnobiological Classification." *Systematic Biology* 43, 293–295.

Stross, B. (1973). Acquisition of botanical terminology by Tzeltal children. In M. Edmonson, ed., *Meaning in Mayan languages*. The Hague: Mouton.

Tanaka, J., and M. Taylor (1991). Object categories and expertise: Is the basic level in the eye of the beholder? *Cognitive Psychology* 23, 457–482.

Tournefort, J. (1694). *Elémens de botanique*. Paris: Imprimerie Royale.

Wallace, A. (1901). *Darwinism*, 3rd ed. London: Macmillan (1st ed. 1889).

Warburton, F. (1967). The purposes of classification. *Systematic Zoology* 16, 241–245.

Zubin, D., and K.-M. Köpcke (1986). Gender and folk taxonomy. In C. Craig, ed., *Noun classes and categorization*. Amsterdam: John Benjamins.

# 10 Species, Stuff, and Patterns of Causation

Frank C. Keil and Daniel C. Richardson

Adults across all cultures seem to think about biological phenomena in distinct ways that are not found in other domains of knowledge. Thought about the living world has its own particular nature that may be different from thought about artifacts and the nonliving natural world. Perhaps the most fundamental unit of that apparently special kind of biological thought is the concept of the species: the basic ontological kind around which so much of folk biology seems to be organized. If this is correct, it is in many ways quite extraordinary, for the phenomenal differences between plants and animals seem so enormous that it is remarkable that they nonetheless might be conceived as vastly more similar to each other than either is to other sorts of kinds. In this essay, we ask how it is that species might be thought of so differently and on what psychological basis.

At the most general level, two, nonmutually exclusive possibilities might explain this phenomenon: (1) there is something in the information about biological species that is structured so differently that it shapes learning differently and results in different kinds of cognitive structures for thinking about species; and (2) humans have certain intrinsic cognitive biases that lead them to think about species in very different ways. The task here is to explore the details of such possibilities and their relative roles. To frame the problem, it is useful to see how each possibility could be the only account.

For example, it might be that the informational patterns associated with biological kinds result in different kinds of knowledge structures being created by completely general learning capacities in humans. The kinds of features seen in living things and the patterns of correlation formed among them over time might result in knowledge of a distinctive type, but that knowledge reflects nothing about any a priori human expectations about living things. A variety of general learning devices, given the kind of information associated with living kinds, would tend to represent that information distinctively and then sequester that information into a coherent domain of biological thought.

Such an approach was implemented in a connectionist model by McRae, de Sa, and Seidenberg (1997). From a large body of feature norms, they found that living kinds had a greater density of intercorrelated features than artifacts and that this distinction could explain the results of several priming studies that revealed latency differences between the two kinds. Furthermore, they claimed that an attractor network that had distributed knowledge of the empirically derived feature norms could simulate the same sort of behavior in priming tasks. Of course, priming results alone are hardly enough reason to suppose that the corresponding living kind concepts were fundamentally different in structure, but the work done by McRae and his colleagues illustrates how such an argument might proceed.[1]

Alternatively, it might be exceedingly difficult to specify any objective informational patterns that set aside plants and animals as a distinct kind with its own rich internal structure; instead, humans may carry with them certain cognitive biases to interpret information about species in highly distinctive and unique ways. Thus, extremely simple perceptual features shared by animals and plants (perhaps a kind of fractal structure seen more commonly in living things) might trigger a cascade of predetermined cognitive biases that make learning and the resulting concepts of biological kinds radically different from other concepts of kinds. There may be some real informational differences as well, such as differing densities of feature clusters, but this view would argue that those differences far underdetermine the nature and kind of specializations seen in biological thought.

We argue that there may be misleading assumptions and misconceptions about the nature of the distinctive information and about the biases that color our notions of what biological thought is like; also, the ensuing misconceptions about biological thought, especially as a kind of intuitive theory, may obscure the true nature of species concepts and how they are embedded within the broader system of biological knowledge. Several new lines of empirical work are suggested.

Much of our discussion examines how biological thought in general and species concepts in particular develop in the child. We take such a developmental perspective because (1) it tells us what sorts of information might be most salient to the naive mind, and (2) it suggests what might be the most fundamental biases that we all have in thought about the living world.

Between the empiricist and nativist extremes there are many gradations and combinations. Our goal here is not to allocate responsibility to these two extremes as much as it is to explore what each may contribute to a more complex interactive model of where species concepts come from and what makes them special. We do so by trying to clarify what informational patterns might be both distinctive to living kinds and salient to humans, and by asking what sorts of cognitive biases might interact with that information, even if those biases themselves are not always reserved exclusively for biological phenomena.

## THE GENERAL NATURE OF FOLKBIOLOGICAL THOUGHT

Any notions of species are embedded within broader systems of belief about biological kinds. One cannot understand what a dog is without also knowing something about animal life cycles, nutritional needs, and the like. For that reason, it is helpful to consider how views of intuitive biological thought have changed in recent years (e.g., Medin and Atran 1999). It has become commonplace to argue that lay people throughout the world possess richly structured beliefs about the living world that might be thought of as intuitive or naive theories about biological processes and systems—such as growth, reproduction, digestion, disease, and death (Medin and Atran 1999). Characterizations of these beliefs as theories or mental models, however, may carry with them somewhat misleading ideas about how that knowledge is represented in the mind of the individual. It might seem that the knowledge must be an explicit set of beliefs connected together in the tightly coherent manner of a formal scientific theory and that the models must contain concrete, imagelike components whose interactions can be clearly visualized. Folk theories would then be said to differ from formal scientific ones only in terms of the particular sets of beliefs they embrace and not so much in terms of their general format. Thus, a belief that demonic possession causes disease contagion might have very much the same kind of mechanistic set of lawful relations as a belief that germs cause disease contagion. (Keil et al. 1999).

But a closer look at intuitive biological theories and perhaps at many theories in the more formal sciences reveals something quite different from an explicit set of propositions all linked together in a tightly connected, logically consistent, and coherent set of inferences. Most people have strikingly little knowledge of the detailed mechanisms at work in their own bodies, let alone in other animals and plants. An exceedingly simple gloss may be all that is known, such as food contains energy and that the body uses that energy as a kind of fuel to power muscles, which make us move. This simple functional schemata may then occasionally get filled with local mechanisms and gradually become interconnected in somewhat larger and more coherent structures, but only the smallest percentage of people in any culture can tell you much of anything about the full causal chain that goes from the ingestion of food to the production of a motor movement based on the energy in that food.

Yet people seem to have far more than a set of functional schemata. They seem to have general and often abstract ways of choosing among classes of explanations about biological phenomena, even though these explanations may be equally satisfactory from a functional point of view. An explanation of mechanisms of digestion in terms of a mechanism in which food particles are converted to light that is routed around the body before being transformed into muscle energy would be vastly less plausible to most adults than a mechanism that invoked transformation of food into a kind of fluid "fuel," even if both mechanisms were ultimately wrong.

There are several sorts of things, above the level of specific mechanisms, that adults and possibly children know and believe about biological kinds and that nonetheless might be distinctive to those kinds and thereby make biological thought different:

1. We know that certain kinds of properties and relations tend to be central to explanations in biology and to the stability of various biological phenomena (cf. Boyd, chapter 6 in this volume). For example, (a) color is likely to play a more important causal role in explaining functions of most living kinds than in explaining functions of most artifacts; (b) size variation might matter less for most living kinds than for most artifacts; (c) difficulty of finding instances may be more important for artifacts; (d) sensitivity to temperature is more important to living kinds, as is (e) their age or stage of development.

2. We know that certain kinds of causal patternings might be distinctive to living kinds and to explanations about their nature. Patterns of causal homeostasis may appear to be richer and more interconnected (Boyd, chapter 6 in this volume). Causal mechanisms may change more dramatically as one goes from "inside" a biological kind to its outside, but not for artifacts and nonliving natural kinds. We may expect the time course of bounded causal events for biological systems to have a certain duration that is distinctive in comparison to the time course for either artifacts or nonliving natural kinds. Artifacts tend to have immediate cause and effects, whereas living things can have a far more delayed reaction to events; for example, a plant may not yield fruit now because it wasn't watered enough a month ago.

3. We know that living kinds may have certain causal powers even if we don't know the mechanisms for these powers or their functions. For example, we might know that humans tend to sneeze when they go rapidly from dim light to bright light, but we may have absolutely no idea how that happens or for what possible reason. The specific causal powers of living things may be quite different in nature and kind from the causal powers of other sorts of things and could thereby give a distinctive structure of biological thought without knowledge of mechanisms.

4. We may also carry distinctive biases about aspects of biological entities and events—biases that may not be correct, but that powerfully constrain our beliefs and explanations about biological phenomena nonetheless. These biases might include the notions that living kinds have fixed inner essences that guide the expression and maintenance of many of their phenomenal properties and that their properties are likely to be present for functional adaptive reasons. The "essentialist bias" for living kinds becomes particularly important later when we examine the developmental course of naive folk biology.

5. We may have notions of how aspects of biological knowledge might be distributed in the minds of others such that we believe that there are people who know certain things about living kinds and thus can answer our questions. Because we cannot understand all the details of most mechanisms, we all learn to divvy up knowledge responsibilities, and part of our understanding of a phenomenon becomes knowing how to access relevant areas of expertise that others have generated. Biological knowledge and most other

areas of scientific understanding have this critical social component (Wilson and Keil 1998). Notice, however, that historically and in most cultures even today, the richness of and social apportionment of knowledge about the biological world far exceeds the richness of knowledge about the rest of the natural world, such as elements and compounds or weather systems or the stars. The cognitive division of knowledge responsibilities may therefore be far and away the most developed in the biological realm.

It is possible that we may have different expectations about how such knowledge is distributed for biological kinds than we do about how it is distributed for other sorts of kinds. For example, it may be that we expect natural kind expertise to be more tied to species or types and artifact expertise to be related to mechanisms or processes. Thus, we have vets, gardeners, pediatricians, and entomologists, on the one hand, and electricians, plumbers, and carpenters, on the other. It seems plausible that this distinction is because knowledge of living kinds has historically begun with the organism and analyzed inward, whereas expertise about nonliving things would consist of grasping a process or mechanism and learning to exploit it for functional means across types. Moreover, as science progresses and we learn more about biological mechanisms, we can analyze them more in terms of processes. This distinction may therefore be dissolving as "cross-type" domains of living kind knowledge develop, such as microbiology or evolutionary biology.

There are many forms of implicit knowledge one could have of living things. Explanatory knowledge of most phenomena can, by necessity, capture only part of the richness and complexity of causal structure in a domain. We therefore cannot ever have full mechanistic knowledge, but at the same time we have more than a collection of surface impressions or skeletal functional schemata. That our knowledge and understanding might be largely organized in these different forms, however, is not always appreciated, and part of the reason it is difficult to see may be that we all tend to have a vividness illusion regarding our own understandings. People may often assume they have complete or extensive mechanistic understanding of a domain when they do not. They might have observed the inner workings of a car engine or a heart or a bicycle derailleur and be convinced that they have an imagelike mental model of how that system works. They might confidently predict their ability to explain exactly what happens in each step of the causal event sequence that characterizes the entity in question; yet when queried about such mechanisms, they might reveal glaring ignorance and inconsistencies. They seem to think that understanding arises from clockwork kinds of vivid concrete steps, and because they have a strong sense of understanding, they assume they have that kind of clockwork knowledge.

Later, we discuss how the vividness illusion may help us understand why knowledge does not always develop from concrete images of interacting components to more abstract notions about property types and their interactions.

Just as we tend to overestimate the centrality of such clockwork mechanisms in our adult understandings, we may tend to overestimate their seminal role in development.

In the end, any notions of species must be powerfully influenced by how we view adult biological knowledge in general. If, for example, we believe that having an essence is a critical part of a species concept, then we must be able to say how we could have a notion of essence without any sort of concrete understanding of what is inside animals and plants. And if we think species are entities with properties that help maintain the survival and integrity of that kind, we must be able to say how we know those patterns of maintenance without knowing many of the details. Almost any way one tries to flesh out the nature of species concepts is powerfully influenced by this view of biological knowledge in general. The species concepts cannot be understood without understanding how it fits into a larger system of folk biology.

## CHILDREN, FOLK, AND SCIENTISTS

Lay people are not children relative to scientists; thus, the study of children's biological concepts is not equivalent to studying biological concepts in people who don't work in the science of biology. All adults have many interactions with the biological world. Even the most nature phobic and jaded urban dweller thinks about the food and drink he consumes and their consequences on his bodily functioning, or about the diseases he encounters in others and how they might influence his own biological state. Adults have spent thousands of more hours having such thoughts than a young child has, and they have surely developed much richer and more elaborate systems of knowledge. As we look across many different cultures, however, we turn to development as highlighting emerging and invariant universal properties of biological thought, for those properties that seem to emerge the earliest might well be the most universal. They may have much more elaborated forms in adults, but examining acquisition may help uncover a core form.

Views of the emergence of biological thought have changed dramatically in recent years. Older accounts described children as going through general stages of cognitive development that would make completely unavailable to them any real notions of biology. Piaget (1954), for example, thought of young children as having "animistic" tendencies in which they endowed a great many things, living and not, with beliefs and desires and explained their properties and actions in such terms. They did not have any notion of plants and animals as forming a common kind of biological things. But these stage views of cognitive development fell from grace in the late 1970s and early 1980s as closer and more systematic examinations of knowledge acquisition did not reveal such stages in which children progressed from having one sort of representational and computational capacity to a qualitatively different one (e.g., Gelman and Baillargeon 1983). Alternative conceptions focused on developmental patterns in specific bounded domains—such as biological thought, theory of mind, naive physical mechanics, and number. This domain specificity approach found a natural affinity with the notion of concepts as embedded in larger systems of explanation or intuitive theories. Despite occasional talk of "theories of everything," the notion of theory tends to imply a bounded domain of phenomena explained by that theory with little reason to think that theoretical knowledge in one domain should necessarily extend to all other domains. Within that framework, two main themes emerged with respect to the emergence of biological thought.

One theme argued that younger children (e.g., before age seven) really did not have an appreciation of biology as a coherent set of phenomena and that they often explained biological phenomena in social and psychological terms (Carey 1985). Thus, they might see sleeping as caused by feeling tired and wanting to sleep, and they might see its function as satisfying those needs. Similarly, they might explain eating in terms of the sensory pleasures of ingesting food and in terms of the social interactions that happen at meal times. They would completely ignore the physiological aspects of these processes. In these accounts, the children would then undergo radical conceptual changes in which an intuitive biology emerged, and now they would explain the same phenomena in completely different terms. An alternative account has emerged from several laboratories (Keil 1992, 1998; Hatano and Inagaki 1996; Wellman and Gelman 1998), however, that suggests that even young preschoolers do have some sense of the domain of biology and have a distinctive mode of biological thought. Their failures on many tasks concerning biological knowledge are attributed to their not knowing specific mechanisms and to sometimes misunderstanding which frame of reference (social versus biological) is being asked for in a task (Gutheil, Vera, and Keil 1998) rather than to a complete lack of an intuitive biology, which is seen as couched in the more implicit forms discussed earlier.

In addition to preferring some mechanisms more than others, young children also often know how knowledge of a specific mechanism is likely to be related to understanding of another mechanism, even when they have no direct knowledge of either mechanism. For example, in one study in progress in our laboratory, children as young as five years of age often respond in an adultlike manner to the following question:

Louise knows all about why kids get a second set of teeth. Cathy knows all about why babies can get afraid of strangers. Who knows why teenagers like to listen to so much music?

Many five-year-old children judge that knowledge about behavioral mechanisms is more likely to "hang together" in the minds of experts (e.g., Cathy is more likely to know about the teenagers predilection for music) than will a mixture of biological and behavioral mechanisms (e.g., where Louise would know more about the teenagers) even though they usually have no explicit notions of the mechanisms whatsoever. Indeed, when asked about the mechanisms, most of them will quickly say they have no idea. Yet, at some level these children are aware that the kinds of processes involved in stranger fear and positive feelings toward music are more similar to each other conceptually than those processes involved in explaining deciduous teeth. The converse also seems to hold—namely, if there are two questions about biological mechanisms and one about behavioral mechanisms, an expert in one of the biological questions is judged more likely to know more about the other biological question than about the behavioral one. These sorts of results raise the question, What do young children know about biology that guides their judgments in such tasks?

More generally, it seems that we do not want to think of young children as only capable of having "concrete" impressions of biological processes wherein specific mechanisms or models are explicitly visualized. Contrary to many decades of claims that younger children are bound to think in concrete terms and can grasp more abstract relations only later, it often seems that knowledge *can* shift from the abstract to the concrete in development (Simons and Keil 1995). We adopt here the view that even preschoolers have an intuitive sense, often at a highly implicit level, of biological phenomena as being distinct from one another, and we ask what that perspective suggests about species concepts and how they develop. One of the most robust findings of that developmental story has been the discovery of very early beliefs in essences for living kinds.

How might a child come to believe more strongly in essences for living kinds than for other sorts of things? We've seen that such a strong belief might arise not only from real informational differences between the causal structures responsible for biological kinds and those structures responsible for other sorts of kinds, but also from cognitive biases that might be related to those differences. Yet another source of information might be parents, but in a surprisingly subtle and implicit manner. Parents do not tell children that things have essences, but they do talk about living kinds quite differently from the way they talk about artifacts and in ways that would seem to suggest a hidden structure with greater richness for living kinds (Gelman et al. 1998). Parents do not provide didactic explanations of hidden properties and their causal consequences. Instead, they seem to indicate in more abstract ways—through different patterns of reference—what sorts of things are kinds and that some kinds are likely to have richer essences. They also provide hints as to what sorts of things are more likely to be taxonomically embedded.

We do not yet know how much influence this subtle pattern of language has on the child's conceptual development, but it does point out a third dimension that might interact with cognitive biases and intrinsic informational differences. We now need to examine the content of this essence concept that adults and children readily attach to living kinds.

## THE ESSENCE OF ESSENCE

All of us may succumb to essentialist biases that compel us to assume and look for those critical aspects of kinds that allow us to make powerful inferences. Gelman and Hirschfeld (1999) contrast three types of "psychological essentialism": sortal, ideal, and causal. *Sortal essence* refers to critical defining features or, in other words, singly necessary and jointly sufficient sets of features for determining category membership—an account that seems to work for only a small set of real-world concepts, however. *Ideal essence* refers to nonexistent perfect cases. The ideal essence of parallel lines has no real counterparts because no physical system can perfectly embody parallelism. Finally, *causal essence* refers to something about a kind that results in its having many of its most typical and stable properties. The nature of that "something" is critical. Most commonly, it seems to be thought of as a fixed inner entity that has multiple causal effects. That entity might be a kind of substance (the essence of gold being atoms with gold's number of protons) or an informational code (the DNA sequence[s] corresponding to tigers) or a process.

Despite much talk about essence in recent years, it still is not clear how most lay people actually conceive of essences or whether the biases are much stronger in younger children and in thought about some sorts of kinds rather than about others. Medin and Ortony (1988) have suggested that we can often believe in an essence without any idea of what the essence actually is—that we have an essence "place holder" concept without the concept of the essence itself. This sense of essence makes clearer how it could be a bias without specific content. But the place holder notion may overlook a related set of beliefs, for even as people commonly believe in essences without knowing any details about those essences, they still might prefer some sorts of future details to others. Thus, there might be a physicalist bias, whereby people prefer essences to be seen as objects or "stuff" rather than kinds of processes. Teaching that photosynthesis is the essence of green plants might be less compelling than teaching that it is a certain DNA sequence, even though the process of photosynthesis may be much more directly connected causally to a far greater range of phenomenal properties of plants.[2]

For species concepts, then, notions of essence may be absolutely essential, yet we have little idea of what sorts of constraints there might be on notions of essence. Some theorists have argued that notions of essence have had extraordinary limiting effects on how we think about species in the context of evolutionary thought. Hull (1965), for example, talks about how Aristotelian essentialism caused a "2,000 year stasis in evolutionary thought" because it assumed that species had fixed essences and thereby could not explain how new species could evolve through natural selection. The notion of species as a probabilistic concept, a distribution of types, seemed to be foreclosed by the essentialist bias even though that notion of distribution is critical to understanding how evolution through natural selection could actually occur. But what notions of essence were involved in Aristotelian essences? Hull's reference to Aristotle suggests something like the sortal essence, yet for much of folk biology, the causal essence seems more appropriate. A causal essence, however, is not on its own incompatible with a probability notion. Indeed, Boyd's notion of causal homeostasis (see chapter 6 in this volume), a process wherein species properties are maintained in stable configurations, fully allows for species themselves to change over time as a consequence of natural selection, yet that kind of process might well be a form of causal essence. The fixedness of essence would seem to arise from a cognitive bias toward not appreciating a process such as causal homeostasis, either because processes in general are not preferred or because any probabilistic components to such processes are not allowed.

The problem of an essence is that it seems indirectly or probabilistically related to the features of an organism, but directly related to its categorization. That is to say, an organism with a leopard "essence" will resemble a leopard *ceteris paribus*, but not if some freak dietary anomaly prevents it from developing a certain pigmentation pattern: it will nevertheless be a leopard under an essentialist conception. The question, then, is what features license this ascription of an essence of a certain type, since an essence is also a basis for ignoring features?

The riddle about causal essence is that unlike both sortal and ideal essences, causal relations in the real world are rarely strictly necessary. Because the entities within a folk theory are usually loosely framed, there can be no necessary causal laws holding between the entities as there can be between entities in a more rigid scientific framework. Therefore, some degree of probability always seems to be associated with causal relations occurring in the real world, even if in some cases that probability is exceedingly high. Are lay notions of causality ignorant of such probabilities, or does probability somehow otherwise get ignored as notions of cause and essence become intertwined? People often seem to know that an event may cause a particular effect, but not always. Eating rotten meat usually causes one to feel sick, but not always. Sexual relations cause the emergence of babies sometime later, but certainly not always. These probabilistic relations are part of how all of us talk every day: why, then, cannot there be a kind of causal essence that is probabilistic? It would seem that the psychological constraints on causal essence strongly discourage certain forms. Perhaps the psychologically appealing sense of causal essence ultimately requires the notion of fixed stuff, rather than a process, as the initial cause, and even if that stuff has only probabilistic causal consequences, its very nature is not at all probabilistic.

Given these considerations, there is the additional problem of the scope of an essence: how it is seen to vary within living kind hierarchies. Is the "essence" a *type*—the DNA sequence varying by "product type and brand" like a supermarket bar code—or is it a *token*, an essence as individual as a fingerprint? Or is it the case that the hypothesized essentialist bias easily admits gradations of similarity between tokens that can be used to categorize them into types? Although it may be more correct to think of a continuum of DNA similarity that is, perhaps arbitrarily, divided up into species, cognitive bias may pull us away from this idea.

In their pioneering studies on prototype concepts, Rosch and Mervis (1975) asked subjects to grade the typicality of various living kinds. For example, a robin was graded as a more typical bird than a duck, and these ratings predicted speed on tasks such as lexical decision. Furthermore, Rips (1975) found that the typicality gave rise to asymmetric judgments: subjects thought it more likely that a duck could catch a disease from a robin than that a robin could catch a disease from a duck. This asymmetry could be interpreted as pointing toward an essentialist bias. Rather than it being the case that robins and ducks simply have a certain degree of similarity, it seems to be the case that we conceive of robins as having a "stronger bird essence" than ducks, so if something affects a robin as a typical bird, it seems more likely to affect a duck than vice versa.

An essentialist folk theory may suggest that biological factors such as susceptibly to a certain disease attach not only to robins, but to their essence as birds; these factors are hence, more likely to extend to a duck. Can the same sense of essence that gives us the concept of species also make one species more central than another in a higher-level category, such as *bird?*

These issues make obvious the need for an extensive set of psychological studies that ask what constraints there might be on different notions of essence. We can further ask how those notions might change over the course of cognitive development in the child—of moving from novice to expert knowledge—and how they vary across kinds. In addition, they may vary across kinds from the earliest points in conceptual development, or they may start as a more common vague notion that gradually differentiates with increasing knowledge in each domain. In short, people do seem to have an essentialist bias and perhaps epecially so for living kinds, a bias that powerfully influences their concepts of species, but we have only begun to understand the real psychological nature of this bias.

## THE VIVID ILLUSION OF SPECIES

So far we have put forward several claims about the nature of our concepts of living kinds, and it seems that there might be a common underlying cognitive explanation. Given that folk-biological thought seems both void of specific mechanisms and inclined toward certain types of explanation, and that we have a strong bias toward essences in living kinds, it now seems plausible to seek an account of the type of illusion discussed earlier, whereby people have a tendency to assume that they have a vivid, clockwork knowledge of certain mechanisms.

This illusion is similar to one that has been repeatedly demonstrated in recent years in studies of visual memory and indeed may arise from common mental sources. In those studies, people look at scenes, often quite simple ones, and assert that they have a clear memory of what it contained. Yet when tested, they can be strikingly ignorant of the details of the scene they just observed. They not only mistakenly remember different colors, textures, and surface patterns on objects, but also often fail to notice in a recognition task when completely different objects are in the scene (e.g., Simons 1996). They do, however, have quite good memories for the spatial layout of the objects in the scene—the general relational topology of objects—even when they forget the details. With dynamic objects and systems, people seem to retain good understandings of the functional "layout" while often losing all the details of particular discrete components in that layout.

The vividness illusion in the mechanisms of folk biology might be seen in terms of a misleading dispositional bias: a variant of the "fundamental attribution error" (Nisbett and Ross 1980) whereby folk tend to think dispositionally about other people, assuming that inner essences, more than situational factors, explain the behaviors of others (Miller 1996). This bias has been shown to extend beyond concepts of people to concepts of other entities in the world, such as a chip of wood in a turbulent stream (Peng and Nisbett forthcoming). It may be that living kinds are far more powerful triggers of the bias than are most other kinds.

A very general cognitive bias may be at work here as well: the tendency to focus on what are known in statistics as main effects and not on interactions. It may be simpler and more cognitively compelling to think of a kind being created by either intrinsic essential properties or environmental forces, rather than by an interaction between the two. Therefore, being aware of some salient endogenous factors may lead to an overzealous assumption of almost exclusively endogenous forces.

The reasons for these illusions or biases are not clear, but they may well have cognitive benefits at some level. For scenes, they perhaps help build the impression of a continuous flow of experience; for systems, they may help build an impression of a continuous chain of understanding without explanatory gaps.

The question with regard to species and essentialism is whether a kind of illusion is created wherein patterns of causal homeostasis result in the relative stability of property clusters, which are then mistakenly assumed to be stable not because of that homeostatic process, but rather because of a fixed physical causal source. That is, causal homeostasis causes stable property clusters, which in turn cause the impression of a fixed physical essence. We know that people succumb to a vividness illusion in several ways. There may be the corresponding assumption that stable property clusters must have stable physical sources. The relative stability of property clusters that emerge through causal homeostasis, as opposed to those clusters that do not, may be so great in relative terms that it leads to the erroneous assumption of absolute stability. Then, the cognitive bias toward essences might consist of positing a stable property for living kinds as a kind of simplifying heuristic.

Many argue that living kinds have a much richer causal structure—that they are causally more complex (Gelman et al. forthcoming). That causal complexity is then thought perhaps to trigger impressions of causal essence. But complexity is notoriously difficult to define and measure, and a little reflection makes one worry about any absolute differences between artifacts and living kinds in terms of causal complexity. To be sure, the vast majority of living kinds have more complex causal internal processes that give rise to surface processes and to activities. When one includes external social and cultural factors that help explain why artifacts are as they are, however, artifacts have vastly more causal complexity connected to such cultural and social properties than most living kinds have.

Again, more subtle and more interesting differences may be at work. The sociocultural causal factors for artifacts may not be nearly as bounded as are the other causal factors for most living kinds. That is, they do not neatly circumscribe the artifact. To know why chairs are the way they are, one has to look at economics, body shapes, and physiological needs in a vastly extended causal network that does not cluster tightly around chairs.

Whereas artifacts may have causal factors that are distributed across sociocultural factors, living kinds seem to have a more bounded, visceral pattern of causal homeostasis. Perhaps living kinds are different in the respect that the causal cluster for each one is more of a dense island in a sea of weaker and less causally complex interconnections. Again, there is a strong intuition here, but it needs to be examined in an experimental manner to see if people see living kinds as forming more dense and distinct clusters (See also Ahn, 1998 on why different features are central for artifacts and natural kinds.).

We all accept that the notion of essence is not unique to thought about living kinds, but its strength and power seem strongest for living kinds even when it may be least correct as a kind of fixed entity in such cases. Thus, at the cognitive level and perhaps also because of the special nature of living things, essences of living kinds seem to have a different character, one that may be more cognitively compelling.

Artifacts are not normally thought of as having essences, but again that notion depends on notions of essence that remain largely unanalyzed. Certain physical constraints make only some two-wheeled pedaled devices physically stable and thereby useful as bicycles (Olson and Kyle 1990). The angle of the front fork can be changed so that the "bicycles" are completely unridable as they start to oscillate in an unstable manner. Thus, a pattern of causal homeostasis makes a stable functional unit, and only certain properties qualify in the pattern for that unit. Are such causal patterns part of the essence of *bicycle*? If not, what makes their case so different from the patterns for some animals? At a psychological level, the difference may lie in the belief that for living kinds, there is some sort of fixed stuff that gives rise to the patterns of homeostasis, a belief that seems counterintuitive for most artifacts.

## SPECIES CONCEPTS AS A DISTINCT KIND OF CATEGORIZATION

A species concept is a kind of categorization. It treats a class of living things as equivalent in important respects, and that equivalence then licenses powerful inductions, which is presumably why species concepts are so useful. But inductive power is said to be a key motivation for almost all cognitively natural categories; what else, beyond essence, is distinctive about species concepts as opposed to other concepts? There appear to be many qualitative distinctions between living kinds and artifacts, including how taxonomies, teleology, and exemplars are construed, yet the question remains whether or not these differences are due to quantitative differences in the spread of causal homeostatic patterns.

Living kinds are said to be much more deeply embedded in taxonomies than are other sorts of kinds, and throughout the world, all peoples seem to realize this taxonomic character very early on in development (Atran, chapter 9 in this volume and 1998). Abundant evidence now shows that this taxonomic assumption is very powerful for living kinds and probably not nearly as powerful for other sorts of kinds (although the latter have not been nearly as carefully or systematically studied). However, what is it about living kinds that leads to this taxonomic assumption? As with essences, it is not so clear what "triggers" the assumption. Thus, it is not so objectively obvious that living kinds occur in more deeply embedded hierarchies. There are, after all, quite deep hierarchies for many naturally occurring compounds (kinds of rocks, soils, gems, etc.). For example, the United States Department of Agriculture gives a four-layered basic taxonomy of soil types—categorizing, to take one instance, from histolsols to fibrists to sphagnofibrists that are pergelic. Therefore, depth of hierarchy is not evidently a cue. Similarly, there are very rich hierarchies for many classes of artifacts. The U.S. Patent Office has more than five hundred classes of artifacts: meandering down the scheme, of these five hundred, the class "Surgery" has forty divisions, within which "Prosthetics" has twenty, within which "Leg" has nine, and of these "Socket" has six categories. Even if the hierarchies of living kinds are generally deeper, it is not at all clear that their depth in itself is responsible for their being seen taxonomically. Thus, there is no evidence to support the intuition that deeper hierarchies are more taxonomically compelling. If depth doesn't predict a sense of taxonomy for the nonliving world, why should it be a factor in the living world?

Do these species-related taxonomies provide the only apparent ways to classify living kinds even though there are multiple ways for other sorts of kinds? No. Animals can be classified as predators or prey, as domesticated or wild, or as edible or not, or they can be classified within theories that focus on phylogenetics, ecology, and so on. It is difficult to know how to count the ways, but it is not obvious that living kinds are different in this respect. Even in folk biology, we see *trees* and *nontrees* as one way of sorting plants that is at intrinsic odds with sorting by more common species names (Dupré 1981).

Do taxonomies for living kinds have more salient properties at each level of the hierarchy? Perhaps that is the case in comparison to nonliving natural kinds, but many artifacts have powerful sets of distinctive features at many levels, and those features license powerful inductions. Perhaps there is a more subtle difference here. Basic-level concepts (Rosch et al. 1976) turn out to be much more robust for some artifacts—such as furniture, clothing, and vehicles—than they do for most living kinds. That is, the basic level for artifacts (e.g., *chair*) represents a vastly richer feature cluster than is present at the superordinate level (e.g., *furniture*); the basic level also offers correspondingly many more inductions. With living kinds, however, the superordinate level, such as *mammal*, is much richer in nature and more full of distinctive properties than it is for most artifact kinds. This difference may suggest more taxonomic continuity between levels of living things because there is less of a jump in feature density as one moves up or down a level. For that reason, the basic level tends to stand out in much sharper relief for artifact categories, with most experimental data demarcating that level more strongly and unambiguously when artifacts are used as stimuli (Rosch et al. 1976).

Another contrast between species categories and categories for other kinds is said to revolve around teleology, the argument being that teleological interpretations work only with living kinds. In a simple form, however, this argument is clearly wrong. Artifacts have purposes, as do their parts. In a more subtle form, however, this view may offer an important contrast. One does not normally ask about the purposes of animals as whole entities, but only about the purposes of aspects of them (the exception is highly domesticated functional species, such as hunting dogs). Such questions are completely normal about artifacts, however. Thus, species categories are different in terms of where teleological questions are directed and when they are legitimate. Why this difference exists and how universal it is remain unclear, however.

Detailed analyses of teleological understandings and some characterization of what it is about living kinds that blocks such holistic teleological interpretations are needed for the contrast between living and nonliving kinds to work (cf. Bloom 1996). With artifacts, human intention guides the holistic teleological account, but why should intention be needed for the holistic account when it is not needed for more local explanations for living kinds (for example, that webbed feet help an animal swim better)?

A final contrast may circle around the nature of the category structure in terms of best exemplars and well definedness. With some natural kinds, the best exemplar would be something like its sortal essence, to use Gelman's terminology: "pure" water is nothing but $H_20$ molecules, for example. The best exemplar of an artifact would be closer to some ideal essence framed in terms of its function, so the exemplar of *computer* would be cheap, fast, easy to use, compatible with all platforms, and so on. However, determining the best exemplar becomes more difficult when the function is not easily framed: What is the one-line functional definition for *church* or *the internet?*

How do we think of the exemplars of living kinds? In some cases, living kinds have functional value for human beings and can be evaluated on that dimension: the best trees are those with the highest fruit yield, for example. Or we might frame the best exemplar as that which fulfills its "purpose": the best butterfly is camouflaged from predators, can extract nectar efficiently, lay a large number of eggs, and so on. But this framework brings in all the problems of teleology discussed above. In addition, how do we evaluate how "doglike" a dog is? It could be the case that we have some prototype or idealized average of all dogs. Does this average equate with a dog essence? Alternatively, we may have some notion that goes beyond the typicality of perceptual features to a representation of a homeostatic cluster of causal properties that are seen to be significant. It becomes difficult, however, to draw a firm distinction between living and nonliving kinds on this criteria of well definedness because there is considerable variation within both classes.

Well definedness in itself may therefore not be the cut between artifacts and living kinds because they both have fuzzy aspects, but the fuzziness may occur in different ways for the two kinds. Furthermore, it seems that when more rigid criteria are applied in the evaluation of living kinds, those kinds are being measured in relation to some functional goal, even the seemingly hollow purpose of winning a dog show. In short, species concepts may be importantly different from other sorts of concepts that categorize kinds, but the dimensions and degrees of difference are still unclear.

It seems, therefore, that species concepts may be distinct in several ways that go beyond notions of essence and that are not closely related to patterns of causal homeostasis. The details of the psychological differences between these ways are just beginning to be uncovered, however.

## CONCLUSIONS

Species concepts seem indeed to reflect a special kind of categorization. Several cognitive biases may interact with some very distinctive informational patterns of living kinds, such as a disposition toward certain types of properties and relations, as well as a tendency to discretize these homeostatic patterns into an essence of some sort. But it is clear that all of these psychological contrasts are just beginning to become apparent. The nature of a species concept is mostly a place holder at present and is framed by only the softest and vaguest of constraints. We therefore do not really know much about how it is that species concepts arise, but this lack of knowledge should not be discouraging because for the first time we are now in a position to learn a great deal more about how those concepts emerge in development and become used by adults. The rapid growth of work on biological thought, especially in cross-cultural and developmental perspectives, has helped set up a framework in which it is now possible to pose highly detailed questions about species concepts—such as how children have them early on, even though they have few concrete details of what an essence might be. We now see more clearly the different ways in which folk biology is likely to be mentally represented and especially how it is not. Even more important, we now see how these issues can be better understood through empirical study of both children and adults. Very different patterns of results are possible in such studies, and those patterns will make a profound difference in our understanding of what it means psychologically to have the concept of a species.

## ACKNOWLEDGMENTS

Preparation of parts of this paper and some of the studies described therein were supported by an National Institutes of Health grant R01-HD23922 to Frank Keil.

## NOTES

1. As an aside, it is interesting to note the similarities between McRae and colleagues' (1997) discussion of an attractor space as a way of describing a cognitive mechanism representing concepts and Goodwin and Webster's (1996) claim that the concept of species should be understood in biology theory in terms of an attractor space or a "morphonogenic field" (see Griffiths, chapter 8 in this volume).

2. Some sort of cognitive bias may be favoring the notion of DNA in lay thinking about living kinds and appropriating it as the essence of living kinds. As argued elsewhere in this volume, DNA is not a sufficient tool with which to divide species because there can be more variability within species than between them. However, it appears as if the scientific term *DNA* has been equated with some intuitive, folk notion of essence, even though essence and DNA divide up the world in very different ways. We discuss the possible psychological grounds for this bias, but for now we might make the observation that when people misunderstand the notion of DNA—as probably all but a few do—the mistakes they make commonly belie an underlying essentialist belief.

## REFERENCES

Ahn, W. (1998). Why are different features central for natural kinds and artifacts? *Cognition*, 69, 135–178.

Atran, S. (1998). Folk biology and the anthropology of science: Cognitive universals and cultural particulars. *Behavioral and Brain Sciences* 21, 547–611.

Bloom, P. (1996). Intention, history, and artifact concepts. *Cognition* 60, 1–29.

Carey, S. (1985). *Conceptual change in childhood.* Cambridge, Mass.: MIT Press.

Dupré, J. (1981). Biological taxa as natural kinds. *Philosophical Review* 90, 66–90.

Gelman, R., and R. Baillargeon (1983). A review of some Piagetian concepts. In J. H. Flavell and E. M. Markman, eds., *Handbook of child psychology*, vol. 3. New York: Wiley.

Gelman, S. A., and L. Hirschfeld (1999). Essentialism and folk biology. In D. Medin and S. Atran, eds., *Folkbiology*. Cambridge, Mass.: MIT Press.

Gelman, S. A., J. D. Coley, K. S. Rosengren, E. Hartman, and A. Pappas (1998). Beyond labeling: The role of maternal input in the acquisition of richly structured categories. *SRCD Monograph Series* 253, 1–114.

Goodwin, B. C., and G. Webster (1996). *Form and Transformation: Generative and relational principles in biology*. Cambridge: Cambridge University Press.

Gutheil, G., A. Vera, and F. C. Keil (1998). Houseflies don't "think": Patterns of induction and biological beliefs in development. *Cognition* 66, 33–49.

Hatano, G., and K. Inagaki (1996). Cognitive and cultural factors in the acquisition of intuitive biology. In D. R. Olson and N. Torrance, eds., *Handbook of education and human development: New models of learning, teaching, and schooling*. Cambridge: Blackwell.

Hull, D. (1965). The effect of essentialism on taxonomy: 2,000 years of stasis. *British Journal for the Philosophy of Science* 15, 314–326, and 16, 1–18.

Keil, F. C. (1992). The origins of an autonomous biology. In M. R. Gunnar and M. Maratsos, eds., *Modularity and constraints in language and cognition: Minnesota symposium on child psychology*, vol. 25. Hillsdale, N.J.: Erlbaum.

Keil, F. C. (1998). Cognitive Science and the origins of thought and knowledge. In R. M. Lerner, ed., *Theoretical models of human development*, vol. 1, *Handbook of Child Psychology*, 5th ed. New York: Wiley.

Keil, F. C., D. T. Levin, B. A. Richman, and G. Gutheil (1999). Mechanism and explanation in the development of biological thought: The case of disease. In D. Medin and S. Atran, eds., *Folkbiology*. Cambridge, Mass.: MIT Press.

McRae, K., de Sa V. R., and M. S. Seidenberg (1997). On the nature and scope of featural representations of word meaning. *Journal of Experimental Psychology General*, 126, 99–130.

Medin, D. L., and A. Ortony (1988). Psychological essentialism. In S. Vosniadou and A. Ortony, eds., *Similarity and analogical reasoning*. New York: Cambridge University Press.

Medin, D., and S. Atran, eds. (1999). *Folkbiology*. Cambridge, Mass.: MIT Press.

Miller, J. G. (1996). Culture as a source of order in social motivation: Comment. *Psychological Inquiry* 7, 240–243.

Nisbett, R., and L. Ross (1980). *Human inference: Strategies and shortcomings of social judgment*. Englewood Cliffs, N.J.: Prentice-Hall.

Olson, J. N., and C. R. Kyle (1990). Bicycle stability. *Bicycling* 31, 134.

Peng, K., and R. E. Nisbett (forthcoming). Cross-cultural similarity and difference in understanding physical causality. In M. Shale, ed., *Culture and Science*. Frankfort Ky.: Kentucky State University Press.

Piaget, J. (1954). *The construction of reality in the child*. New York: Basic Books.

Rips, L. (1975). Inductive judgments about natural categories. *Journal of Verbal Learning and Verbal Behavior* 14, 665–681.

Rosch, E., and C. B. Mervis (1975). Family resemblances: Studies in the internal structure of categories. *Cognitive Psychology* 7, 573–605.

Rosch, E., C. B. Mervis, W. D. Gray, D. Johnson, and P. Boyes-Braem (1976). Basic objects in natural categories. *Cognitive Psychology* 8, 382–439.

Simons, D. J. (1996). In sight, out of mind: When object representations fail. *Psychological Science* 7, 301–305.

Simons, D. J., and F. C. Keil (1995). An abstract to concrete shift in the development of biological thought: The "insides" story. *Cognition* 56, 129–163.

Wellman, H. M., and S. A. Gelman (1998). Knowledge acquisition in foundational domains. In D. Kuhn and R. Siegler, eds., *Cognition, perception and language,* vol 2, *Handbook of Child Psychology,* 5th ed. New York: Wiley.

Wilson, R. A., and F. C. Keil (1998). The shadows and shallows of explanation. *Minds and Machines* 8, 137–159.

# V Species Begone!

# 11 Species and the Linnaean Hierarchy

Marc Ereshefsky

Prior to the eighteenth century, biological taxonomy was a chaotic discipline marked by miscommunication and misunderstanding. Biologists disagreed on the categories of classification, how to assign taxa to those categories, and even how to name taxa (Heywood 1985). Fortunately for biology, Linnaeus saw it as his divinely inspired mission to bring order to taxonomy. The system he introduced offered clear and simple rules for constructing classifications. It also contained rules of nomenclature that greatly enhanced the ability of taxonomists to communicate. Linnaeus's system of classification was widely accepted by the end of the eighteenth century. That acceptance brought order to a previously disorganized discipline, and it laid the foundation for "the unprecedented flowering of taxonomic research" of the late eighteenth and early nineteenth centuries (Mayr 1982, 173).

In the last two hundred years, the theoretical landscape of biology has changed drastically, however. Species and other taxa are not the result of divine creation, but the products of evolution. Taxa are not static, timeless classes of organisms, but evolving and temporary entities. The theoretical assumptions of the Linnaean system have been replaced by those of Darwinism and the Modern Synthesis. In light of these changes, one might wonder if the Linnaean system remains a practical system for constructing classifications. Some authors maintain that it is not (Griffiths 1974, 1976; de Queiroz and Gauthier 1992, 1994; Ereshefsky 1994). The process of evolution and the resultant diversity of biota, they argue, render the Linnaean rules of classification and nomenclature seriously flawed. Some detractors of the Linnaean system even suggest that it should be replaced (Hull 1966, Hennig 1969, Griffiths 1976, de Queiroz and Gauthier 1992, 1994). Defenders of the Linnaean system respond that the Linnaean system is worth saving, though they agree that some revision is necessary (Wiley 1979, Eldredge and Cracraft 1980, chapter 5).

The focus of this chapter is not the entire Linnaean hierarchy, but a particularly troublesome portion of it: the species category. Two types of problems face the species category: ontological ones and pragmatic ones. For Linnaeus, as well as for the architects of the Modern Synthesis, the divide between the lower and higher ranks of the Linnaean hierarchy was supposed to reflect an important ontological divide in nature. As we have discovered, the existence of that divide is suspect. In addition, Linnaeus and the architects of the Modern Synthesis thought that all species taxa are comparable in some important respect. That assumption is problematic as well. If there is no clear distinction between species and higher taxa, and species taxa are noncomparable entities, then we have grounds for doubting the existence of the species category.

Theoretical problems grade over to pragmatic ones. The Linnaean system requires that the names of some taxa indicate their rank. Species are given binomials, higher taxa have uninomials, and subspecies are assigned trinomials. So a species' binomial designation indicates its rank. But if the existence of the species category is suspect, then so too is the practice of using a taxon's name to indicate that it is a species, for if there is no species category, then no taxa should be designated as "species." This is one potential problem with the Linnaean rules of nomenclature. There are other problems with those rules, and they arise regardless of whether one is a skeptic of the species category.

The bulk of this chapter outlines the problems facing the species category and the rules of nomenclature governing the naming of species. Those problems are serious enough to consider abandoning the entire Linnaean system. Replacing the Linnaean system with an alternative system is no small task, however, nor should it be done lightly. The Linnaean system is firmly entrenched in biology, not to mention in popular culture. Biologists and philosophers must build a persuasive case for rejecting the Linnaean system, and they must develop viable alternatives. The last section of this chapter introduces some non-Linnaean systems worthy of consideration. In doing so, it offers a glimpse of what might be the future of biological nomenclature.

## THE DISTINCTION BETWEEN SPECIES AND HIGHER TAXA

The assumption that there is a fundamental ontological divide between lower taxa and higher taxa is an old one, dating back to Aristotle. This section examines several prominent criteria for distinguishing lower from higher taxa, starting with Linnaeus's and ending with two contemporary suggestions. The criteria for distinguishing lower from higher taxa encounter a number of difficulties. This negative result casts a shadow on the distinction between species and higher taxa as well as on the existence of the species category (cf. Boyd, chapter 6 in this volume).

### Linnaean Species

For Linnaeus, the fundamental divide in his hierarchy lies between genera and all other higher taxa. Classifications of species and genera reflect real groups in nature, whereas classifications of classes and orders are artificial (Cain 1958, 148, 152–153). This distinction stems from a central tenet of

Linnaeus's biological theory: Aristotelian essentialism. According to Linnaeus, the members of species and genera are endowed with essential natures created by God. The job of the biologist is to discover those essences and their associated species and genera (Ereshefsky 1997). Orders and classes, on the other hand, are constructed on pragmatic grounds and are artificial constructs. "An order is a subdivision of classes needed to avoid placing together more genera than the mind can follow" (Linnaeus, quoted in Mayr 1982, 175). According to Linnaeus, species and genera have mind-independent essences, whereas orders and classes owe their existence to us.

When we dive a bit deeper into Linnaeus's biological theory, in particular his sexual system, we see more clearly why he thought that species and genera, but not orders and classes, have essences. Following Cesalpino, Linnaeus believed that plants have two vital functions: nutrition, which preserves the individual, and reproduction, which preserves the kind (Larson 1971, 146). The function of reproduction in plants is found in their "fructification structures"—namely, their flowers and fruits. Fructification structures are found at the level of genera, and are the essences of genera, according to Linnaeus (Larson 1971, 74 ff.). Moving down a level, a species' essence is a combination of its genus's fructification structure and those parts involved in the function of nutrition (Larson 1971, 115 ff.). The essences of species and genera are particularly important for Linnaeus because they are responsible for the continued existence of taxa beyond God's original creation. Fructification structures allow the members of a species or a genus to reproduce and thus allow taxa to continue. Classes and orders, on the other hand, are not defined by fructification structures. They are merely aggregates of organisms containing different fructification structures.

In the last hundred years, Linnaeus's divide between genera and higher taxa has fallen on hard times. One reason is the rejection of essentialism in Darwinian biology. Linnaeus thought that species and genera, but not orders and classes, had essences; thus, the former are real, but not the latter. In Darwinian biology, no taxon of any rank has a taxon-specific essence. (See Hull 1965 and Sober 1980 for arguments against taxa, especially species, having essences; cf. Wilson, chapter 7 in this volume.) The relevance here is that species and genera no more have essences than do orders and classes, so Linnaeus's essentialist ground for holding the distinction between lower and higher taxa no longer applies.

Linnaeus also based his distinction on the assumption that only species and genera have generic-specific fructification structures. Orders and classes are artificial groups of genera containing different fructification structures. Biologists no longer believe that fructification structures are responsible for the existence of taxa. Instead, many hold that taxa are the result of interbreeding among conspecific organisms (see, for example, Mayr 1970, 373–374, Eldredge and Cracraft 1980, 89–90). Species are populations of organisms that exchange genetic material through interbreeding. That process causes the local populations of a species to evolve as a unit. In contrast, the species of a higher taxon do not exchange genetic material, and the evolution of a higher taxon is merely the by-product of evolution occurring within its species. Thus, instead of Linnaeus's ontological divide between genera and other higher taxa, the authors of the Modern Synthesis, as well as many contemporary biologists, hold that the ontological divide lies between species and all higher taxa.[1]

## The Units of Evolution

In place of Linnaeus's distinction we have the species/higher taxa distinction of the Modern Synthesis. Species are actively evolving entities, often referred to as *the* units of evolution. Higher taxa, including genera, are merely artifacts of processes occurring at lower levels of organization. This distinction is widely held by contemporary biologists (for example, Mayr 1970, 373–374; Eldredge and Cracraft 1980, 89–90, 249; Ghiselin 1987, 141). Typically, two sorts of arguments are given to support this distinction, each highlighting a process that is supposed to be common and unique to species taxa. According to the first argument, gene flow is an essential process for the continued existence of species. New traits that arise in one local population are spread by gene flow to the other local populations of a species. As a result, a species evolves as an evolutionary unit rather than an aggregate of local populations. No comparable mechanism to gene flow exists among the species of a higher taxon. The evolution of a higher taxon is merely a historical by-product of evolution occurring in its species. Species, therefore, are active agents of evolution, while higher taxa are merely passive results of evolution.

Though this basis for the species/higher taxa distinction is quite popular, it has been called into question. A number of biologists argue that many species lack the integrating force of gene flow. If they are correct, then many species are akin to higher taxa: they are merely aggregates of processes working at lower levels of biological organization (Ehrlich and Raven 1969, Grant 1980, Mishler and Donoghue 1982). Suppose, as many biologists do, that asexual organisms form species taxa. The members of such species are not bound by interbreeding, but by such processes as selection, genetic homeostasis, or developmental canalization (see Templeton 1989 for a discussion of these processes). Such processes cause a group of organisms to belong to a single species without requiring any causal interaction among those organisms.[2] Selection, for instance, can maintain the unity of a species by affecting individual members in a similar fashion. The same goes for the actions of genetic homeostasis and developmental canalization. If asexual organisms form species, then such species are both structurally and causally akin to higher taxa.

One need not posit the existence of asexual species to find problems with the species/higher taxa distinction. Ehrlich and Raven (1969) and Templeton (1989) suggest that many species of sexual organisms contain local populations that exchange little or no genetic material. Here, the claim is not that the local populations momentarily fail to exchange genetic material, but that they fail do so for a significant amount of time. If these biologists are right, then many species of sexual organisms are bound by processes working at lower levels of biological organization than at the level of the entire species (cf. Sterelny, chapter 5 in this volume). The unity of such species may be the result of interbreeding within local populations, or their unity may be due to processes that independently affect organisms, such as selection or genetic homeostasis. Either way, some species of sexual organisms are akin to higher taxa: they are bound by causal mechanisms acting at lower levels of biological organization. If such sexual species exist, or if there are asexual species, then the process of gene flow does not universally distinguish species from higher taxa.

Another argument for the species/higher taxa distinction highlights the process of speciation. This argument can be found in Eldredge and Cracraft (1980, 327) and Mayr (1982, 296), and it runs as follows. Speciation is the primary cause of change in evolution. It occurs in species but not higher taxa. Therefore, species, but not higher taxa, are the units of evolution. Once again, species are the active agents of evolution, whereas higher taxa are merely passive results. The trouble with this line of reasoning is that the locus of speciation is neither the species nor the higher taxon, but the founder population. According to Mayr's (1970) allopatric model, speciation begins when a small population of organisms becomes isolated and is exposed to new selection pressures. The population undergoes a "genetic revolution" and becomes the founding population of a new species.

The point here is that the process of speciation occurs in founding populations, not in entire species. So if higher taxa are not evolutionary units because speciation occurs in only a portion of them (founder populations), then by parallel reasoning, species are not evolutionary units either.

Now one might counter that it is inaccurate to assert that speciation occurs in founder populations rather than entire species because there is nothing more to an incipient species than its founder population. Suppose we grant that point. Then it applies to incipient higher taxa as well: at some stage in their development they are nothing more than founder populations. Thus, if species are evolutionary units because speciation occurs in their incipient forms, then higher taxa are evolutionary units as well. In brief, the process of speciation does not separate species from higher taxa.

Stepping back, we see that the distinction between species and higher taxa is problematic. Linnaeus drew the distinction along the lines of essential natures and fructification systems. Modern biology has rejected essentialism and Linnaeus's sexual system. Alternatively, the authors of the Modern Synthesis and many contemporary authors base their distinction between species and higher taxa on the processes of speciation and interbreeding, but those processes do not adequately distinguish species from higher taxa, either.

Perhaps some yet to be discovered process or feature will clearly distinguish species from higher taxa. That certainly is a possibility. But until we know of such a process, we should maintain a healthy skepticism toward the species/higher taxa distinction.

Doubt concerning the species/higher taxa distinction affects our confidence in the reality of the species category. A minimal requirement for believing in the existence of a category is having sufficient grounds for distinguishing it from its neighboring categories. If no biological criteria adequately distinguish species taxa from higher taxa, then we lack grounds for believing that species, genera, and orders are ontologically distinct categories. It is worth emphasizing that the argument here is merely against the existence of the Linnaean categories, especially the species category. Nothing I have said casts doubt on the reality of particular taxa. We can remain confident in the existence of such taxa as *Homo* and *Homo sapiens* even if the Linnaean categories go by the wayside (cf. Mishler, chapter 12 in this volume).

## SPECIES PLURALISM

Reasons for doubting the existence of the species category come from other quarters as well. The current taxonomic literature contains no less than a dozen species concepts (see Ereshefsky 1992b and recent issues of *Systematic Biology*). Biologists and philosophers have responded to this wealth of concepts in two ways. Monists believe that biologists should settle on a single correct concept. Pluralists suggest that a number of species concepts should be accepted as legitimate (cf. Dupré and Hull, chapters 1 and 2 in this volume).[3]

Undoubtedly, a number of currently proposed species concepts will be found wanting and relegated to the history of science. However, two major approaches to species—the interbreeding and phylogenetic—are currently well entrenched in biology for good theoretical and empirical reasons (de Queiroz and Donoghue 1988, Ereshefsky 1992a). The interbreeding and phylogenetic approaches highlight noncomparable types of species taxa, so if one accepts these two approaches, then there is no single unitary species category, but a heterogeneous collection of base taxa referred to by the term *species*. Species pluralism, in other words, poses a threat to the existence of the species category. The aim of this section is to display that threat and to highlight the disunity of the species category. But before getting to that, we need a quick introduction to the interbreeding and phylogenetic approaches.

The *interbreeding approach* is typified by Mayr's (1970) biological species concept. Species are gene pools held together by interbreeding and protected by various reproductive isolating mechanisms. Examples of reproductive isolating mechanisms include courtship behavior that prevents the mating of two interspecific organisms and hybrid inviability if such mating does occur. Other species concepts that fall under the general interbreeding approach include Ghiselin's (1974) reproductive competition concept and Paterson's (1985) mate recognition concept. Though these concepts differ in important respects, they agree that a species is "a field for genetic recombination" (Carson 1957).

The *phylogenetic approach* is found in the work of cladists (for example, Cracraft 1983, Mishler and Brandon 1987, Ridley 1989). Propinquity of descent is the operative notion here. A taxon must contain a single ancestral species as well as all and only its descendant species. Any taxon meeting that requirement is *monophyletic*. The founder of cladism, Willi Hennig, did not intend the notion of monophyly to apply to species, but reserved it for higher taxa. Recent cladists have extended its use, however. Mishler and Brandon (1987, 46), for example, define a species as "the least inclusive taxon recognized in a classification into which organisms are grouped because of degree of monophyly" (see also Mishler, chapter 12 in this volume). Phylogenetic species are base monophyletic taxa maintained by a number of forces—including selection, interbreeding, genetic homeostasis, and developmental canalization.

Given these two approaches to species—the interbreeding and the phylogenetic—one might wonder what feature unifies phylogenetic and interbreeding species into a single category (cf. de Queiroz, chapter 3 in this volume). A simple requirement for the existence of a category is that its entities share a feature that distinguishes them from entities in other categories. If phylogenetic and interbreeding species are both species, then they should share some common and distinctive feature. If they lack such a feature, then the species category consists of noncomparable entities.[4] The remainder of this section examines those features that might render interbreeding and phylogenetic species comparable.

A good place to start is with the processes that maintain the existence of interbreeding and phylogenetic species. In interbreeding species, gene flow is the primary unifying force. How much and how often genetic material must be exchanged varies from interbreeding species to interbreeding species (Templeton 1989, 165). For many phylogenetic species, the situation is quite different. Some phylogenetic species contain sexual organisms living in isolated populations. In such species, selection, genetic homeostasis, or developmental canalization are the primary unifying forces, not gene flow. Similarly, phylogenetic species consisting of asexual organisms are bound by forces other than interbreeding, so different types of species are unified by different types of processes. Consequently, no single unifying process (or set of processes) serves as the common feature of all species taxa.[5]

Perhaps we would be better off looking at the *structures* of interbreeding and phylogenetic species. Perhaps a significant similarity can be found there. Some interbreeding species consist of a single local population. Other interbreeding species consist of a number of local populations connected by interbreeding. Either way, the process of interbreeding causes such species to have the structure of causally integrated entities. Of course, interbreeding among the members of a species is not a continuous affair, but contains temporal gaps. Moreover, many members of a species may not breed at all (though they are the result of interbreeding). Nonetheless, the members of such species are bound into single species by the interactive force of interbreeding.

The structure of many phylogenetic species is different. Some are base monophyletic taxa of asexual organisms. The forces that maintain the existence of such species are selection, genetic homeostasis, and developmental canalization. These forces can maintain a group of organisms as a species without requiring causal interaction among those organisms. Selection, for example, can maintain the unity of a species by individually affecting its organisms. Similarly, homeostatic genotypes separately cause organisms to have certain traits, and developmental canalization independently constrains the ontogeny of each organism. Such phylogenetic species are not causally integrated entities. They are the result of independent forces acting at the level of the organism. The same observation can be made of phylogenetic species consisting of sexual organisms living in isolated populations. In such species, interbreeding may preserve unity at the level of local populations, but the preservation of entire species is due to selection, genetic homeostasis, or developmental canalization (Ehrlich and Raven 1969, Eldredge and Gould 1972, Mishler and Donoghue 1982).

Stepping back, we see that interbreeding species are causally integrated entities, whereas many phylogenetic species are not. Interbreeding species are bound by interactive processes that occur at the level of entire species; many phylogenetic species are the result of noninteractive processes working at lower levels of biological organization. These considerations lead to the conclusion that interbreeding species and many phylogenetic species have different ontological structures. In other words, the species category is an ontologically mixed bag of entities.

So far we have seen that species taxa are bound by different types of processes and that species taxa have very different ontological structures. The heterogeneous nature of the species category is brought into sharper focus when we see that lineages that are considered species in one approach often fail to be species in the other. For example, some interbreeding species fail to form phylogenetic species because they do not contain all the descendants of a common ancestor. The freshwater fish group *Xiphophorus* contains a series of populations (figure 11.1). The members of C and F successfully interbreed and are reproductively isolated from the members of the other populations. Thus $C + F$ forms a single interbreeding species (Rosen 1979, 275–279). Yet on the phylogenetic approach, $C + F$ cannot be a single species because it does not contain all the descendants of the common ancestor X. In the phylogenetic approach, C and F are two distinct species.

This type of discrepancy is far from unusual. Consider ancestral species. Suppose an interbreeding species A spawns a new species C; yet A continues to exist as B (figure 11.2). In the interbreeding approach, $A + B$ forms one species, but C forms another. In the phylogenetic approach, $A + B$ cannot be

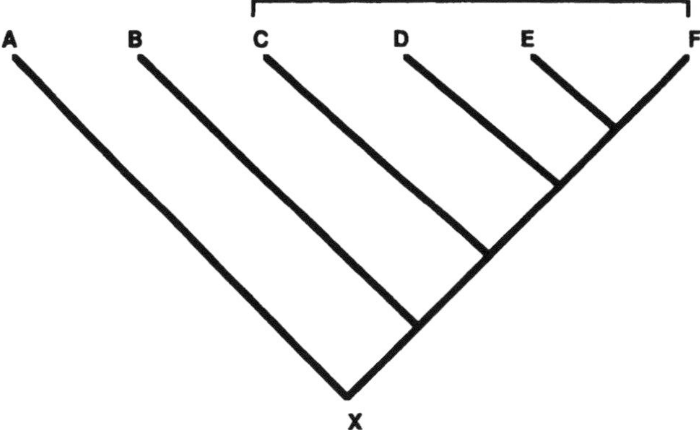


**Figure 11.1** A simplified cladogram of the swordfish group *Xiphophorus* (after Rosen 1979, 276). $C+F$ forms a species on the interbreeding approach but not on the phylogenetic approach.

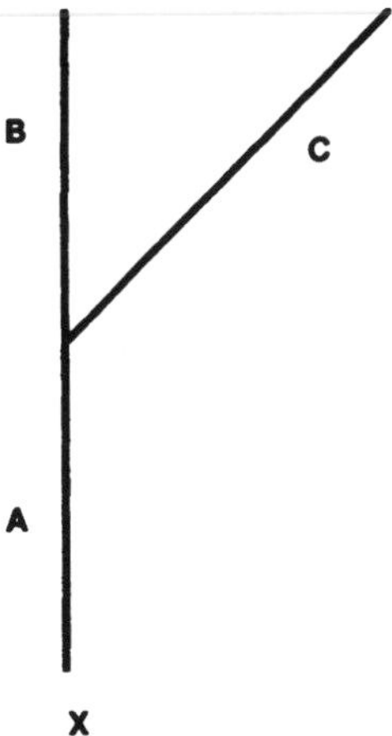


**Figure 11.2** On the interbreeding approach, $A+B$ forms one species, whereas $C$ forms another species. On the phylogenetic approach, $A$, $B$, and $C$ are three different species.

a species because it does not contain all the descendants of the common ancestor X : $A+B$ is missing the organisms of C. So, for many cladists, A goes extinct at the time of speciation and two new species, B and C, take its place.

Phylogenetic species often fail to be interbreeding species as well. Some proponents of the phylogenetic approach believe that asexual organisms form species (Mishler and Brandon 1987). Supporters of the interbreeding approach disagree and argue that asexual organisms do not belong to any species (Ghiselin 1987, Hull 1987). For my part, I see no reason why the term *species* should be reserved for only sexual organisms (see Ereshefsky 1992b and 1998 for arguments). If asexual organisms do form phylogenetic species, then many phylogenetic species fail to be interbreeding species. This sort of problem does not depend on the assumption that asexual organisms form species. It also occurs for some groups of sexual organisms. As we have seen, some phylogenetic species consist of sexual organisms living in isolated populations. Given the lack of sufficient internal gene flow, such species also fail to be interbreeding species.

The above examples illustrate that the phylogenetic and the interbreeding approaches carve the tree of life in different ways. Often, lineages considered species in one approach fail to be species in the other. Some phylogenetic species are higher taxa in the interbreeding approach, as in the case of *Xiphophorus*. Those phylogenetic species consisting of asexual organisms do not form any species in the interbreeding approach. The discrepancy between interbreeding and phylogenetic species is not the result of one or two exceptions, but of widespread cleavages in the organic world. For example, much of life reproduces asexually (Templeton 1989, 164), so it forms phylogenetic but not interbreeding species. Conversely, numerous ancestral species form interbreeding but not phylogenetic species. The heterogeneous nature of the species category is extensive and therefore does not stem from a few isolated cases.

In summary, the forces of evolution segment the tree of life into noncomparable base lineages. Species are maintained by different types of processes, and species have varying ontological structures. Furthermore, interbreeding and phylogenetic species cross-classify the world's organisms. Of course, this picture of the organic world could be wrong. Perhaps some of the forces discussed earlier lack the ability to produce stable taxonomic entities, so the species category is not as multifarious as suggested. Perhaps we will end up with a monistic species category. Yet, as things now stand in biology, we have no reason to believe that a monistic definition is on the horizon. In fact, we have every reason to believe that the species category will remain heterogeneous.

## IMPLICATIONS FOR THE SPECIES CATEGORY

The results of the last two sections cast a shadow on the reality of the species category. Minimal constraints on the existence of the species category are twofold: species taxa should differ significantly from other taxa, and species taxa should be comparable in a theoretically important manner. The fulfillment of either constraint is doubtful. As illustrated in the first section of this chapter, the species/higher taxa distinction has been drawn along several lines. Unfortunately, all are problematic. Linnaeus's essentialist approach for distinguishing lower taxa from higher taxa has been undermined by Darwinism, and its evolutionary replacement—that species, but not higher taxa, are *the* units of evolution—has also fallen on hard times. We should put stock in the existence of a category only if we have reason to believe that it contains entities that are distinctive from entities in other categories. "No entity without identity" is the Quinean refrain with which philosophers are familiar. At this stage in biology, it is not clear what divides species taxa from other taxa. As a result, it is not clear whether a distinctive species category exists.

Problems for the species category come from another angle. The entities of an existing category should have a theoretically important commonality. The species category seems to fail that requirement. The problem is not simply that species taxa vary, but that they lack a common and distinctive biological feature. Species taxa are maintained by different processes, they have different ontological structures, and they carve the tree of life in different ways. This disunity is not caused by a few species taxa lacking a biological feature found in most species taxa, but by vast numbers of species taxa forming one *type* of species but not another.

We now have two separate lines of reasoning against the existence of the species category. Together, they provide a strong case for doubting the reality of the category.[6] Once again, the case here is merely against the existence of the species category, not against the reality of those taxa we call "species." None of the arguments given thus far should cause us to doubt the existence of such taxa as *Homo sapiens* or *Drosophila melanogaster*.

The idea that the species category does not exist is far from new. According to Ghiselin (1969, 93 ff.) and Beatty (1985, 277 ff.), Darwin held this view. In the *Origin of Species*, Darwin writes, "I look at the term species as one arbitrarily given for the sake of convenience to a set of individuals closely resembling each other, and that it does not essentially differ from the term variety" (1859, 52). In a letter to Joseph Hooker, he goes even further:

> It is really laughable to see what different ideas are prominent in various naturalists' minds, when they speak of "species"; in some, resemblance is everything and descent of little weight—in some, resemblance seems to go for nothing, and Creation the reigning idea—in some, sterility an unfailing test, with others it is not worth a farthing. It all comes, I believe, from trying to define the indefinable. (December 24, 1856; in Darwin 1887, vol. 2, 88)

Of course, Darwin did believe in the existence of evolving lineages, including those called "species." He just doubted the reality of the species category and the other categories of the Linnaean hierarchy.

If the species category does not exist, then how should we treat the term *species?* Should it be dropped from the discourse of current science and relegated to the history of science? Perhaps in a world of purely rational agents, this choice would be appropriate. We could, for instance, disambiguate the language of biology by replacing *species* with terms that distinguish the various taxa we call "species." Grant (1981, 36) suggests calling interbreeding species "biospecies." We could add the term *phylospecies* for phylogenetic species. In the abstract, this suggestion is attractive, but there are practical matters worth considering. The term *species* is well entrenched in contemporary biology and everyday life, so eliminating it might cause more problems than simply keeping it. Whether we should or can eliminate the term *species* is not an easy question to answer. I return to it in the last section of the chapter, but before getting to that, I examine another aspect of the Linnaean system: the rules for naming species.

## BINOMIAL NAMES

Linnaeus's system of classification consists of three parts: a hierarchy of categorical ranks, a set of rules for sorting organisms into taxa, and a list of rules for naming taxa. Though his rules of nomenclature do not have the theoretical significance of the other aspects of his system, they have practical importance. Prior to Linnaeus, biologists widely disagreed on the methods for naming taxa and, as a result, often assigned different names to the same taxon. In an effort to provide unity and clarity to biological nomenclature, Linnaeus introduced his own rules for naming taxa. This section focuses on Linnaeus's most famous rule: the prescription that every species be assigned a binomial—a prescription that has turned out to be another problematic aspect of the Linnaean system.

Species names contain two parts: a generic name that indicates a species' genus and a specific name that distinguishes a species from others in its genus. Linnaeus had several motivations for assigning binomials to species (1737, sec. 286). One motivation was the metaphysically significant role of genera. Fructification structures occur at the level of genera, and such structures are responsible for the continued existence of species and genera, so Linnaeus believed that the names of species should make reference to their genera. The introduction of binomials was also motivated by practical reasons. Binomials indicate the taxonomic placement of a species by citing its genus. Furthermore, the generic name of a species can guide a biologist to the taxonomic position of a species in a kingdom, assuming she knows the order and class of that genus.

Binomials serve one other function as well. One requirement of the Linnaean rules is that the ranks of various taxa, especially species, be encoded in their names. Thus, all species taxa, and only species taxa, are assigned binomials. Higher taxa are given uninomials, and subspecies are assigned trinomials. A more recent addition to the Linnaean rules assigns rank-specific suffixes to the names of many higher taxa. In the animal kingdom, for instance, all names of tribes have the suffix *–ini*, and all names of families have the suffix *–idae*. The rules for assigning suffixes to the names of higher taxa are set out in the contemporary rules of nomenclature, and those suffixes vary from kingdom to kingdom.

As useful as these rules might have been in the past, they are far from optimal these days. The discussion of the previous sections should cause us to wonder if the rank of a taxon should be indicated in its name. If the heterogeneous nature of the species category implies that no species category exists in nature, then it is misleading to distinguish some taxa as species by their names. Furthermore, if no workable ontological divide between species and higher taxa exists, then it is inappropriate to incorporate that distinction in the names of taxa. The first problem of the Linnaean rules governing the names of species is therefore a semantic one: the designation of a taxon's rank through the use of binomials may mark no distinction in nature. Similar observations apply to the use of suffixes in the names of higher taxa and the use of trinomials for subspecies (Ereshefsky 1994, 1997).

There are other problems with the use of binomials, and they arise regardless of whether one is a skeptic of the species category. One of Linnaeus's motivations for introducing binomials was his belief that a biologist should memorize the taxonomic positions of all the species within the kingdom he studies (Cain 1958, 156 ff.). Linnaeus thought there were too many species to do that, more than ten thousand plant species by his estimation (Atran 1990, 171). But he believed that the number of genera in the world was sufficiently small—approximately three hundred plant genera and three hundred animal genera (Mayr 1969, 344)—and that the number of genera would not increase significantly. With not too much difficulty, a biologist could memorize the names and positions of all the genera in a kingdom. Thus, binomials served as handy guides for memorizing the taxonomic positions of all the species in a kingdom.

However, Linnaeus did not envision a world where evolution continually gives rise to new species and genera. As a result, he grossly underestimated the diversity of the organic world. He believed that there were 312 animal genera; more recent estimates cite more than 50,000. As Mayr (1969, 344) observes, "a generic name no longer tells much to a zoologist except in a few popular groups of animals." Matters are even worse when we turn to plants. Linnaeus thought there were, at most, 6,000 species of plants. More recent studies estimate hundreds of thousands. Because there are too many generic names and taxonomic positions to memorize, Linnaeus's original motivation for assigning binomial names has been lost. Nevertheless, the names of species must still include a generic name. Indeed, the genus of a species must first be determined before a species can even be named, which may seem like an innocuous requirement. According to Cain (1959, 242), however, "the necessity of putting a species into a genus before it can be named at all is responsible for the fact that a great deal of uncertainty is wholly cloaked and concealed in modern classifications." Simply put, some species are placed in genera without adequate empirical information—a practice that is the result of an outdated rule of nomenclature.

Another problem with binomials involves taxonomic revision. Classifications of the organic world are constantly revised. For example, a species assigned to one genus may be reassigned to a different genus as the result of a new DNA analysis. The assignment of binomials further complicates such revisions by requiring that a species be given a new name when it is assigned to a different genus. Consider a simple case. When the species *Cobitis heteroclita* was found to belong to the genus *Fundulus*, its name was changed to *Fundulus heteroclita* (Wiley 1981, 399). As a result, all previous classifications of that species became outdated. Taxonomic revision is an epistemological problem that cannot be eliminated from biological taxonomy: new discoveries are made all the time. Yet the instability of a species' name is the result of the Linnaean requirement that the name cite the species' genus. That instability is avoidable.

A more problematic case occurs when a species is assigned to a genus that already has a species with that specific name. When this occurs, both a new generic name and a new specific name are needed. For example, the bee species *Paracolletes franki* was found to belong to the genus *Leioproctus*, yet a species named *Leioproctus franki* already existed. Consequently, *Paracolletes franki* was given an entirely new name to avoid the existence of a homonym (Michener 1964, 183–184, 188). These cases are far from isolated. As Mayr (1969, 344) notes, "In these examples we are not dealing with the results of excessive splitting or any sort of arbitrariness, but with a serious weakness of the entire system of binomial nomenclature."

Another difficulty with binomials arises when biologists disagree on the taxonomic placement of a taxon—usually regarding either the rank of a taxon or the genus of a species. Suppose one biologist adopts the interbreeding approach to species and assigns a group of organisms the rank of species, but another biologist adopts the phylogenetic approach and believes that the same group is a genus. According to the Linnaean system, the first biologist must assign the group a binomial, whereas the second biologist must assign it a uninomial. In other words, these biologists must assign different names to what they agree is the very same group of organisms.[7] The same problem occurs when biologists disagree over the genus of a species. According to the Linnaean system, they must assign that species different generic and hence binomial names. In all of these disagreements, the requirement that the name of a species reflect its rank and position is a hindrance rather than a help.

Linnaeus's introduction of binomial names seemed like a good idea in his day, but the advent of evolutionary theory has undermined their effectiveness. His original motivation for binomials has been lost because there are too many generic names to memorize. The binomial rule causes unnecessary instability in taxonomy by requiring that a species' name be changed in revision. And, the binomial rule requires the assignment of two or more names when biologists disagree on a taxon's rank or placement. In other words, the traditional use of binomials goes against two virtues that taxonomists prize: taxonomic stability and uniqueness of names. Finally, if the species category lacks an ontological foundation, as suggested earlier, then it is inappropriate to designate certain taxa as species by their names.[8]

The binomial rule is flawed on pragmatic grounds, and as we have seen, the ontological status of species category is itself suspect. Taken together, these problems cast a shadow on the continued use of the species category and its associated rules of nomenclature—and thereby on the continued use of the entire Linnaean system of classification. This chapter has discussed only those problems concerning species, yet many other aspects of the Linnaean system are problematic as well (Griffiths 1976, Ereshefsky 1994, de Queiroz and Gauthier 1994). Given the problems facing the system, perhaps we should consider replacing it. The next and final section of this chapter briefly introduces some alternative systems.

## ALTERNATIVES TO THE LINNAEAN SYSTEM

A comprehensive introduction to alternative systems of classifications, not to mention a proper comparison between them and the Linnaean system, is beyond the scope of this section. Its aim is more modest: (1) to show that alternative systems of classification do exist, which means that the abandonment of the Linnaean system would not leave biological taxonomy in a vacuum, and (2) to show that there are promising non-Linnaean systems worthy of attention and further development.

One aspect of the Linnaean system that should be replaced is its hierarchy of categorical ranks. The challenge here is not to the assumption that life is hierarchically arranged, though some theorists have questioned that assumption (Hull 1964). The challenge is to the continued use of the Linnaean categories to capture that hierarchy. For Linnaeus, taxa of a particular rank should be comparable. All species taxa, for example, should have some common feature that distinguishes them from all other types of taxa. This assumption was also held by the proponents of the Modern Synthesis and is held by many contemporary biologists. Yet species taxa may be noncomparable units: some are interbreeding units, others are phylogenetic (see the second section of this chapter). If such taxa are noncomparable, then it is not very meaningful—indeed, it is misleading—to say that they belong to a single, unified category. In addition, the distinction between species taxa and genera has been called into question (see the first section). If that distinction lacks a basis in nature, then it is misleading to designate one taxon as a species and another as a genus.

For these reasons, and many others, some biologists have suggested that the categorical ranks of the Linnaean hierarchy lack an ontological basis (Hennig 1969, Griffiths 1976, Eldredge and Cracraft 1980, 168). Some have even suggested that we completely abandon the Linnaean categories because not only do those categories lack independent existence, but their continued use misleads biologists into thinking that there are essential differences among taxa of different Linnaean ranks (Griffiths 1976, Ax 1987, chap. K). Two non-Linnaean systems for indicating the subordination of taxa have been offered: an *indentation system* and a *numerical system.*

The indentation system indicates the relative taxonomic positions of taxa by indenting subordinate taxa below their higher taxa (Farris 1976, Ax 1987, Gauthier et al. 1988). Sister taxa have the same indentation. Here is an example from Gauthier and colleagues (1988, 65):

Amniota
  Synapsida
  Reptilia
    Anapsida
    Diapsida

Indented classifications have several advantages over Linnaean ones. They represent the hierarchical relations of taxa without the use of ontologically suspect Linnaean categories. Taxonomic revision can be performed by merely moving the names of taxa because there are no categorical ranks to be altered, and biologists need not quarrel over the rank of a taxon. Finally, indented classifications free biologists from the need to create more and more Linnaean ranks. As some biologists have pointed out, the standard twenty-one ranks of the Linnaean hierarchy are woefully insufficient for representing life's complexity (Patterson and Rosen 1977).

Not all biologists are happy with the indentation system, however. Wiley (1981, 203–204) objects that it is difficult to compare the hierarchical positions of taxa listed on different pages of an indented classification. A ruler, he suggests, is needed to compare their positions. Ax (1987, 244) responds that such objections are "purely technical scruples" overshadowed by the benefits of the indentation system.

The other non-Linnaean system for representing subordination employs numerical notation (Hull 1966, Hennig 1969, Griffiths 1974). A numerical prefix indicates the hierarchical position of a taxon within a particular classification, and the last digit of a prefix can be used to indicate sister relations. Here is an example from Hennig (1969, 30):

2.2.2.2.4.6    Mecopteroidea
2.2.2.2.4.6.1    Amphiesmenoptera
2.2.2.2.4.6.1.1    Trichoptera
2.2.2.2.4.6.1.2    Lepidoptera

The use of numerical prefixes avoids Wiley's objection that the indentations of different taxa are difficult to compare, and like indented classifications, numerical ones indicate hierarchical relations without using Linnaean ranks. Supporters of the Linnaean system, however, suggest that the numerical alternative is inferior because "numerical prefixes are not the languages of ordinary use and are foreign to our efforts to communicate" (Wiley 1981, 202; see also Eldredge and Cracraft 1980, 224–225). In addition, the numerical prefixes for lower taxa in large classifications may be inordinately long. Anyone who desires a non-Linnaean system can avoid these problems by adopting the indented approach. Alternatively, they can use the numerical approach and argue that its inconveniences are slight compared to the costs of using the Linnaean system.

Another aspect of the Linnaean system worth replacing is the incorporation of a taxon's rank and taxonomic placement within its name. As we have seen, species are distinguished from other taxa by having binomial names, and the names of many higher taxa have rank-specific suffixes. If the Linnaean categories should be abandoned, then so too should the practice of incorporating those categories in the names of taxa. Furthermore, the practice of incorporating a taxon's rank within its name causes a number of practical problems. If a taxonomic revision reveals that a species belongs to a different genus than previously thought, then its generic name must be altered. If two biologists disagree on the genus of a species, they must give that species different binomial names. These problems arise from binomials serving a dual role as both a taxon's name and an indicator of its taxonomic placement. Separate these functions, and such problems will not arise. More specifically, a taxon's name should be simply that; it should not be used as a device for indicating a taxon's rank or placement. Given this suggestion, how should base taxa be named?

First, it should be noted that if one adopts either the indentation or numerical system, there is no need to incorporate the rank of a taxon within its name. Its hierarchical position within a classification is indicated by either indentation or a numerical prefix. Thus, the indentation and numerical systems nicely divorce the function of naming a taxon from that of classifying it. Still, questions remain. Should we preserve current binomial designations or replace binomials with uninomials? As Cain (1959, 242–243) and Griffiths (1976, 172) point out, replacing current binomials with uninomials would cause too much instability in taxonomy: biologists would need to relearn too many new names. Alternatively, Griffiths suggests that binomials be preserved, but that generic names be considered *forenames*. Cain (1959) and Michener (1963, 1964), on the other hand, suggest that the two parts of a binomial be hyphenated to form a single name. In either proposal, the first name of a taxon (its forename) will no longer refer to a genus, but will merely be a part of a taxon's name. Moreover, such binomial or hyphenated designations will not indicate the rank of a taxon.

Both suggestions address the question of what to do with taxa that already have binomials. But how should biologists name newly discovered base taxa? Should such taxa be given binomials? Both Griffiths (1976, 172) and Michener (1963, 165) think that they should. There are so many base taxa, they argue, that the assignment of forenames to newly discovered base taxa would help avoid the occurrence of homonymy. But then the question arises, How should forenames be chosen? For example, should the forename of a taxon be the name of its next inclusive taxon? If forenames are assigned in that manner, then it is important to keep in mind that they are merely names and not means for designating the taxonomic position of a taxon. Alternatively, forenames could be assigned in the same way that specific names are chosen, with no reference to other taxa.

Undoubtedly, these alternative systems of nomenclature are in need of further development, but they are nevertheless an improvement because they avoid a number of problems facing the Linnaean system by divorcing the task of naming a taxon from that of indicating its taxonomic position. In particular, they preserve the stability of a taxon's name during taxonomic revision, and they assign a unique name to a taxon even when its placement is controversial. Furthermore, these alternative systems free the names of taxa from incorporating meaningless categorical designations. The alternative systems surveyed here are not the panacea for all the problems facing the Linnaean system—probably no system is—but they are a step in the right direction. They promote stability within biological taxonomy and at the same time rid it of false metaphysical connotations. At the very least, they are worthy of further consideration.

The introduction of the Linnaean system was a watershed event for biological taxonomy. It brought order to a chaotic discipline by providing practical and theoretically sound rules for naming and sorting taxa. Much has happened in biology since the eighteenth century, however. The theoretical foundations of Linnaeus's sorting rules—essentialism and his sexual system—have been rejected. The existence of the species category has come under attack on a number of fronts, and the Linnaean rules for naming taxa have become more of a hindrance than a help. Perhaps these developments warrant the replacement of the Linnaean system with another system. At the very least, they warrant a serious review of the continued use of the species category and its associated rules of nomenclature. Biological theory has changed drastically in the last two hundred years. Perhaps it is time we changed the way we represent the organic world.

## ACKNOWLEDGMENTS

I thank David Hull, Jay Odenbaugh, Tony Russell, and Rob Wilson for their helpful comments on earlier drafts of this chapter. Financial support was generously provided by the Killam Foundation.

## NOTES

1. One might wonder whether Linnaeus's genera/higher taxa distinction, and the species/higher taxa distinction of the Modern Synthesis are one and the same distinction. After all, Linnaeus's fructification systems are the reproductive organs of plants, whereas the species of the Modern Synthesis are groups of organisms that can successfully interbreed. These distinctions, however, are quite different. Linnaeus's sexual system classifies organisms according to qualitative similarities among fructification structures, but with the advent of Darwinism, qualitative similarity took a backseat to genealogical connectedness. Furthermore, many characteristics of fructification structures (numbers of stamens and pistils, for example) are of little functional significance in conspecific reproduction (Mayr 1982, 178).

2. I am not denying that the members of asexual species are causally connected in some manner; after all, they are connected through parent-offspring relations. Of concern here, however, are

intragenerational connections, such as interbreeding, which are supposed to distinguish species from higher taxa but are missing in asexual species.

3. Species pluralism is a growth industry in the philosophy of biology, with several versions on the market. The type of pluralism adopted in this chapter differs from other prominent versions in two ways. Dupré's (1993) and Kitcher's (1984) forms of pluralism allow species to be spatiotemporally unrestricted classes of organisms; the pluralism used here requires that species form genealogical entities. Mishler and Brandon's (1987) pluralism requires that an organism belong to only one species; the pluralism adopted here permits an organism to belong to two different species at the same time. For full illustrations of the version of pluralism used here, see Ereshefsky 1992a and 1998.

4. One could avoid this conclusion by denying that both phylogenetic and interbreeding species are species, a response that is considered in Ereshefsky 1992a and 1998.

5. One might respond that inheritance is a unifying process that all species taxa share, which is correct, but it is not a process that distinguishes species from other types of taxa. Recall that we are looking for a unifying process that occurs in all and only species.

6. One might try to save the species category by employing Boyd's (1991) "causal homeostasis" account of natural kinds. However, I don't think that this particular philosophical approach will help. Consider Griffiths's (1997, chapter 8 in this volume) application of the causal homeostasis approach to taxa. The two arguments offered against the species category in this chapter remain unaffected. The natural kind traits Griffiths cites—homologies—do not distinguish species from higher taxa; thus, no new grounds for distinguishing species from higher taxa are offered. Furthermore, Griffiths buys into a pluralistic account of species very similar to the one articulated in this chapter, so the disunity of the species category remains.

7. For examples of discrepancies between the interbreeding and phylogenetic approaches to species, see the second section. For examples where taxonomic disagreements cause biologists to assign different names to the same taxon, see Michener 1964 and Ereshefsky 1994.

8. For additional problems with the binomial rule and the other Linnaean rules of nomenclature, see Ereshefsky 1994.

## REFERENCES


Atran, S. (1990). *Cognitive foundations of natural history: Towards an anthropology of science*. Cambridge: Cambridge University Press.

Ax, P. (1987). *The phylogenetic system*. New York: Wiley and Sons.

Beatty, J. (1985). Speaking of species: Darwin's strategy. In D. Kohn, ed., *The Darwinian heritage*. Princeton: Princeton University Press.

Boyd, R. (1991). Realism, anti-foundationalism, and the enthusiasm for natural kinds. *Philosophical Studies* 61, 127–148.

Cain, A. (1958). Logic and memory in Linnaeus's system of taxonomy. *Proceedings of the Linnaean Society of London* 169, 144–163.

Cain, A. (1959). Taxonomic concepts. *Ibis* 101, 302–318.

Carson, L. (1957). The species as a field for genetic recombination. In E. Mayr, ed., *The species problem*. Washington, D.C.: American Association for the Advancement of Science.

Cracraft, J. (1983). Species concepts and speciation analysis. In R. Johnston, ed., *Current ornithology*. New York: Plenum Press.

Darwin, C. (1859, facsimile edition 1964). *On the origin of species: A facsimile of the first edition*. Cambridge, Mass.: Harvard University Press.

Darwin, F., ed. (1887). *The life and letters of Charles Darwin, including an autobiographical chapter*, 2 vols. 3rd ed. London: John Murry.

de Queiroz, K., and M. Donoghue (1988). Phylogenetic systematics and the species problem. *Cladistics* 4, 317–338.

de Queiroz, K., and J. Gauthier (1992). Phylogenetic taxonomy. *Annual review of ecology and systematics* 23, 449–480.

de Queiroz, K., and J. Gauthier (1994). Toward a phylogenetic system of biological nomenclature. *Trends in Ecology and Evolution* 9, 27–31.

Dupré, J. (1993). *The disorder of things: Metaphysical foundations of the disunity of science*. Cambridge, Mass.: Harvard University Press.

Ehrlich, P., and P. Raven (1969). Differentiation of populations. *Science* 165, 1228–1232.

Eldredge, N., and J. Cracraft (1980). *Phylogenetic patterns and the evolutionary process*. New York: Columbia University Press.

Eldredge, N., and S. Gould (1972). Punctuated equilibria: An alternative to phyletic gradualism. In T. Schopf, ed., *Models in paleobiology*. San Francisco: Freeman Cooper.

Ereshefsky, M. (1992a). Eliminative pluralism. *Philosophy of Science* 59, 671–690.

Ereshefsky, M. (1994). Some problems with the Linnaean hierarchy. *Philosophy of Science* 61, 186–205.

Ereshefsky, M. (1997). The evolution of the Linnaean hierarchy. *Philosophy and Biology* 12, 493–519.

Ereshefsky, M. (1998). Species pluralism and anti-realism. *Philosophy of Science* 65, 103–120.

Ereshefsky, M., ed. (1992b). *The units of evolution: Essays on the nature of species*. Cambridge, Mass.: MIT Press.

Farris, J. (1976). Phylogenetic classification of fossils with recent species. *Systematic Zoology* 25, 271–282.

Gauthier, J., R. Estes, and K. De Queiroz (1988). A phylogenetic analysis of lepidosauromorpha. In R. Estes and G. Pregill, eds., *Phylogenetic relationships of lizard families*. Palo Alto, Calif.: Stanford University Press.

Ghiselin, M. (1969). *The triumph of the Darwinian method*. Chicago: University of Chicago Press.

Ghiselin, M. (1974). A radical solution to the species problem. *Systematic Zoology* 23, 536–544.

Ghiselin, M. (1987). Species concepts, individuality, and objectivity. *Biology and Philosophy* 2, 127–143.

Ghiselin, M. (1989). Sex and the individuality of species: A reply to Mishler and Brandon. *Biology and Philosophy* 4, 77–80.

Griffiths, G. (1974). On the foundations of biological systematics. *Acta Biotheoretica* 3–4, 85–131.

Griffiths, G. (1976). The future of Linnaean nomenclature. *Systematic Zoology* 25, 168–173.

Griffiths, P. (1997). *What emotions really are*. Chicago: Chicago University Press.

Grant, V. (1980). Gene flow and the homogeneity of species populations. *Biologisches Zentralbaltt* 99, 157–169.

Grant, V. (1981). *Plant speciation*, 2nd ed. New York: Columbia University Press.

Hennig, W. (1969, English edition 1981). *Insect phylogeny*. Translated by A. C. Pont. New York: John Wiley Press. [Originally published as *Die Stammesgeschichte der Insekten*. Frankfurt: Waldemar Kramer.]

Heywood, V. (1985). Linnaeus—the conflict between science and scholasticism. In J. Weinstock, ed., *Contemporary perspectives on Linnaeus*. Lanham, Md.: University Press of America.

Hull, D. (1964). Consistency and monophyly. *Systematic Zoology* 13, 1–11.

Hull, D. (1965). The effect of essentialism on taxonomy: Two thousand years of stasis. *British Journal for the Philosophy of Science* 15, 314–326, and 16, 1–18.

Hull, D. (1966). Phylogenetic numericlature. *Systematic Zoology* 15, 14–17.

Hull, D. (1987). Genealogical actors in ecological roles. *Biology and Philosophy* 2, 168–183.

Kitcher, P. (1984). Species. *Philosophy of Science* 51, 308–333.

Larson, J. (1971). *Reason and experience: The representation of natural order in the work of Carl von Linne*. Berkeley: University of California Press.

Linnaeus, C. (1737, English edition 1938). *The critica botanica*. Translated by A. Hort. London: Ray Society.

Mayr, E. (1969). *Principles of systematic zoology*. Cambridge, Mass.: Harvard University Press.

Mayr, E. (1970). *Populations, species, and evolution*. Cambridge, Mass.: Harvard University Press.

Mayr, E. (1982). *The growth of biological thought*. Cambridge, Mass.: Harvard University Press.

Michener, C. (1963). Some future developments in taxonomy. *Systematic Zoology* 12, 151–172.

Michener, C. (1964). The possible use of uninominal nomenclature to increase the stability of names in biology. *Systematic Zoology* 13, 182–190.

Mishler, B., and R. Brandon (1987). Individuality, pluralism, and the phylogenetic species concept. *Biology and Philosophy* 2, 397–414.

Mishler, B., and M. Donoghue (1982). Species concepts: A case for pluralism. *Systematic Zoology* 31, 491–503.

Paterson, H. (1985). The recognition concept of species. In E. Vrba, ed., *Species and speciation*. Pretoria: Transvall Museum. 21–29.

Patterson, C., and D. Rosen (1977). Review of ichthyodectiform and other mesozoic teleost fishes and the theory and practice of classifying fossils. *Bulletin of the American Museum of Natural History* 158, 81–172.

Ridley, M. (1989). The cladistic solution to the species problem. *Biology and Philosophy* 4, 1–16.

Rosen, D. (1979). Vicariant patterns and historical explanations in biogeography. *Systematic Zoology* 27, 159–188.

Sober, E. (1980). Evolution, population thinking and essentialism. *Philosophy of Science* 47, 350–383.

Templeton, A. (1989). The meaning of species and speciation: A genetic perspective. In D. Otte and J. Endler, eds., *Speciation and its consequences*. Sunderland, Mass.: Sinauer. Reprinted in M. Ereshefsky, ed. (1992). *The units of evolution: Essays on the nature of species*. Cambridge, Mass.: MIT Press.

Wiley, E. (1979). The annotated Linnaean hierarchy, with comments on natural taxa and competing systems. *Systematic Zoology* 28, 308–337.

Wiley, E. (1981). *Phylogenetics: The theory and practice of phylogenetic systematics*. New York: Wiley and Sons.

# 12 Getting Rid of Species?

Brent D. Mishler

The debate about species concepts over the last twenty years follows a curious pattern. Rather than moving toward some kind of consensus, as one might expect, the trend has been toward an ever-increasing proliferation of concepts. Starting with the widely accepted species concept that took precedence in the 1940s and 1950s as a result of the Modern Synthesis—the biological species concept (BSC)—we heard calls for change from botanists, behaviorists, and others. Despite the babel of new concepts, the BSC continues to have fervent advocates (Avise and Ball 1990, Avise and Wollenberg 1997) and has itself spawned several new variants. A recent paper by Mayden (1997) lists no fewer than twenty-two prevailing concepts! We can't seem to eliminate any existing concept, only produce new ones.

Why? The obvious conclusion one might draw—that biologists are contrarians who want to make their own personal marks in a debate and thus coin their own personal concepts to defend—is really not the case; this is no debate about semantics. The conceptual divisions are major and real. In my opinion, the plethora of ways in which different workers want to use the species category reflects an underlying plethora of valid ways of looking at biological diversity. The way forward is to recognize this view and face its implications: the basis of the confusion over species concepts is a result of heroic but doomed attempts to shoehorn all this variation into an outdated and misguided classification system, the ranked Linnaean hierarchy. Most of the confusion can be eliminated simply by removing the ranks. The issues that remain can then be dealt with by carefully considering what we want formal classification to represent as the general reference system and then by carefully specifying criteria for grouping organisms into these formal classifications.

To develop this argument, I first make the case for generalizing the species problem as a special case of the taxon problem. For a consistent, general reference classification system, all taxa must be of the same type; species should be regarded as simply the least-inclusive taxon in the system. Then, I review the reasons why phylogeny provides the best basis for the general purpose classification: species should be considered as just another phylogenetically based taxon. Next, I address the recent calls for rank-free classification in general and pursue the central thesis of this paper: *the species rank must disappear along with all the other ranks.* Finally, I explore the practical implications of eliminating the rank of species for such areas as ecology, evolution, and conservation.

## SPECIES AS JUST ANOTHER TAXON

In their particular theories of systematics, many authors have made a firm distinction between species and higher taxa (e.g., Wiley 1981, Nelson and Platnick 1981, Nixon and Wheeler 1990; see also the discussion by de Queiroz, chapter 3 in this volume). The idea is that somehow species are units directly participating in the evolutionary process, whereas higher taxa are at most lineages resulting from past evolutionary events. However nicely drawn this distinction is in theory, these arguments have resulted more from wishful thinking than from empirical observations. When anyone has looked closely for an empirical criterion to distinguish the species rank uniquely and universally from all others, the attempt has failed.

One early suggestion was phenetic: a species is a cluster of organisms in Euclidean space separated from other such clusters by some distinct and comparable gap (e.g., Levin 1979). This idea has been clearly shown to be mistaken: phenotypic clusters are actually nested inside each other with continuously varying gap sizes. Current entities ranked as species are not comparable either in the amount of phenotypic space they occupy or in the size of the "moat" around them, nor can they be made to be comparable through any massive realignment of current usage.

Another suggestion for a unique ranking criterion for species is expressed in the biological species concept: a species is a reproductive community separated by a major barrier from crossing with other such communities (Mayr 1982). Like the phenetic gap, this view (nice in theory perhaps) fails when looking at real organisms (cf. Nanney, chapter 4 in this volume). Despite the publication of many conceptual diagrams that depict a distinct break between reticulating and divergent relationships at some level (Nixon and Wheeler 1990, Roth 1991, Graybeal 1995), actual data suggests that in most groups, the probability of intercrossability decreases gradually as more and more inclusive groups are compared (Mishler and Donoghue 1982, Maddison 1997). There usually is no distinct point at which the possibility of reticulation drops precipitously to zero.

Similar suggestions have been made based on ecological criteria: a species is a group of organisms occupying some specific and unitary ecological niche (Van Valen 1976). Maybe species "can define themselves"; we just need to see whether two organisms treat each other as belonging to the same or different species. Again, actual studies show no such distinctive level where ecological interactions change abruptly from "within kind" to "between kind." Cryptic, ecologically distinct groups can be found below the species level, and large guilds of organisms from divergent species can act as one group ecologically in some situations.

Finally, there have been attempts to distinguish species from all other taxa phylogenetically (Nixon and Wheeler 1990, Graybeal 1995, Baum 1992). In this view, species are the smallest divergent lineage, inside of which there is no recoverable divergent phylogenetic structure (only reticulation). Again, nice in theory, but unsound empirically, at least as a general principle (Mishler and Donoghue 1982, Mishler and Theriot 1999). Some biological situations fit the model well (e.g., in organisms with complex and well-defined sexual mate recognition systems and no mode of asexual propagation). However, in many clonal groups (e.g., aspen trees, bracken fern) discernible lineages go down to the within-organism level (the problem of "too little sex"; see Templeton 1989). On the other hand, occasional horizontal transfer events ("reticulations") occur between very divergent lineages (the problem of "too much sex"; see Templeton 1989). In all such cases, a large gray area exists between strictly diverging patterns of gene genealogies and strictly recombining ones (cf. Avise and Wollenberg 1997).

To sum up, we have no and are unlikely to have any criterion for distinguishing species from other ranks in the Linnaean hierarchy, which is not to say that particular species taxa are unreal. They are real, but only in the sense that taxa at all levels are real. Species are not special.

## THE NECESSITY FOR PHYLOGENETIC CLASSIFICATIONS

The debate over classification has a long and checkered history, but this essay is not the place to detail the history fully (see Stevens 1994 and Ereshevsky, chapter 11 in this volume). I want to begin with the conceptual upheaval in the 1970s and 1980s that resulted in the ascension of Hennigian phylogenetic systematics (for a detailed treatment, see the masterful book by Hull 1988). Many issues were at stake in that era, foremost of which was the nature of taxa. Are they just convenient groupings of organisms with similar features, or are they lineages, marked by homologies? A general, if not completely universal, consensus has been reached that taxa are (or at least should be) the latter (Hennig 1966, Nelson 1973, Farris 1983, Sober 1988).

A full review of the arguments for why formal taxonomic names should be used solely to represent phylogenetic groups is beyond the scope of this paper, but they can be summarized as follows. Evolution is the single most powerful and general process underlying biological diversity. The major outcome of the evolutionary process is the production of an ever-branching phylogenetic tree, through descent with modification along the branches. This results in life being organized as a hierarchy of nested monophyletic groups. Because the most effective and natural classification systems are those that "capture" the entities resulting from processes that generate the things being classified, the general biological classification system should be used to reflect the tree of life.

The German entomologist Willi Hennig codified the meaning of these evolutionary outcomes for systematics in what has been called the *Hennig Principle* (Hennig 1965, 1966). Hennig's seminal contribution was to note that in a system evolving via descent with modification and splitting of lineages, characters that changed state along a particular lineage can serve to indicate the prior existence of that lineage, even after further splitting occurs. The Hennig Principle follows from that conclusion: homologous similarities[1] among organisms come in two basic kinds, *synapomorphies* due to immediate shared ancestry (i.e., a common ancestor at a specific phylogenetic level) and *symplesiomorphies* due to more distant ancestry. Only the former are useful for reconstructing the relative order of branching events in phylogeny. A corollary of the Hennig Principle is that classification should reflect reconstructed branching order; only *monophyletic groups*[2] should be formally named. Phylogenetic taxa will thus be "natural" in the sense of being the result of the evolutionary process.

This isn't to say that phylogeny is the only important organizing principle in biology. There are many ways of classifying organisms into a hierarchy because of the many biological processes impinging on organisms. Many kinds of nonphylogenetic biological groupings are unquestionably useful for special purposes (e.g., producers, rain forests, hummingbird pollinated plants, bacteria). However, it is generally agreed that there should be one consistent, general reference system, for which the Linnaean hierarchy should be reserved. Phylogeny is the best criterion for the general-purpose classification, both theoretically (the tree of life is the single universal outcome of the evolutionary process) and practically (phylogenetic relationship is the best criterion for summarizing known data about attributes of organisms and predicting unknown attributes). The other possible ways to classify can, of course, be used simultaneously, but should be regarded as special purpose classifications and clearly distinguished from phylogenetic formal taxa.

## THE ADVANTAGES OF A RANK-FREE TAXONOMY

A number of calls have been made recently for the reformation of the Linnaean hierarchy (e.g., de Queiroz and Gauthier 1992). These authors have emphasized that the roots of the Linnaean system are to be found in a non-evolutionary worldview—a specially created worldview. Perhaps the idea of fixed ranks made some sense under that view, but not under an evolutionary worldview. Most aspects of the current code, including priority, revolve around the ranks, which leads to instability of usage. For example, when a change in relationships is discovered, several names often need to be changed to adjust, including the names of groups whose circumscription has not changed. Authors often frivolously change the rank of a group even though there is no change in postulated relationships. Although practicing systematists know that groups given the same rank across biology are not comparable in any way (i.e., in age, size, amount of divergence, internal diversity, etc.), many users of the system do not know this. For example, ecologists and macroevolutionists often count numbers of taxa at a particular rank as an erroneous measure of "biodiversity." The nonequivalence of ranks means that at best (to those who are knowledgeable) they are a meaningless formality and perhaps not more than a hindrance. At worst, formal ranks lead to bad science in the hands of a user of classifications who naively assumes that groups at the same rank are comparable in some way.

It is not completely clear at this point how exactly a new code of nomenclature should be written, but the basics are clear. Such a new code should maintain the principle of priority (the first name for a clade should be followed) and other aspects of the current code that promote effective communication of new names to the community. However, the major change would be that the Linnaean ranks (e.g. phylum, family) should be abandoned for more efficient and accurate representation of phylogenetic relationships. Instead, names of clades should be hierarchically nested uninomials regarded as proper names. A clade would retain its name regardless of where new knowledge might change its phylogenetic position, thus increasing nomenclatorial stability. Furthermore, because clade names would be presented to the community without attached ranks, users would be encouraged to look at the actual attributes of the clades they compare, thus improving research in comparative biology.

It is important to emphasize that despite misrepresentations to the contrary, theorists who advocate getting rid of Linnaean ranks do not at all advocate getting rid of the hierarchy in biological classification. Nesting of groups within groups is essential because of the treelike nature of phylogenetic organization. Think of a nonsystematic example: a grocer might classify table salt as a spice, and group spices together under the category *food items*. This simple hierarchy is clear, but requires no named ranks to be understood. In fact, all human thought is organized into hierarchies, and becoming educated in a field essentially means learning the hierarchical arrangement of concepts in that field. Taxonomy is unusual in the assigning of named ranks to its hierarchies; they are superfluous to true understanding.

## GETTING RID OF THE SPECIES RANK

Curiously, so far in this debate, even the advocates of rank-free phylogenetic classification have retained the species rank as a special case. All other ranks are to be abandoned, but the species rank is to be kept, probably because the species concept is so ingrained and comfortable in current thinking. However, all the arguments that can be massed against Linnaean ranked classification in general can be brought to bear against the species rank as well. As difficult as it is to overthrow ingrained habits of thinking, logical consistency demands that all levels in the classification should be treated alike.

Given the background developed in the previous three sections, the conclusion seems inescapable: *the species rank must go the way of all others.* We must end the bickering over how this rank should be applied and instead get rid of the rank itself. This solution is truly the "radical solution to the species problem" sought unsuccessfully by Ghiselin (1974). Biological classification should be a set of nested, named groups for internested clades. Not all clades need be named, but those that are should be named on the basis of evidence for monophyly (see further discussion of the meaning of monophyly in Mishler and Brandon 1987). We stop naming groups at some point approaching the tips of the phylogeny because we don't have solid evidence for monophyly at the present stage of knowledge. This may be due to rampant reticulation going on below some point or simply to a lack of good markers for distinguishing finer clades. We shouldn't pretend, however, that the smallest clades named at a particular time are ontologically different from other, more inclusive named clades. Further research could easily result in subdividing these groups or lumping several of them into one if the original evidence that supported them is discovered to be faulty.

Given the redundancy now present in species epithets (e.g., *californica* is used in many genera), there needs to be a way to uniquely place each smallest named clade in the classification. My recommendation for nomenclature at the least inclusive level under a totally rank-free classification would be to regard names in a similar way as personal names are regarded in an Arabic culture. Each clade, including the least inclusive one named, has its own uninomial name; however, the genealogical relationships of a clade are preserved in a polynomial giving the lineage of that clade in higher and higher groups. Therefore, the familiar binomial, which does after all present some grouping information to the user, could be retained, but should be inverted. Our own short clade name thus should be *Sapiens Homo*. The full name for our terminal clade should be regarded as a polynominal that gives the names of the more and more inclusive clades all the way back. To use the human example, this full name would be something like: *Sapiens Homo Homidae Primate Mammalia Vertebrata Metazoa Eucaryota Life.*[3] Again, as in a traditional Arabic name, this formal and complete name would be used only rarely and for the most formal purposes (although it would be very useful behind the scenes for data-basing purposes); the everyday name of the clade would be *Sapiens Homo*.

## PRACTICAL IMPLICATIONS

"Getting rid of species" has another, all too ominous meaning in today's world. Named species are being driven to (and over) the brink of extinction at a rapid rate. What will be the implications of the view of taxa advocated in this paper? If we get rid of the species rank, with all its problems, will we hamstring conservation efforts? I tend to think not; scientific honesty seems the best policy here as elsewhere. The rather mindless approach followed in conservation—that if a lineage is ranked as a species, it is worth saving, but if it is not considered a species, it is not worth considering—is misguided in many ways. It is wrong scientifically; the species rank is a human judgment rather than any objective point along the trajectory of diverging lineages. It is also wrong ethically; any recognizable lineage is worth conservation consideration. Not all lineages need be conserved, or at least be given the same conservation priority, but such judgments should be made on a case by case basis.

All biologists are concerned about defining biodiversity and about its current plight; thus, the radical move suggested here (i.e., getting rid of the species rank) will no doubt worry many. People who want to characterize and conserve biodiversity commonly complain that "without species we will have no way of quantifying biodiversity or of convincing people to preserve it." This viewpoint, although expressing a commendable and important concern, is ultimately misguided, both in theoretical and practical terms. There may a comfortable self-deception going on to the contrary, but only a moment of thoughtful reflection is enough to remind us that species are not comparable in any important sense and cannot be made so.

However, the recognition that a count of species is not a good measure of biodiversity does not mean that biodiversity cannot be quantified. All named species are unique, with their own properties and features, and they represent only the tip of the underlying iceberg of biodiversity. We must face these facts and move to develop valid measures of the diversity of lineages, taking into account their actual properties and phylogenetic significance. A number of workers have suggested quantitative measures for phylogenetic biodiversity, which take into account the number of branch points and possibly branch lengths separating the tips of the tree (Vane-Wright, Humphries, and Williams 1991; Faith 1992a, 1992b).

Many macroevolutionary studies are framed in terms of comparing diversity patterns at some particular rank (e.g., families of marine invertebrates, phyla of animals). The adoption of rank-free classification would (fortunately) make such studies impossible, but would it make all studies of macroevolution impossible? Of course not: comparisons among clades would still be quite feasible, but it would be up to the investigator to establish that the clades being compared were the same with respect to the necessary properties (i.e., equivalent age or disparity, and so on). Similar arguments could be made with respect to the many ecological studies that compare numbers of species in different regions or communities. The bottom line is that rank-free classification would lead to much more accurate research in ecology and evolution because, investigators would be encouraged to use cladograms directly in their comparative studies instead of relying on equivalence in taxonomic rank as a (very) crude proxy for comparability of lineages. Given the rapid progress in development of quantitative comparative methods (Funk and Brooks 1990, Brooks and McLennan 1991, Harvey and Pagel

1991, Martins 1996) and the rapid proliferation of ever-improving cladograms for most groups of organisms, this change can only be for the best.

Species, RIP.

## NOTES

1. In Hennigian phylogenetic systematics, *homology* is defined historically as a feature shared by two organisms because of descent from a common ancestor that had that same feature.

2. A strictly monophyletic group (a clade) is one that contains all and only descendants of a common ancestor. A paraphyletic group is one the excludes some of the descendants of the common ancestor.

3. Note that some of the nested clades will have formal suffixes indicating their previous rank (e.g., *-idae* for family). Although these endings would be retained for existing clade names in order to avoid confusion, there would be no meaning attached to them, and newly proposed clade names would have no particular suffix requirement.

## REFERENCES


Avise, J. C., and R. M. Ball (1990). Principles of genealogical concordance in species concepts and biological taxonomy. *Oxford Surveys in Evolutionary Biology* 7, 45–67.

Avise, J. C., and K. Wollenberg (1997). Phylogenetics and the origin of species. *Proceedings of the National Academy of Science*, (USA) 94, 7748–7755.

Baum, D. (1992). Phylogenetic species concepts. *Trends in Ecology and Evolution* 7, 1–2.

Brooks, D. R., and D. A. McLennan (1991). *Phylogeny, ecology, and behavior.* Chicago: University of Chicago Press.

de Queiroz, K., and J. Gauthier (1992). Phylogenetic taxonomy. *Annual Review of Ecology and Systematics* 23, 449–480.

Faith, D. P. (1992a). Conservation evaluation and phylogenetic diversity. *Biological Conservation* 61, 1–10.

Faith, D. P. (1992b). Systematics and conservation: On predicting the feature diversity of subsets of taxa. *Cladistics* 8, 361–373.

Farris, J. S. (1983). The logical basis of phylogenetic analysis. In N. Platnick and V. Funk, eds., *Advances in Cladistics*, vol. 2. New York: Columbia University Press.

Funk, V. A., and D. R. Brooks (1990). *Phylogenetic systematics as the basis of comparative biology.* Washington, D.C.: Smithsonian Institution Press.

Ghiselin, M. T. 1974. A radical solution to the species problem. *Systematic Zoology* 23, 536–544.

Harvey, P. H., and M. D. Pagel (1991). *The comparative method in evolutionary biology.* Oxford: Oxford University Press.

Hennig, W. (1965). Phylogenetic systematics. *Annual Review of Entomology* 10, 97–116.

Hennig, W. (1966). *Phylogenetic systematics.* Urbana: University of Illinois Press.

Hull, D. L. (1988). *Science as a process: An evolutionary account of the social and conceptual development of science.* Chicago: University Chicago Press.

Levin, D. A. (1979). The nature of plant species. *Science* 204, 381–384.

Maddison, W. P. (1997). Gene trees in species trees. *Systematic Biology* 46, 523–536.

Martins, E. P. (1996). Phylogenies, spatial autoregression, and the comparative method: A computer simulation test. *Evolution* 50, 1750–1765.

Mayden, R. L. (1997). A hierarchy of species concepts: The denouement in the saga of the species problem. In M. F. Claridge, H. A. Dawah, and M. R. Wilson, eds., *Species: The units of biodiversity*. London: Chapman and Hall.

Mayr, E. (1982). *The growth of biological thought*. Cambridge, Mass.: Harvard University Press.

Mishler, B. D., and R. N. Brandon (1987). Individuality, pluralism, and the phylogenetic species concept. *Biology and Philosophy* 2, 397–414.

Mishler, B. D., and M. J. Donoghue (1982). Species concepts: A case for pluralism. *Systematic Zoology* 31, 491–503.

Mishler, B. D., and E. Theriot (1999). Monophyly, apomorphy, and phylogenetic species concepts. In Q. D. Wheeler and R. Meier, eds., *Species concepts and phylogenetic theory: A debate*. New York: Columbia University Press.

Nelson, G. (1973). Classification as an expression of phylogenetic relationships. *Systematic Zoology* 22, 344–359.

Nelson, G., and N. Platnick (1981). *Systematics and biogeography: Cladistics and vicariance*. New York: Columbia University Press.

Nixon, K. C., and Q. D. Wheeler (1990). An amplification of the phylogenetic species concept. *Cladistics* 6, 211–223.

Roth, V. L. (1991). Homology and hierarchies: Problems solved and unresolved. *Journal of Evolutionary Biology* 4, 167–194.

Sober, E. (1988). *Reconstructing the past*. Cambridge, Mass.: MIT Press.

Stevens, P. F. (1994). *The development of biological systematics*. New York: Columbia University Press.

Templeton, A. R. (1989). The meaning of species and speciation: A genetic perspective. In D. Otte and J. A. Endler, eds., *Speciation and its consequences*. Sunderland, Mass.: Sinauer.

Vane-Wright, R. I., C. J. Humphries, and P. H. Williams (1991). What to protect?—Systematics and the agony of choice. *Biological Conservation* 55, 235–254.

Van Valen, L. (1976). Ecological species, multi-species, and oaks. *Taxon* 25, 233–239.

Wiley, E. O. (1981). *Phylogenetics: The theory and practice of phylogenetic systematics*. New York: Wiley and Sons.

# Index

Printed in Great Britain
by Amazon

56571782R00199